Advanced Researches in Materials Science and Engineering

Advanced Researches in Materials Science and Engineering

Edited by **Andrew Green**

WILLFORD PRESS

New York

Published by Willford Press,
118-35 Queens Blvd., Suite 400,
Forest Hills, NY 11375, USA
www.willfordpress.com

Advanced Researches in Materials Science and Engineering
Edited by Andrew Green

International Standard Book Number: 978-1-68285-123-4 (Hardback)

Printed in the United States of America.

Contents

Preface

Materials science as a discipline has its prime focus on the study of new materials, especially solid materials. This book explores the mechanical, optical, chemical properties of various materials along with their applications. The topics covered in this book include crystallography, polymers, composites, characterization techniques like calorimetry, neutron diffraction, etc. It also focuses on the industrial use of materials. Those in search of information to further their knowledge in this discipline will be greatly assisted by this book. This text is apt for students specializing in materials science and allied branches of science and engineering.

This book unites the global concepts and researches in an organized manner for a comprehensive understanding of the subject. It is a ripe text for all researchers, students, scientists or anyone else who is interested in acquiring a better knowledge of this dynamic field.

I extend my sincere thanks to the contributors for such eloquent research chapters. Finally, I thank my family for being a source of support and help.

Editor

Effect of Layer and Film Thickness and Temperature on the Mechanical Property of Micro- and Nano-Layered PC/PMMA Films Subjected to Thermal Aging

Ahmed Abdel-Mohti [1], Alison N. Garbash [2], Saad Almagahwi [2] and Hui Shen [2,*]

[1] Civil Engineering Department, Ohio Northern University, Ada, OH 45810, USA;
    E-Mail: a-abdel-mohti@onu.edu

[2] Mechanical Engineering Department, Ohio Northern University, Ada, OH 45810, USA;
    E-Mails: agarbash@gmail.com (A.N.G.); s-almagahwi@onu.edu (S.A.)

* Author to whom correspondence should be addressed; E-Mail: h-shen@onu.edu;

Academic Editor: Giorgio Biasiol

Abstract: Multilayered polymer films with biomimicking, layered structures have unique microstructures and many potential applications. However, a major limitation of polymer films is the deterioration of mechanical properties in working environments. To facilitate the design and development of multilayered polymer films, the impact of thermal aging on the mechanical behavior of micro- and nano-layered polymer films has been investigated experimentally. The composition of the polymer films that have been studied is 50 vol% polycarbonate (PC) and 50 vol% poly(methyl methacrylate) (PMMA). The current study focuses on the effect of film and layer thickness and temperature on the mechanical properties of the materials subjected to thermal aging. To study the effect of film and layer thickness, films with the same thickness, but various layer thicknesses, and films with the same layer thickness, but various film thicknesses, were thermally aged at 100 °C in a constant temperature oven for up to six weeks. The results show that as the layer thickness decreases to 31 nm, the film has a higher stiffness and strength, and the trend of the mechanical properties is relatively stable over aging. The ductility of all of the films decreases with aging time. To study the effect of temperature, the films with 4,096 layers (31 nm thick for each layer) were aged at 100 °C, 115 °C and 125 °C for up to four weeks. While the 100 °C aging results in a slight increase of the stiffness and strength of the films, the higher aging temperature caused a decrease of the stiffness and strength of the films.

The ductility decreases with the aging time for all of the temperatures. The films become more brittle for higher aging temperatures.

Keywords: multilayered polymer film; thermal aging; film thickness; layer thickness; aging temperature; mechanical properties

---

1. Introduction

Polymers have a wide range of applications due to their inherent properties, such light-weight, strength, resistance to chemicals, thermal resistance, *etc.* Many processing techniques have been developed to advance the material properties and extend the range of applications. Multilayered polymers have been developed by coextrusion of polymeric systems [1]. The coextrusion technology was first developed by the Dow Chemical Company in the 1970s [2–6]. Multilayered polymers include microlayered and nanolayered polymer depending on the scale of the thickness of layers. Multilayered polymers have been reported to have improved mechanical properties, such as ductility and impact strength, as the layer thickness is reduced [7–9]. The microlayered polymer films have been successfully used in industrial applications, such as food packaging and coating [10–12]. The 3M Company adopted multilayered polymers for light management in mirrors and the screens of laptop computers. These types of polymer films have some unique properties in their applications. For example, the films can retain the aroma, flavor and freshness of food when used in food packaging. As the coextrusion techniques have advanced in recent years, innovative nanolayered polymer films have been developed, which have more complex hierarchical systems and a truly biomimetic nature [13,14]. There are many potential applications for the nanolayered films, such as gas barrier materials, next-generation flat-panel displays and spherical gradient refractive index (GRIN) lenses [15,16]. To mimic the structures of eyes in nature, polymer lenses made of nanolayered polymer film with GRIN distributions have been developed in recent years. The GRIN crystalline lenses in biological eyes, such as human being's eyes, typically contain approximately 22,000 layers [17,18]. Layered polymeric optical lenses have been reported to have better optical properties than glass lenses and have been used to replace the traditional glass lens [17,18]. Some researchers have studied the physical and mechanical properties, such as transparency and flexibility, of the novel thin films, although the reports are relatively few [19–22].

Due to the intrinsic structure of polymers, the mechanical properties of the polymer films can vary widely depending on the material formulation, environmental temperature and time. To facilitate the application of polymers, the deterioration of the material over time, especially at elevated temperature, needs to be studied. We have previously carried out a preliminary investigation to experimentally study the impact of thermal aging on the mechanical behavior of the micro- and nano-layered polymer films [23]. However, in the previous study, all of the films have various layer thicknesses and film thicknesses. This makes it hard to study the effect of layer thickness and film thickness on the change of properties after aging separately. In the current study, the films have either the same film thickness with various layer thicknesses or the same layer thickness with various film thicknesses. This way, the effect of film and layer thickness on the aging of the film can be studied independently. The effect of aging temperature on a specific nanolayered film has also been studied.

The composition of the polymer films under study is 50 vol% polycarbonate (PC) and 50 vol% poly(methyl methacrylate) (PMMA). The layer thickness ranges from 31 nm to 1,984 nm with a 5-μm film thickness and a film thickness of 31.8 μm to 508 μm with a 496-nm layer thickness. These films were thermally aged at 100 °C in a constant temperature oven for up to six weeks. Their mechanical properties, including the modulus of elasticity, tensile strength and ductility, were compared. It has been observed that the thermal aging temperature and aging time have significant effects on the overall character of the stress-strain responses, and layer and film thickness play important roles. The films with 4,096 layers (31 nm thick for each layer) were aged at 100 °C, 115 °C and 125 °C for up to four weeks to study the effect of aging temperature. It has been observed that the film with a 31-nm layer has relatively stable mechanical properties. The microstructural scale level contributes to the material mechanical properties, such as the mechanical stability and durability. The microstructural features start to play a role as the layer thickness reduces to a certain level, which is very important for applications, such as in GRIN lens designs.

2. Materials and Tests

The multi-layered polymer films under study were fabricated with the unique layer-multiplying coextrusion process at Case Western Reserve University. The film sample is shown in Figure 1. The process combines 50 vol% PC and 50 vol% PMMA as perfectly alternating layered systems. The films with various layer thicknesses are shown in Table 1 and films with various film thicknesses in Table 2. The room temperature modulus of elasticity, tensile strength and fracture strain (as a measurement of ductility) of the films were studied with static tension tests using an Instron 4467 instrument in tension mode fitted with film tension grips. The film samples were prepared and tested as directed by ASTM D882. Samples for the tension test were approximately 6 in (152 mm) long and 0.2 in (5 mm) wide. The gage length was approximately 4 in (102 mm). To study the effect of thermal aging, these films were aged 100 °C in a constant temperature oven for up to six weeks. Samples were removed from the oven at aging times of 3 days, 1 week, 2 weeks, 3 weeks, 4 weeks, 5 weeks and 6 weeks (42 days) and tested. The test matrix and sample numbers for each test are listed in Table 3. The test matrix was designed to test about 11 samples for each type of film for aging times up to 42 days. Film samples with 0 days of aging are pristine samples without any aging. Type 4 films were aged to 115 °C and 125 °C. Five samples were removed from the oven at each aging time of 3 days, 1 week, 2 weeks, 3 weeks, 4 weeks, 5 weeks and 6 weeks (42 days) and tested. Note that films with an aging temperature of 125 °C and an aging time of 2 weeks or longer could not be used for the tension test. These films became warped and very brittle after removing from the oven.

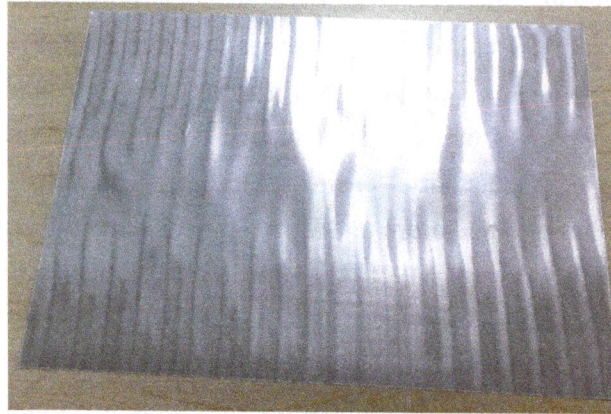

Figure 1. The sample of the multilayered film.

Table 1. Films with various layer thicknesses.

Type	Number of Layers	Film Thickness (μm)	Layer Thickness (nm)
1	64	127 (5 mil *)	1984
2	256	127 (5 mil)	496
3	1,024	127 (5 mil)	124
4	4,096	127 (5 mil)	31

* mil is a thousandth of an inch.

Table 2. Films with various film thicknesses.

Type	Number of Layers	Film Thickness (μm)	Layer Thickness (nm)
5	64	31.75 (1.25 mil)	496
2	256	127 (5 mil)	496
6	1,024	508 (20 mil)	496

Table 3. Test matrix showing sample numbers at each aging time for testing for each sample type at 100 °C aging temperature.

Aging Time (day)	Multi-Layer Polymer PC/PMMA Films					
	Type 1	Type 2	Type 3	Type 4	Type 5	Type 6
0	11	13	11	13	11	11
3	8	8	8	9	8	9
7	8	8	8	8	8	9
14	8	8	8	8	8	8
21	8	8	8	8	8	8
28	8	8	8	8	8	10
35	8	8	8	9	8	8
42	8	7	8	8	8	8

The modulus of elasticity was determined from the early linear portion of the stress-strain curves; tensile strength was determined from the maximum stress of the curves; and the fracture strain was the strain when the films finally broke. Film samples in Table 3 were tested, and the data are reported in the next section.

3. Results and Discussion

3.1. The Effect of the Layer Thickness

For each test in the test matrix, the results were compiled and the average values were compared. The data for the modulus of elasticity *vs.* aging time at 100 °C are summarized in Figure 2. It is observed that aging has an effect on the modulus of elasticity of films. The significance of this effect changes with the layer thickness. The modulus increases with the aging time to reach about 10%, increases at 42 days of aging for the films with a layer thickness of 1,984 nm and 496-nm films, but decreases for the one with a 124-nm layer thickness. The trend is stable for the films with a 31-nm layer thickness. The modulus value of the 31-nm layer film is the highest among the four types of films. This is because as the layer thickness reduced to 31 nm, the molecular chains of PMMA and PC are more aligned.

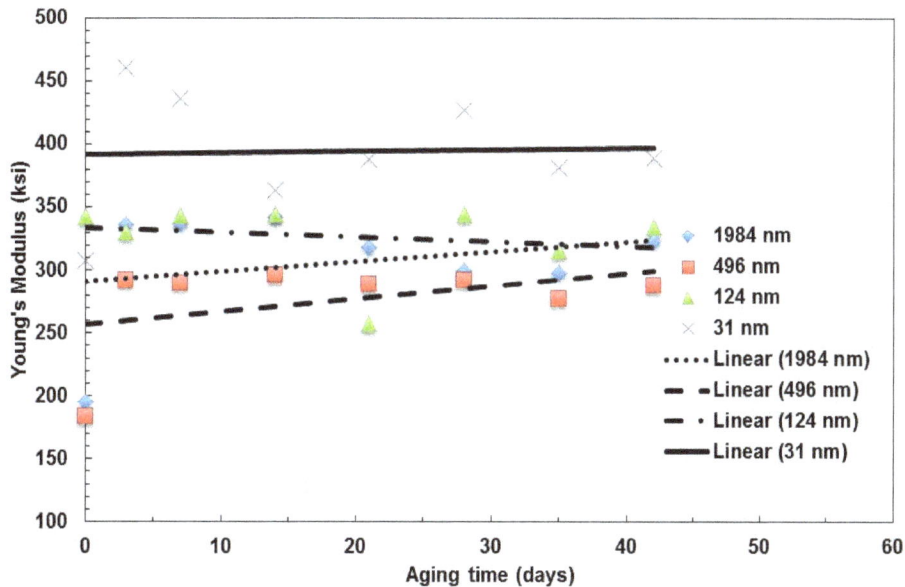

Figure 2. Summary of the change in Young's modulus with aging time for films with various layer thicknesses aged at 100 °C.

The data for the tensile strength *vs.* aging time at 100 °C are summarized in Figure 3. The trend of strength does not change with aging time for the film with a 31-nm layer thickness, but it increases with the aging time for the other three types of films. Similar to the modulus of elasticity, the strength of the 31-nm layer film is the highest among the four types of films.

The data for the fracture strain *vs.* aging time at 100 °C are summarized in Figure 4. The trend of ductility decreases with aging time for all of the films. The ductility was measured by the amount of strain that the material can sustain before it fractures. The thermal aging conditions destroy the covalent bonds in the molecular chains of the polymers. The 31-nm layer film has the lowest ductility among the four types of films. As the film layer reduced to a certain level, the molecular chains of the polymers are aligned and confined in the nanolayer without much ability to be stretched.

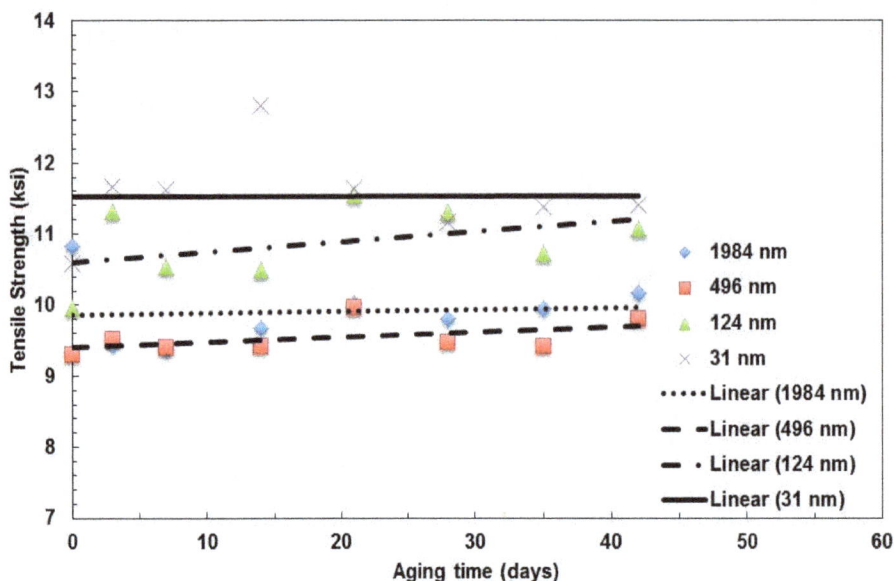

Figure 3. Summary of the change in tensile strength with aging time for films with various layer thicknesses aged at 100 °C.

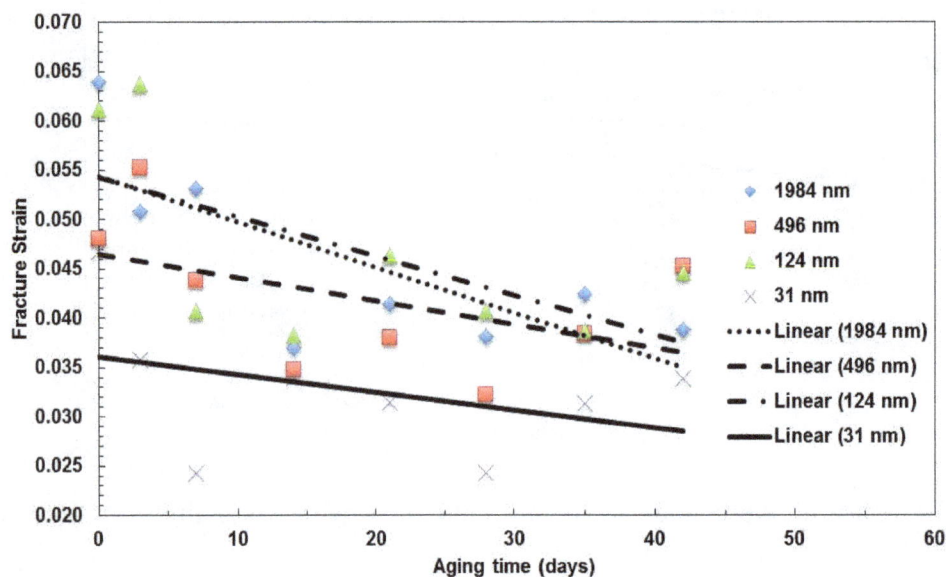

Figure 4. Summary of the change in fracture strain with aging time for films with various layer thicknesses aged at 100 °C.

3.2. The Effect of the Film Thickness

The three types of films with a 496-nm film thickness have been tested to investigate the effect of the film thickness. The moduli of elasticity of films with the same layer thickness, but various film thicknesses, are shown in Figure 5. It seems that for the relatively thick films (5 mil and 20 mil thickness films) the pristine film (zero days of aging) have relatively low moduli, which increases with the aging time. The values become relatively stable after aging over three days. For the relatively thin film (1.25 mil), the modulus slightly decreases with aging time.

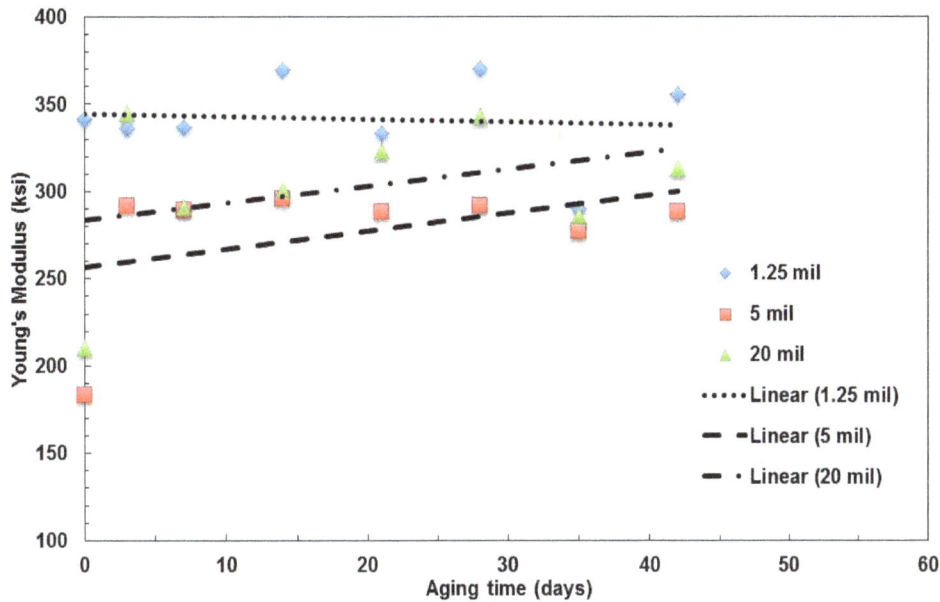

Figure 5. Summary of the change in Young's modulus with aging time for films with various film thicknesses aged at 100 °C.

The strength values of the three types of films are shown in Figure 6. The trend of the thin film (1.25 mil thick) is different from the other two types of films. The strength decreases with aging time, but the other two slightly increase with the aging time.

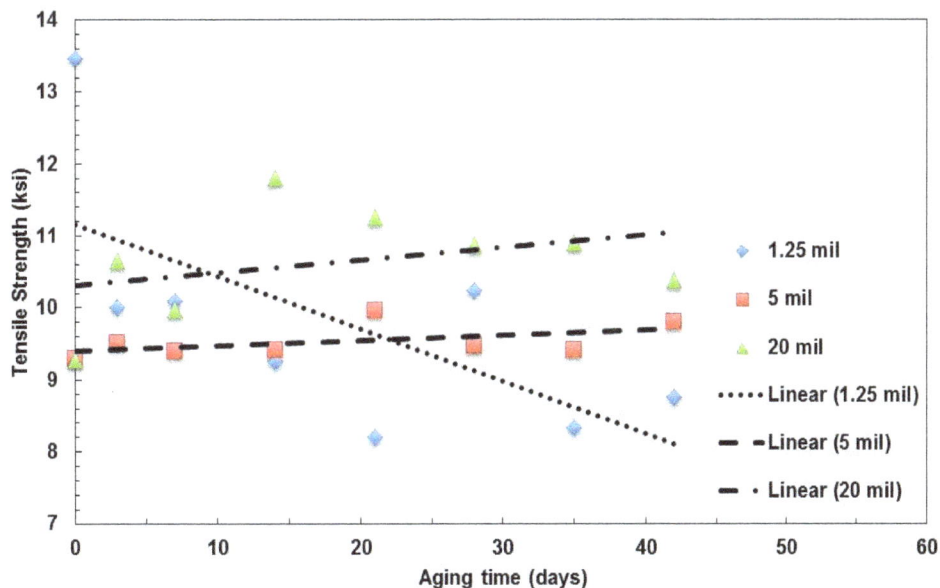

Figure 6. Summary of the change in tensile strength with aging time for films with various film thicknesses aged at 100 °C.

The fracture strains of the films are shown in Figure 7. The ductility of the three films decreasing with aging time shows the degradation of the polymer films.

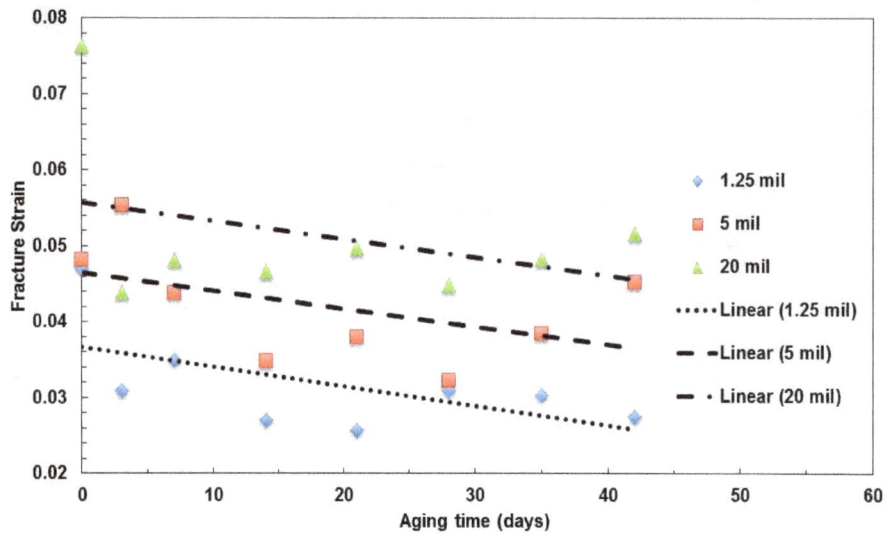

Figure 7. Summary of the change in fracture strain with aging time for films with various film thicknesses aged at 100 °C.

3.3. The Effect of the Aging Temperature

The effect of temperature was studied for the nanolayer film with 4,094 layers of 31-nm thick (Type 4 film in Table 2) at temperatures of 100 °C, 115 °C and 125 °C. While the samples for the aging tests at 100 °C are not from the same batch as the sample for the 115 °C and 125 °C tests, the mechanical properties are not the same for the pristine (original) films. To compare the trend of films from different batches, all of the data were normalized by the pristine material properties. That is, all of the pristine properties are one for zero days without aging. The moduli of elasticity of the aging at three temperatures are shown in Figure 8, from which it can be observed that the trend for the stiffness stays relatively stable over aging time for all three temperatures. The aging at 100 °C has a higher average moduli than the other two.

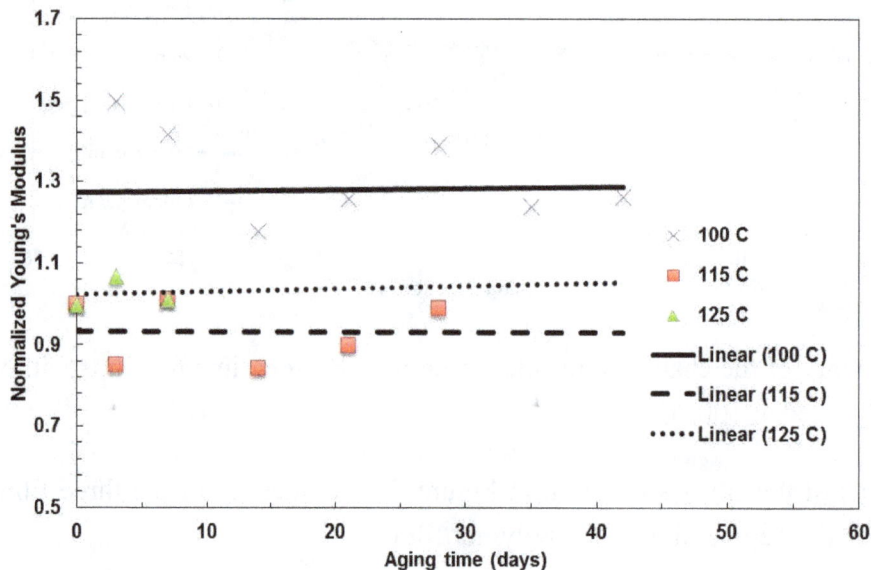

Figure 8. Summary of the change in Young's modulus with aging time for films with a 31-nm layer thickness aged at various temperatures.

The tensile strengths of the films are shown in the Figure 9. While the aging resulted in a noticeable decrease for the 125 °C aging, there is no obvious change for the 100 °C aging for the film. The 115 °C aging cause a slight decrease of the property.

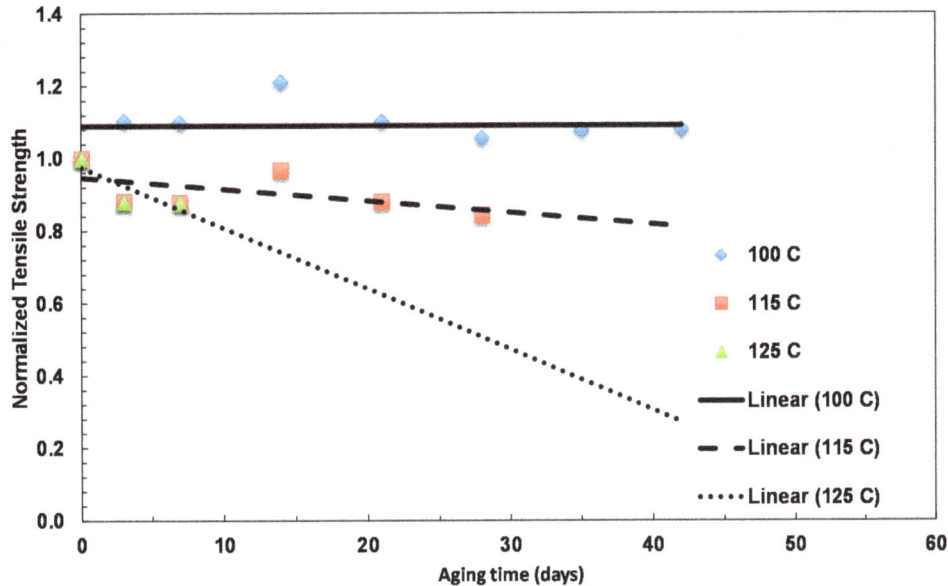

Figure 9. Summary of the change in tensile strength with aging time for films with a 31-nm layer thickness aged at various temperatures.

The fracture strains for all three aging temperatures are shown in Figure 10. As all of the films lost ductility during aging, the higher the temperature and the longer the aging time, the more brittle the films become over aging, which indicated the greater degradation of the nanolayer films.

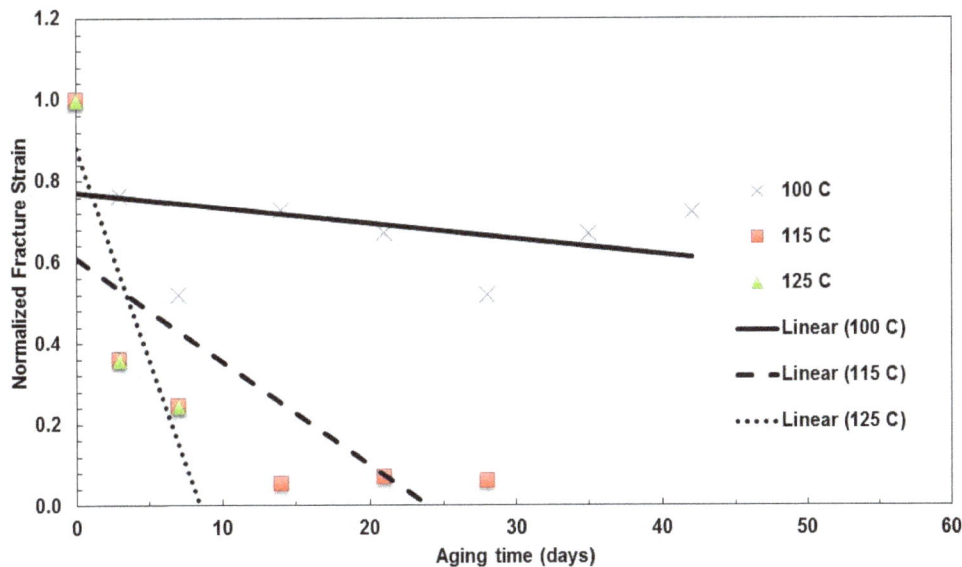

Figure 10. Summary of the change in fracture strain with aging time for films with a 31-nm layer thickness aged at various temperature.

4. Conclusions

In this paper, the effects of layer and film thicknesses and thermal aging on the PC/PMMA micro- and nano-layered films have been studied experimentally. It is observed from the results that the material layered structure has a great effect on the mechanical properties and aging. For the films with the same film thickness, but various layer thicknesses, as the layer thickness goes down to 31 nm, the film has higher stiffness and strength than other films with thicker layers. Meanwhile, the stiffness and strength stay stable with the aging time. As the layer becomes very thin, down to the nanoscale, the confinement from the layer boundaries holds the molecules, makes the molecules more tightly packed and, finally, aligned. The aligned molecules have more inter-molecule van der Waals force, which makes them lock tighter to make the material deteriorate more slowly. As the layer thickness stays the same, the stiffness and strength of thinner films decrease with aging time, but for the relatively thicker films, stiffness and strength increase with the aging time. The aging temperature has a great effect on the strength and ductility for the 31-nm nanolayered film. For all of the films, the thermal aging results in a noticeable reduction in the ductility of the films. The higher temperature results in a high loss of the ductility. The findings of this research work demonstrate the breakthrough of the technology to make nanolayered polymer materials not only increase their stiffness and strength, but also increase their stability and durability. While a major limitation of polymer films is the deterioration of mechanical properties in working environments, this finding would be important for the application of the polymer films.

Acknowledgments

The authors acknowledge the financial support of the National Science Foundation through Grant Number DMR-0423914 and the materials provided by Eric Baer at Case Western Reserve University.

Author Contributions

Ahmed Abdel-Mohti analyzed experimental data and wrote part of the paper. Alison N. Garbash performed all the experiments. Saad Almagahwi compiled experimental data. Hui Shen designed the experiment procedures and wrote part of the paper. Ahmed Abdel-Mohti and Hui Shen read and approved final version of manuscript to be submitted.

Appendix

In this Appendix, testing data for some tests are shown to illustrate the data processing procedures. The data for all of the plots in Figures 2–10 are shown in the Tables A1–A9 for reference.

The static tensile tests at room temperature were done for all of the pristine and thermally-aged PC/PMMA multi-layered films. A tensile curve of a Type 4 film sample (31 nm thick, 4,096 layers) is shown in Figure A1. This curve just serves as an example of all of the tests. The values for the modulus of elasticity, tensile strength and fracture strain of each film are obtained from the tensile curves and compiled.

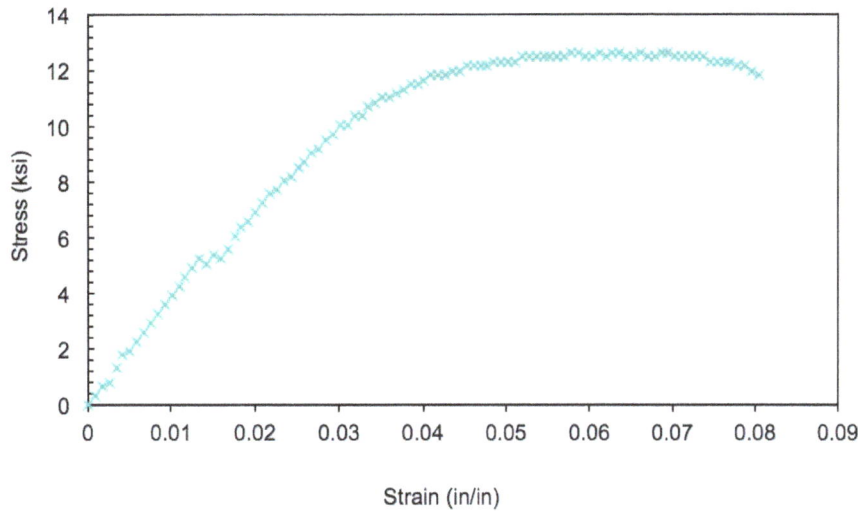

Figure A1. Stress-strain curve.

Table A1 shows the data values for the modulus of elasticity, tensile strength and fracture strain for all types of films at various aging times and temperatures. These are the data for Figures 1–9.

Table A1. The modulus of elasticity (unit: ksi) *vs.* aging time at a 100 °C aging temperature for film Types 1–4, which have the same film thickness (5 mil), but various layer thicknesses.

Aging Time (days)	0	3	7	14	21	28	35	42
Film Type 1	195	336	337	342	318	300	297	323
Film Type 2	184	292	289	296	289	292	277	288
Film Type 3	343	330	344	344	258	345	316	334
Film Type 4	308	460	436	363	388	427	382	389

Table A2. Tensile strength (unit: ksi) *vs.* aging time at a 100 °C aging temperature for film Types 1–4, which have the same film thickness (5 mil), but various layer thicknesses.

Aging Time (days)	0	3	7	14	21	28	35	42
Film Type 1	10.83	9.45	9.37	9.68	10.02	9.81	9.95	10.17
Film Type 2	9.30	9.51	9.41	9.42	9.97	9.47	9.42	9.81
Film Type 3	9.96	11.32	10.53	10.50	11.55	11.31	10.73	11.08
Film Type 4	10.58	11.66	11.62	12.80	11.64	11.16	11.38	11.41

Table A3. Fracture strain *vs.* aging time at a 100 °C aging temperature for film Types 1–4, which have the same film thickness (5 mil), but various layer thicknesses.

Aging Time (days)	0	3	7	14	21	28	35	42
Film Type 2	0.064	0.051	0.053	0.037	0.041	0.038	0.042	0.039
Film Type 3	0.048	0.055	0.044	0.035	0.038	0.032	0.038	0.045
Film Type 4	0.061	0.064	0.041	0.038	0.046	0.041	0.039	0.045
Film Type 6	0.047	0.036	0.024	0.034	0.031	0.024	0.031	0.034

Table A4. modulus of elasticity (unit: ksi) *vs.* aging time at a 100 °C aging temperature for films with the same layer thickness, but various film thicknesses.

Aging Time (days)	0	3	7	14	21	28	35	42
Film Type 5	341	336	337	370	333	370	290	356
Film Type 2	184	292	289	296	289	292	277	288
Film Type 6	210	345	291	301	324	343	286	314

Table A5. Tensile strength (unit: ksi) *vs.* aging time at a 100 °C aging temperature for films with the same layer thickness, but various film thickness.

Aging Time (days)	0	3	7	14	21	28	35	42
Film Type 5	13.47	10.01	10.09	9.26	8.20	10.24	8.33	8.75
Film Type 2	9.30	9.51	9.41	9.42	9.97	9.47	9.42	9.81
Film Type 6	9.28	10.66	9.98	11.81	11.26	10.87	10.90	10.38

Table A6. Fracture strain *vs.* aging time at a 100 °C aging temperature for films with the same layer thickness, but various film thicknesses.

Aging Time (days)	0	3	7	14	21	28	35	42
Film Type 5	0.047	0.031	0.035	0.027	0.026	0.031	0.030	0.028
Film Type 2	0.048	0.055	0.044	0.035	0.038	0.032	0.038	0.045
Film Type 6	0.076	0.044	0.048	0.047	0.050	0.045	0.048	0.052

Table A7. Modulus of elasticity of the nanolayered film with 4,094 layers of 31 nm thick (Type 4 film in Table 2) aged at temperatures of 100 °C, 115 °C and 125 °C. The data were normalized by the pristine material properties.

Aging Time (days)	0	3	7	14	21	28	35	42
100 °C	1.00	1.50	1.42	1.18	1.26	1.39	1.24	1.26
115 °C	1.00	0.85	1.01	0.84	0.90	0.99		
125 °C	1.00	1.07	1.01					

Table A8. Tensile strength of the nanolayered film with 4,094 layers of 31-nm thick (Type 4 film in Table 2) aged at temperatures of 100 °C, 115 °C and 125 °C. The data were normalized by the pristine material properties.

Aging Time (days)	0	3	7	14	21	28	35	42
100 °C	1.00	1.10	1.10	1.21	1.10	1.05	1.08	1.08
115 °C	1.00	0.88	0.88	0.97	0.88	0.84		
125 °C	1.00	0.88	0.88					

Table A9. Fracture strain of the nanolayered film with 4,094 layers of 31 nm thick (Type 4 film in Table 2) aged at temperatures of 100 °C, 115 °C and 125 °C. The data were normalized by the pristine material properties.

Aging Time (days)	0	3	7	14	21	28	35	42
100 °C	1	0.76	0.52	0.73	0.67	0.52	0.67	0.72
115 °C	1	0.36	0.24	0.05	0.07	0.06		
125 °C	1	0.36	0.24					

Conflicts of Interest

The authors declare no conflict of interest.

References

1. Ponting, M.; Hiltner, A.; Baer, E. Polymer nanostructures by forced assembly: Process, structure, and properties. *Macromol. Symp.* **2010**, *294*, 19–32.
2. Schrenk, W.J.; Chisholm, D.S.; Cleereman, K.J.; Alfrey, T., Jr. Method of Preparing Multilayer Plastic Articles. U.S. Patent 3,565,985, 23 February 1971.
3. Alfrey, T.; Schrenk, W. Highly Reflective Thermoplastic Bodies for Infrared, Visible or Ultraviolet Light. U.S. Patent 3,711,176, 16 January 1973.
4. Schrenk, W.J. Apparatus for Multilayer Coextrusion of Sheet or film. U.S. Patent 3,884,606, 20 May 1975.
5. Schrenk, W.J.; Shastri, R.K.; Ayres, R.F.; Gosen, D.J. Interfacial Surface Generator. U.S. Patent 5,094,788, 10 March 1992.
6. Ramanathan, R.; Schrenk, W.J.; Wheatley, J.A. Coextrusion of Multilayer Articles Using Protective Boundary Layers and Apparatus Therefor. U.S. Patent 5,269,995, 14 December 1993.
7. Gregory, B.; Hiltner, A.; Baer, E.; Im, J. Dynamic mechanical behavior of continuous multilayer composites. *Polym. Eng. Sci.* **1987**, *27*, 568–572.
8. Gregory, B.; Siegmann, A.; Im, J.; Hiltner, A.; Baer, E. Deformation behavior of coextruded multilayer composites with polycarbonate and poly(styrene-acrylonitrile). *J. Mater. Sci.* **1987**, *22*, 532–538.
9. Im, J.; Baer, E.; Hiltner, A. *High Performance Polymers*; Baer, E., Moet, A., Eds.; Hanser Publishers: New York, NY, USA, 1991; p. 175.
10. Cheng, W.; Gomopoulos, N.; Fytas, G.; Gorishnyy, T.; Walish, J.; Thomas, E.L.; Hiltner, A.; Baer, E. Phonon dispersion and nanomechanical properties of periodic 1D multilayer polymer films. *Nanoletters* **2008**, *8*, 1423–1428.
11. Wang, H.; Keum, J.K.; Hiltner, A.; Baer, E.; Freeman, B.; Rozanski, A.; Galeski, A. Confined crystallization of polyethylene oxide in nanolayer assemblies. *Science* **2009**, *323*, 757–760.
12. Guillorya, P.; Deschainesa, T.; Hensona, P. Analysis of multi-layer polymer films. *Materialstoday* **2009**, *12*, 38–39.
13. Ania, F.; Baltá-Calleja, F.J.; Henning, S.; Khariwala, D.; Hiltner, A.; Baer, E. Study of the multilayered nanostructure and thermal stability of PMMA/PS amorphous films. *Polymer* **2010**, *51*, 1805–1811.
14. Ania, F.; Puente, O.I.; Baltá-Calleja, F.J.; Roth, S.; Khariwala, D.; Hiltner, A.; Baer, E.; Roth, S.V. Ultra-small-angle X-ray scattering study of PET/PC nanolayers and comparison to AFM results. *Macromol. Chem. Phys.* **2008**, *209*, 1367–1373.
15. Baer, E.; Hiltner, A.; Shirk, J.S. Multilayer Polymer Gradient Index (GRIN) Lenses. U.S. Patent 7,002,754, 21 February 2006.

16. Gupta, M.; Lin, Y.; Deans, T.; Baer, E.; Hiltner, A.; Schiraldi, D.A. Structure and gas barrier properties of poly (propylene-graft-maleicanhydride)/phosphate glass composites prepared by microlayer coextrusion. *Macromolecules* **2010**, *43*, 4230–4239.

17. Song, H.; Singer, K.; Wu, Y.; Zhou, J.; Lott, J.; Andrews, J.; Hiltner, A.; Baer, E.; Weder, C.; Bunch, R.; *et al.* Layered polymeric optical systems using continuous coextrusion. In Proceedings of the SPIE7467 Nanophotonics and Macrophotonics for Space Environments III, San Diego, CA, USA, 2 August 2009.

18. Beadie, G.; Shirk, J.S.; Rosenberg, A.; Lane, P.A.; Fleet, E.; Kamdar, A.R.; Jin, Y.; Ponting, M.; Kazmierczak, T.; Yang, Y.; *et al.* Optical properties of a bio-inspired gradient refractive index polymer lens. *Opt. Express* **2008**, *16*, 11540–11547.

19. Ebina, T.; Mizukami, F. Flexible transparent clay films with heat-resistant and high gas-barrier properties. *Adv. Mater.* **2007**, *19*, 2450–2453.

20. Tetsuka, H.; Ebina, T.; Tsunoda, T.; Nanjo, H.; Mizukami, F. Highly transparent flexible clay films modified with organic polymer: Structural characterization and intercalation properties. *J. Mater. Chem.* **2007**, *17*, 3545–3550.

21. Podsiadlo, P.; Kaushik, A.K.; Arruda, E.M.; Waas, A.M.; Shim, B.S.; Xu, J.; Nandivada, H.; Pumplin, B.G.; Lahann, J.; Ramamoorthy, A.; *et al.* Ultrastrong and stiff layered polymer nanocomposites. *Science* **2007**, *318*, 80–83.

22. Bonderer, L.J.; Studart, A.R.; Gauckler, L.J. Bioinspired design and assembly of platelet reinforced polymer films. *Science* **2008**, *319*, 1069–1073.

23. Shen, H.; Gannon, N.D. Analysis of the mechanical properties of pristine and thermally aged PC/PMMA micro and nanolayered films. In Proceedings of the ASME 2010 International Mechanical Engineering Congress & Exposition IMECE2010, Vancouver, BC, Canada, 12–18 November 2010.

Glycopolymeric Materials for Advanced Applications

Alexandra Muñoz-Bonilla [1,*] and Marta Fernández-García [2,*]

[1] Departamento de Química Física Aplicada, Facultad de Ciencias, Universidad Autónoma de Madrid (UAM), C/ Francisco Tomás y Valiente 7, Cantoblanco, 28049 Madrid, Spain

[2] Instituto de Ciencia y Tecnología de Polímeros (ICTP-CSIC), C/ Juan de la Cierva 3, 28006 Madrid, Spain

* Authors to whom correspondence should be addressed;
E-Mails: alexandra.munnoz@uam.es (A.M.-B.); martafg@ictp.csic.es (M.F.-G.);

Academic Editor: Klara Hernadi

Abstract: In recent years, glycopolymers have particularly revolutionized the world of macromolecular chemistry and materials in general. Nevertheless, it has been in this century when scientists realize that these materials present great versatility in biosensing, biorecognition, and biomedicine among other areas. This article highlights most relevant glycopolymeric materials, considering that they are only a small example of the research done in this emerging field. The examples described here are selected on the base of novelty, innovation and implementation of glycopolymeric materials. In addition, the future perspectives of this topic will be commented on.

Keywords: glycopolymers; biorecognition; anti-adhesion therapy; bioactive delivery; vaccines; sensors; imaging; diagnostics

1. Introduction

It is well known that carbohydrates located in the cell membrane as glycoproteins, glycolipids or glycans play a crucial role in many physiological as well as pathological events, including cellular proliferation and cancer metastasis, intercellular communication, recognition processes, adhesion, *etc.* This is mainly as a result of specific interactions of the carbohydrates with protein receptors such as enzymes or lectins that triggering the above mentioned biological functions. This molecular recognition ability is greatly enhanced by multivalency or the so-called glycocluster effect [1,2], commonly found in natural glycoproteins or glycolipids clusters that strongly recognize sugar-binding proteins.

Numerous investigations have been focused on the preparation of synthetic glycopolymers with pendant sugar moieties along the polymer backbone or at the end of the chain, which are able to interact with lectins through multivalent interactions, mimicking natural biomolecules. Especially interesting are the controlled polymerization techniques and efficient chemical reactions such as click chemistry to tune the architecture of the glycopolymers and optimize the recognition process [3–5]. The design of the glycopolymer is essential, because the strength of the binding strongly depends on the type of sugars, anomeric status, and linkage position and the linker that connects the carbohydrate to the polymer backbone, and also on the density of sugars, degree of polymerization and branching [6]. For example, by increasing the polymer length, then the valence, polymers can access to multiple binding sites in proteins, thereby increasing their affinity. Nevertheless, when all accessible binding sites are occupied, further increase in polymer length will not yield enhancements in their interaction. Moreover, more flexible backbones and linkers allow polymer to adopt a conformation/orientation that leads to effective interactions. It is important to mention that the interaction between glycopolymer and lectin is exclusive for each binomial team; that is, the best polymer structure for a particular lectin is not necessarily the optimum structure for other lectins, even those with the same carbohydrate specificity.

Synthetic polymeric techniques such as reversible addition-fragmentation chain transfer polymerization (RAFT) [7] and atom transfer radical polymerization (ATRP) [8] allow the development of well-defined glycopolymers of a variety of composition and topologies, including homopolymers, statistical and block copolymers.

In the last years, efforts were mainly dedicated to the synthesis of new glycopolymer structures and their use as biomimetic model to fundamentally investigate the specific carbohydrate-protein interactions. Nowadays, besides the synthetic development, glycopolymeric nanostructures, both polymer glyconanoparticles and hybrid nanoparticles are receiving more and more attention. Nanoparticles due to their size and high surface area have shown great potential in different fields, especially in biomedicine. Recently, research has demonstrated that the size and shape of the nanoparticles significantly influence their interactions with cells. A variety of approaches, such as grafting from or grafting on, have been used to prepare glyconanoparticles with different compositions, sizes and shapes to study and understand the interactions between particles and proteins and optimize their applicability [9].

Based on the extensive investigations of the biological activity of the glycopolymers and how their structure affects the binding with lectins the focus of research has moved more to the pure applications of glycopolymers, which also contribute to amplify the knowledge in the recognition process. Lately, advancements in the applications of glycopolymers have shown an explosion, with the consolidation and advance of several uses, such as target delivery systems and also emerging new applications. This article will examine the most common and recent applications of glycopolymers particularly focused in biomedical and biological uses.

2. Biomimetic Model to Investigate Carbohydrate-Protein Interactions

As mentioned above, glycopolymers can strongly interact with specific proteins, lectins, by the multivalent glycocluster effect. There are many fundamental examples that investigate glycopolymer-lectin recognition [10–23], especially with Concanavalin A, Con A, and *Ricinus Communis* Agglutinin 120, RCA_{120}, mainly by UV-Vis or fluorescence spectroscopies; in the latter when either the lectin or the glycopolymer are fluorescence-labeled. It is well-known that Con A, which is a tetramer with four carbohydrate binding sites, specifically binds to glycopolymers containing mannosyl and glucosyl residues while RCA_{120} lectin interacts to those containing galactosyl residues. These residues have to be present in a specific form, otherwise there will be no interactions. This is the case of gluconolactone derivatives, where the glucose moieties are in an open-form and the interactions are not observed either with RCA_{120} or with Con A [24,25].

Moreover, the binding abilities of glycopolymers have been also measured by microarray and surface plasmon resonance (SPR) techniques [26–31]. Microarray technique of biomolecules, also called biochip, is a collection of microscopic spots of these molecules attached to a solid surface [32,33]. Meanwhile, surface plasmon resonance sensing is also a powerful and quantitative probe of the interactions of a variety of biopolymers with various ligands. It provides a means not only for identifying these interactions and quantifying their equilibrium constants, kinetic constants and underlying energetics, but also for employing them in very sensitive biochemical assays [34,35].

In this context, the glycopolymers can be efficiently attached to surfaces, where they approximate physiological cell–cell and cell–extracellular matrix interactions and retain the ability to engage proteins. This is the case of chondroitin sulfate glycomimetic polymers that present sulfated disaccharide units, which mimic the multivalent architecture of native glycosaminoglycans chains [27]. Glycopolymers were attached to microarray surfaces with a high-precision contact-printing robot delivering nanoliter volumes of the biotin-labeled glycopolymers to streptavidin-coated slides, yielding spots of approximately 200 mm in diameter (see Figure 1).

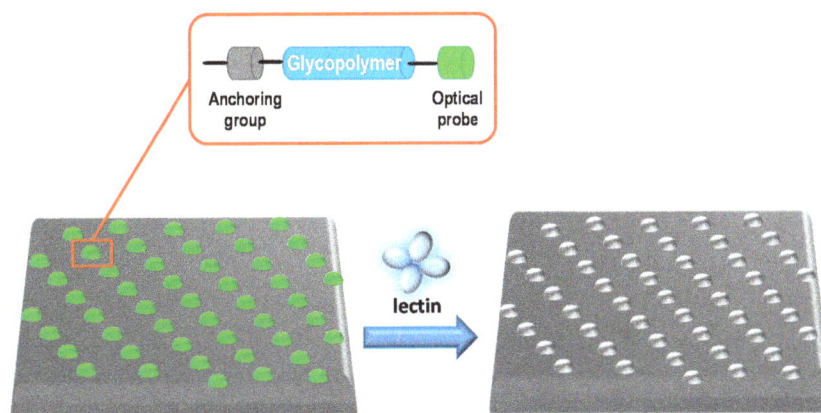

Figure 1. Schematic representation of glycopolymers anchored on chips and their interaction.

Bertozzi group has used the microcontact printing (mCP) technique to covalently array the glycopolymers on azide-functionalized biochips in well-defined patterns using a poly(dimethylsiloxane) stamp with approximately 2 mm circular structures [31]. They tested the binding effectiveness of *Helix Pomatia* Agglutinin (HPA) lectin, proving the specific interaction.

Narain group has studied the recognition ability of glyconanoparticles with immobilize lectins, *i.e.*, *N*-acetyl β-D-glucosaminoside decorated gold nanoparticles, which specifically interact with wheat germ agglutinin (WGA) lectin, whereas α-D-mannoside derivatives specifically interact with Con A lectin. The glycopolymer structure gives the capability of lectin interaction, but can also infer other properties. This is the case of glycopolymers obtained by RAFT process, where the polymer is conjugated with tripeptide reduced glutathione (GSH) by thiol-disulfide exchange reaction [36]. The resulting disulfide-linked GSH bioconjugate polymer (Figure 2) possesses antioxidant character, with prolonged radical scavenging activity. This is because of GSH, which can protect cells from oxidative stress through scavenging singlet oxygen, hydroxyl radicals, and superoxide radicals. Moreover, the specific recognition with Con A lectin is enhanced about three times when it is conjugated with a GSH, other than on its own.

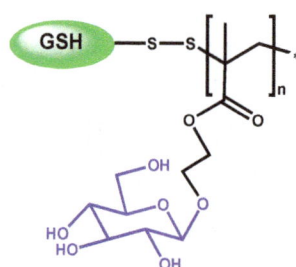

Figure 2. Tripeptide reduced glutathione-bioconjugate glycopolymer structure.

Stimuli-responsive structures enlarge the properties of glycopolymers and make the binding properties more versatile. For example, thermoreversible glycopolymers used to control lectin interaction or bacterial aggregation [37,38], where the binding between glycopolymer and lectin or bacteria is regulated by the polymer thermoreversibility. In the first case, the random glycopolymer based on *N*-isopropyl acrylamide (NIPAAm) and *N*-acryloyl glucosamine (AGA) is grafted from the honeycomb surface. Below the lower critical solution temperature (LCST) of the surface, the conjugation is switched off, while above the LCST the surface grafted glucose moieties bind strongly to Con A (Figure 3). Contrary effect is observed in the second example in which the block glycopolymer based also on NIPAAm is in solution. In this situation, when temperature is below its LCST, the chains are extended and display glucose moieties in solutions, then it is able to bind to the cell of *Escherichia coli* bacteria, whereas above LCST, the glycopolymer is presented in a globular state, which hides the carbohydrate moieties (Figure 4).

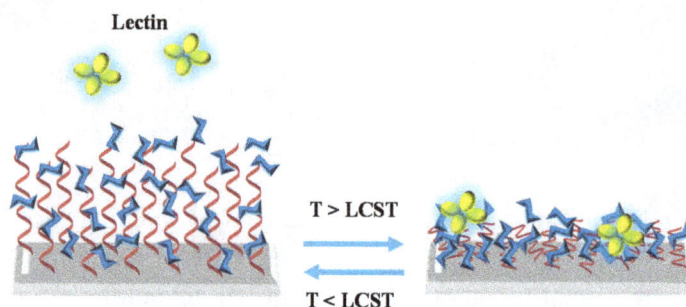

Figure 3. Representation of interaction between random thermoreversible glycopolymers attached to surface and lectin.

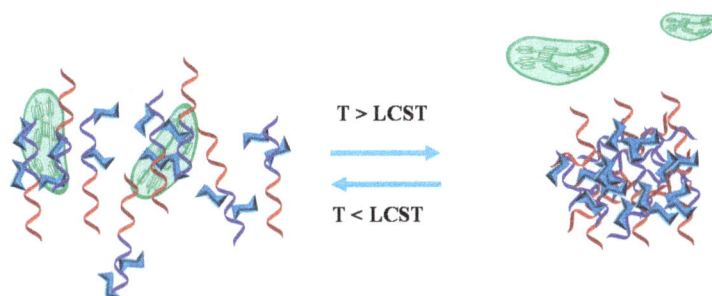

Figure 4. Representation of interaction between block thermoreversible glycopolymers and bacteria in water solution.

3. Anti-Adhesion Therapy

Currently, one of the most attracting and emerging applications of glycopolymers is anti-adhesion therapy in which glycopolymers behave as antagonists to interfere with the binding of lectins from a pathogen to the host organism and, therefore, prevent infection from occurring. Glycopolymers have been designed to act against infections by different AB₅ toxins, including the cholera toxin (CT), Shiga toxins, and the heat-labile enterotoxin (LT). Different degrees of carbohydrate incorporation have been obtained by post-modification of poly(L-glutamic acid) (PGA) backbone with *N*-(ε-aminocaproyl)-β-D-galactosylamine over the β-D-galactosylamine. The results indicate that higher accessibility of galactose in *N*-(ε-aminocaproyl)-β-D-galactosylamine improves the inhibition of binding [39,40].

Moreover, glycopolymers bearing globotriose or lactose and acrylamide have been used to evaluate the inhibitory effects on cytotoxicity of Shiga toxins (Stxs; Stx1 and Stx2). The results showed that glycopolymers having globotriose units exhibited inhibitory effect on cytotoxicity of Stx1, while glycopolymers having small amount of acrylamide exhibited inhibitory effect on cytotoxicity of Stx2. Also, the glycopolymer, having lactose, was able to bind Stx1, showing inhibitory effect on its cytotoxicity [41]. The potential of glycopolymer containing mannose to interact with human DC-SIGN, thereby inhibiting the interactions between DC-SIGN and the HIV envelope glycoprotein gp120 was also reported [42]. They prepared a library of glycopolymers and demonstrated that simple structures effectively prevent infections. This study has a considerably importance because the DC-SIGN is also implicated in the action mechanism of other viruses including Ebola and hepatitis C.

Following the same strategy, different glycopolymers were synthetized by post-modification to inhibit bacterial toxins such as cholera [43–45]. They proved the importance of the glycopolymer structure in terms of linker length and branching, carbohydrate density and molecular weight on the inhibition of the bacterial toxins. Specifically, longer linkers increase inhibition of the B subunit of cholera toxin, which is attributed to the depth of the binding pocket. Moreover, Gibson group also demonstrated that branched side chains into the linker in galacto-terminal polymers increase binding affinity to their corresponding lectins, relative to simple monosaccharides.

4. Bioactive Delivery Systems

Drug delivery systems are one of the main applications of glycopolymers, since, theoretically, they do not produce undesirable side effects but can easily reach their targets and increase the solubility of the system. Most of the glycopolymer systems used in drug delivery are nanostructures, in particular micellar systems adequate for oral and intravenous delivery route of poorly water-soluble drugs (Figure 5).

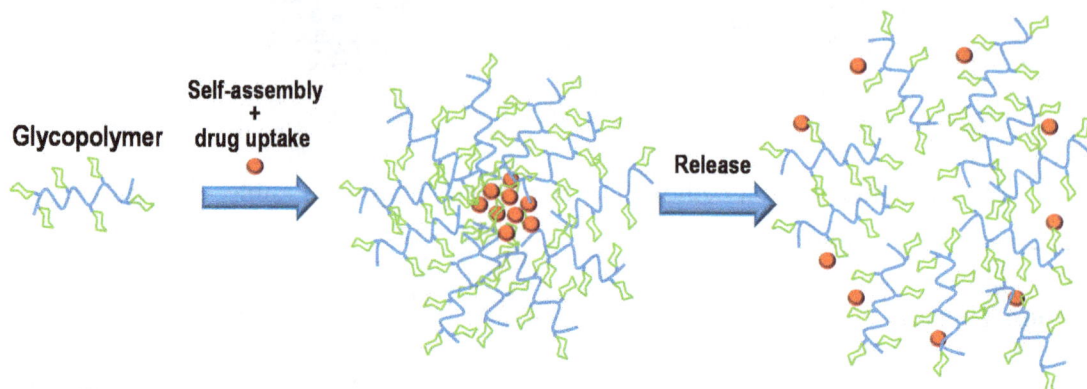

Figure 5. Scheme of glycopolymer self-assembly along with bioactive drug uptake and posterior release of the compound.

Especially interesting are stimuli responsive systems that allow smart strategies of delivery. For instance block glycocopolymer based on poly(azobenzene methacrylate) (PMAzo) and poly(3-O-4-vinylbenzoyl-D-glucopyranose) (PBG) was reported (Figure 6). Although this glycopolymer has not been used for bioactive delivery, they are able to encapsulate, release, and re-encapsulate water-insoluble Nile Red as model compound [46]. PMAzo-*b*-PBG glycopolymer self-assembles into spherical aggregates in water but UV irradiation destabilizes the aggregates as a result of the *trans–cis* isomerization of the MAzo units. The dissociated aggregates reunite to form tubular aggregates when the solution is irradiated at 450 nm.

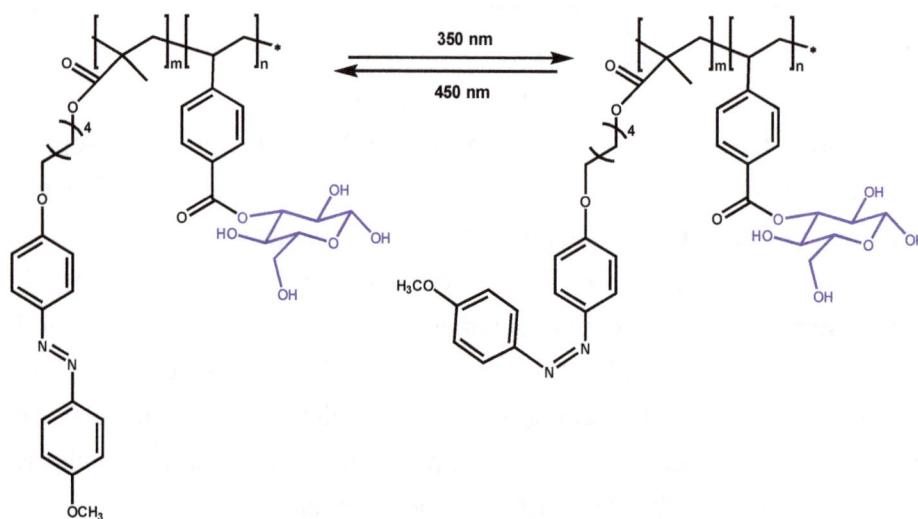

Figure 6. Light-induced isomerization of block copolymers poly(azobenzene methacrylate)-*b*-poly(3-O-4-vinylbenzoyl-D-glucopyranose), under alternate ultraviolet and visible irradiation for encapsulate-release process.

Other glycopolymers have been used to form micelles through their self-aggregation to encapsulate gold nanoparticles [47]. In detail, maltoheptaose-*b*-polystyrene (MH-*b*-PS) can incorporate PS surface-functionalized gold nanoparticles. In this article, the influence of the preparation method of micelles on the particle size is described. Nanoprecipitation of the block copolymer solution, where the block copolymer molecules exist as well-swollen single chains into a large amount of selective solvent for carbohydrate segment, produces micelle nanoparticles of ~30 nm while the reverse procedure produces random aggregations and formed large micelles of ~80 nm.

Suriano *et al.* [48] have synthetized amphiphilic block glycopolymers containing D-glucose, D-galactose and D-mannose by metal-free organocatalyzed ring-opening polymerization. These glycopolymers were able to form micelles with particle sizes of 100 nm and low polydispersities, which are not toxic to asialoglycoprotein receptors (ASGP-R), positive HepG2 (human liver hepatocellular carcinoma) cells or ASGP-R negative HEK293 cells. These authors demonstrated the selectivity of galactose-containing micelles that deliver doxorubicin (DOX) more efficiently into HepG2 cells than in HEK293 cells. Moreover, the cytotoxicity of DOX against HepG2 cells was significantly increased when delivered using the galactose-containing micelles as compared to the free DOX formulation and the glucose-containing micelles. It is important to mention that the regioisomerism is crucial for binding properties. Glycopolymers carrying different constitutional isomers of the pendent sugar species, *i.e.* the glycosidic linkage of the galactose unit from the anomeric position, position 1, and position 6 shows different interaction. The first has strong binding to lectins of peanut (*Arachis hypogea*) agglutinin (PNA) and *Erythrina cristagalli* agglutinin (ECA), while the later does not. More significantly, they show binding behavior similar to the ASGP-R, but different internalization pathways in the HepG2 cell after ASGP-R mediated endocytosis [49].

Glycopolymers can also be used as nasal drug delivery systems for delivery of proteins and peptides, since they can avoid degradation in gastrointestinal tract and metabolism by liver enzymes and are non-toxic. However, nasal delivery has to overcome the self-defense mechanisms, such as mucocillary clearance, ciliary beat frequency and inflammation, and enzymatic barrier, which affect the permeability of drugs through the nasal mucosa. The boronic acid and its derivatives are known to possess the ability to reversibly interact with diols, sugars, and glycoproteins, and transport saccharide across lipid bilayers. However, the main drawback of these materials is the cytotoxicity. In this sense, glycopolymers have been proposed to increase the biocompatibility and hydrophilicity of these compounds. Jin *et al.* [50] described the synthesis of amphiphilic copolymers based on 3-acrylamidophenylboronic acid (AAPBA) with maleimide-glucosamine (MAGA) at different ratios (Figure 7).

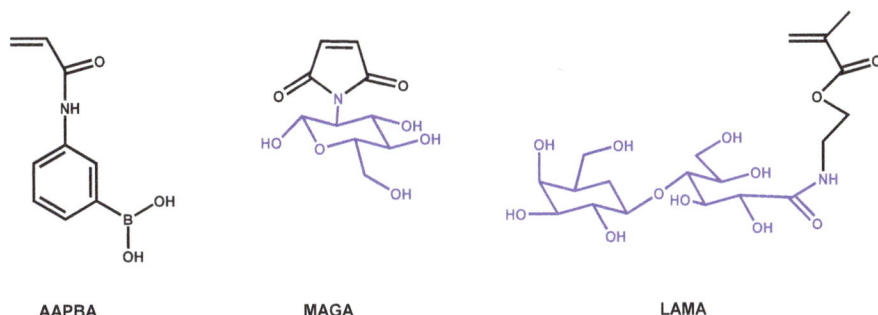

Figure 7. Different monomers used for copolymerization.

They demonstrated the biocompatibility and, therefore, the copolymers are innocuous and do not have impact on cell proliferation against NIH 3T3 fibroblast cell line. In addition, insulin was loaded into the glycopolymer particles due to both hydrophilic and hydrophobic interactions, with loading capacity and encapsulation efficiency of 9–11 and 60%–80% approximately, depending on the composition. Afterwards, this group [51] also synthesized a random amphiphilic glycopolymer of AAPBA with 2-lactobionamidoethyl methacrylate (LAMA). [51] The self-assembly of this copolymer was able to highly load insulin (loading capacity and encapsulation efficiency of 12% and 90%, respectively). In addition, they have proved that after nasal administration to diabetic rats, it can significantly decrease the glucose levels.

Stenzel group has synthesized homo and block copolymers with glycomonomer bearing protected thiosugar units. These systems can then be converted by post-modification into polymeric gold(I) complexes (Figure 8). The formed micelles were tested measuring the proliferation OVCAR-3 human ovarian carcinoma cells. [52]. The activity of the block glycopolymer complex is comparable to deacetylated auranofin or auranofin, which are potent antitumor drug.

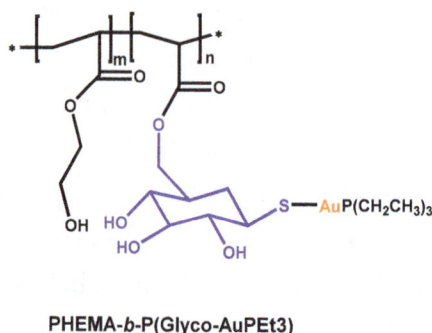

PHEMA-*b*-P(Glyco-AuPEt3)

Figure 8. Block copolymer complex structure.

Likewise, Narain group also synthesized statistical glycopolymers of 3-gluconamidopropyl methacrylamide (GAPMA) (Figure 9) with 3-aminopropyl methacrylamide (APMA). These were subsequently modified with gold (I) phosphine to form the polymeric complexes [53]. The copolymer with lower molecular weight showed higher activity toward different cell lines (Hep G2, HEK 293T, MCF-7, and human dermal fibroblasts), especially toward MCF-7 cells. This polymeric complex presented higher accumulation and cytotoxicity in cancer cells under hypoxic conditions, which were used to mimic the therapy-resistant conditions in cancer diseases, in comparison to the normoxic conditions. Posterior, this group also synthesized statistical glycopolymers of 2-gluconamidoethyl methacrylamide (GAEMA), 2-lactobionamidoethyl methacrylamide (LAEMA) (see also Figure 9), and glycidyl methacrylamide (GMA) with different molecular weights [54]. These glycopolymers are decorated with dithiocarbamate to prepare conjugated glyconanoparticles with the anticancer drug, gold (I) triphenylphosphine, which were then used for targeted delivery to ASGP-R overexpressing HepG2 cells.

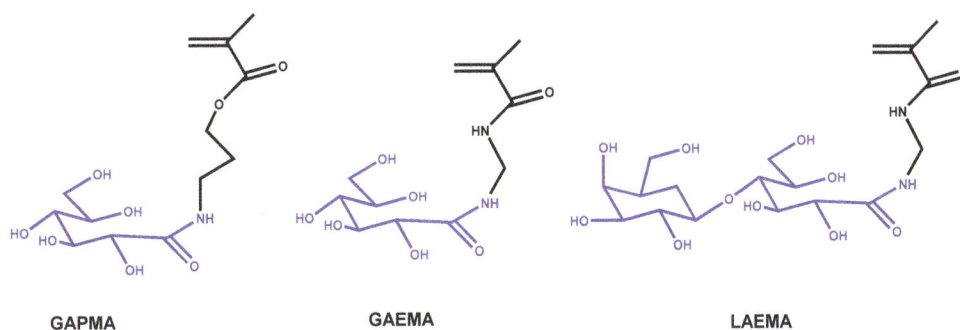

Figure 9. Different glycomonomer structures.

Fleming *et al.* [55] prepared a carbohydrate-antioxidant hybrid polymer using a different approach in order to deliver vitamin E, α-tocopherol, to porcine spermatozoa. The results demonstrated the interaction between the polymer and the galactose-binding protein presented on the mammalian spermatozoa, followed by the endocytosis and release of antioxidant. Consequently, an increase of fertilization rates by a diminishment of the degradation process is produced.

Series of diblock glycopolycations of 2-deoxy-2-methacrylamido glucopyranose (MAG), *N*-[3-(*N*,*N*-dimethylamino) propyl] methacrylamide (DMAPMA), and *N*-(2-aminoethyl) methacrylamide (AEMA) have been developed by aqueous RAFT [56]. These glycopolymers are all able to complex plasmid DNA into polyplex structures and to prevent colloidal aggregation of polyplexes in physiological salt conditions. Moreover, they are able to transfect in HeLa (human cervix adenocarcinoma) cells and HepG2 cells, being much higher in those with AEMA primary amine blocks and in HepG2 cells than in HeLa cells. Therefore, the glyco-block benefits the hepatocyte transfection and DNA delivery.

A series of cyclodextrin-based mannose and fucose glycoconjugates, including glycoclusters, star glycopolymers and star diblock glycopolymers, have also been synthesized via combination of copper(I)-catalyzed Azide-Alkyne Cycloaddition (CuAAC) Huisgen coupling and ATRP. These glycoconjugates showed high affinity binding to the human transmembrane lectin DC-SIGN and act as inhibitors to prevent the binding of human immunodeficiency virus (HIV) envelope protein gp120 to DC-SIGN at nanomolar concentrations as demonstrated before. In addition, the star diblock glycopolymers showed high loading capacity of hydrophobic 1,4-dihydroxyanthraquinone and Saquinavir mesylate as anticancer and anti-HIV drugs, respectively. [57]

5. Vaccines

Glycopolymers can also be good candidates to develop vaccines and to provide active acquired immunity against diseases [58,59]. A polymerizable version of the Tn-antigen glycan has been prepared and converted into well-defined glycopolymers by RAFT polymerization [60]. These glycopolymers were then conjugated to gold nanoparticles with narrow size distribution of diameter between 5 and 20 nm. Immunologically, these nanomaterials generated strong and long-lasting production of antibodies that are selective to the Tn-antigen glycan and cross-reactive toward mucin proteins displaying Tn.

Very recently, a block glycopolymer has been prepared by cyanoxyl-mediated free radical polymerization followed by conjugation with a tumor associated carbohydrate antigens (TACA) Tn antigen and a mouse T helper cells (Th cells) peptide epitope derived from polio virus to afford the vaccine construct (Figure 10) [61].

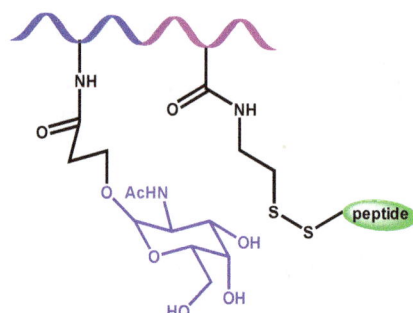

Figure 10. Glycopolymer vaccine structure.

The glycopolymer vaccine stimulated an anti-Tn immune response with significant titers of IgG antibodies, which recognized Tn-expressing tumor cells. Although glycopolymeric platforms are still discreet, they offer great flexibilities to adjust antibody generator density, valence and the ratio of tumor associated carbohydrate antigens (TACA) against Th epitope. In addition, the immunogenicity of the glycopolymer backbone is not elevated, which probably will not compete significantly with the desired TACA for B cell activation.

6. Sensors

Due to the binding capacity of glycopolymers towards protein and bacteria, they have been extensively investigated as fundamental part of biosensors systems, mostly of them based on gold nanoparticles/surfaces and in their plasmonic properties. As commented above bacterial infections are today one of the people major concerns and glycopolymers can help in the solution as bacterial sensor. Gold nanoparticles with carbohydrates directly immobilized on their surface by Au-S bond formation showed very rapid responses to both lectins and lectin-expressing bacteria. In this sense, Richard *et al.* [62] have described a system capable of discriminating between different strains of *Escherichia coli* using this approach; that is, gold nanoparticles functionalized with poly(ethylene glycol) (PEG) with galactose and mannose-ended (Figure 11). The use of the PEG chains increases the saline stability retaining their biorecognition properties. The optical properties of nanoparticles change when bind to protein *Fim*H positive bacteria, enabling the identification of bacteria strain.

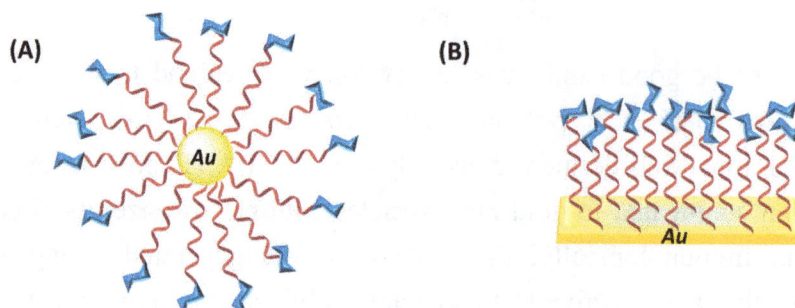

Figure 11. Schematic representation of glycopolymers attached to (**A**) gold particles and (**B**) flat gold surface, e.g., quartz crystal microbalance.

Other statistical glycopolymers having pendant thioglycosides have been synthetized by RAFT and then immobilized on gold nanoparticles as well as on quartz crystal microbalance (QCM). They showed strong and specific binding with their corresponding lectins, with estimated association constant (K_a) values of around 10^7 M^{-1} [63]. This value is reached with only 10% carbohydrate units, which gives an idea of the strong glycocluster effect (the interaction between lectin and free saccharide molecule is around 10^3 M^{-1}).

p-Acrylamidophenyl-α-D-mannopyranoside homopolymers and mannose-incorporating nanogel particles have been immobilized on nano-imprinted cyclo-olefin polymer films [64]. The strong and specific interactions between the mannose-incorporating polymers and Con A lectin were analyzed by monitoring the changes in the reflection intensity of the film. The nanogel particles showed a larger binding capacity compared with the homopolymers, and are able to detect low protein concentrations (with a detection limit of 6.0 ng/mL). Furthermore, the detection limit of the developed biosensor was lower than that of surface plasmon resonance sensor (1.43 mg/mL).

Very recently, amphiphilic block glycopolymers based on 2-O-methacryloyloxyethyl-(β-D-lactose) (Lac) and 4-pyridilmethyl methacrylate (PyMA) have been synthesized by RAFT process. Through the pyridine groups those can be chemo-adsorbed onto gold surface of QCM [65]. The constant value obtained for the specific binding with RCA$_{120}$ is 6.26×10^6 M^{-1}. This approach was also employed to modify gold nanorods. A very small amount of lectin (100 pg/mL = 8.3×10^{-13} M) was detected by the aggregation of these glyco-gold nanorods, which gives idea of the potential of these systems [66].

Statistical glycopolymers of LAEMA and cationic 2-aminoethyl methacrylamide hydrochloride (AEMA) monomers obtained by RAFT polymerization were immobilized on a sensor surface for studies of bacterial adhesion by QCM with dissipation [67]. Galactose specific lectin RCA$_{120}$ was used to test the QCM bacterial adhesion. Then, significantly higher amount of *Pseudomonas aeruginosa* PAO1 adhere on the glycopolymer surface with strong contact point stiffness as compared to *E. coli* K-12, since *P. aeruginosa* has galactose-specific binding while *E. coli* has mannose-specific one. Moreover, RAFT glycopolymers of acrylamide and glycomonomer bearing triazole-linked sialyloligosaccharides have been also immobilized on a gold-coated sensor of QCM. These glycopolymers strongly bind to both human and avian influenza A viruses [68].

Layer-by-layer (LbL) is a widely used technique for thin film preparation. It consists of the alternate deposition of interacting species on a substrate with an intervening rinsing step following each deposition [69]. Usawa *et al.* [70] employed LbL adsorption methodology for the assembly of glycochips by using polyanionic glycopolymers (Figure 12). They prepared three glycochips carrying globobioside (Gb2), β-lactoside (β-Lac), or α-D-mannoside (α-Man) residues for the detection of Shiga toxins, Stx1 and Stx2, by surface plasmon resonance (SPR). Stx1 and Stx2 toxins show binding specificity for the Gb2 glycochip and a weak affinity for the β-Lac glycochip. Therefore, they give different SPR response allowing discriminating between the two toxins.

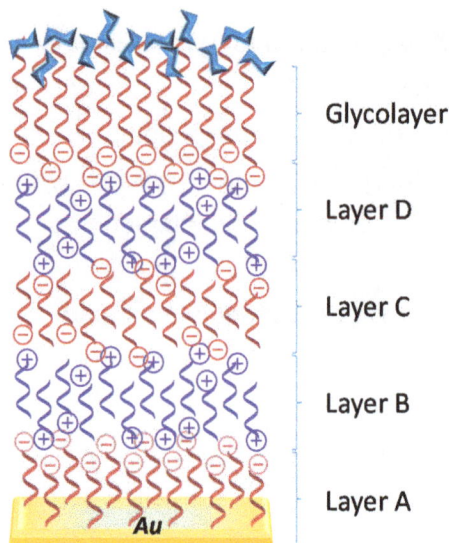

Figure 12. Schematic representation of layer-by-layer preparation of glycochips.

7. Imaging and Diagnostic

7.1. Glycopolymers

The early detection of tumors or other illnesses are attracting ongoing attention. Nowadays, the development of systems with high cellular internalization efficiency and specificity, as well as low toxicity and good stability is of urgent need. Related to this, the conjugation of glycopolymers with contrast agent probes is a very promising alternative to achieve these purposes [71]. Wang *et al.* [72] have terpolymerized glucose- and lactose-containing methacrylate (GEMA and MAMA) monomers with radiopaque 2-(2'-iodobenzoyl)-ethyl methacrylate (2-IEMA) in presence of methyl methacrylate (MMA) as target imaging carrier materials for controlled drug release. These materials are thermally stable and a 0.50 mm thick terpolymer disk exhibits radiopacity similar to a 0.88 mm thick aluminum plate. Both glycopolymeric systems are radiopaque but the most of them is the one based on LAMA with a molar fraction of 0.5. Computed Tomography (CT) showed that aqueous solution of terpolymers with glycopolymer compositions of 0.5, especially lactose derivative, enhances visibility in the liver and kidney organs. CT values demonstrated that glycopolymer stays in kidney organs for at least one hour without an appreciable loss of contrast.

Nevertheless, the conjugation with fluorescent probe is probably the most investigated strategy. Mannose-derivative of 2-hydroxyethyl methacrylate (MEMA) glycomonomer was copolymerized at different molar fractions with HEMA to analyze the ability to interact with Con A [73]. Besides, it was copolymerized with fluorescence 7-(*N*-(4-vinylbenzyl)amino)-4-trifluoromethylcoumarin to visualize the cellular internalization via endocytosis without apparent damage to the HeLa cells. The no toxicity of the glycopolymer was demonstrated on the basis of the maintenance of NADH dehydrogenase activity of cells after incubation for 24 hours.

Another fluorescence glycopolymer has been synthesized by simultaneous atom transfer radical polymerization (ATRP) and click chemistry using as fluorescent initiator N,N'-bis{2-[2-[(2-bromo-2-methylpropanoyl)oxy]ethoxy]ethyl}perylene-3,4,9,10-tetracarboxylic acid bisimide [74]. The glycopolymers with different length emit strong green fluorescence and exhibit good biocompatibility with 3T3 fibroblasts, macrophages, and KB cells. Therefore, they can also be used as cell labeling agents.

In addition, glycopolymers with near infrared fluorescence have been synthesized employing a combination of ring opening polymerization and RAFT polymerization, followed by post-functionalization with an aminocyanine molecule [75]. These materials after incubation with HepG2 cells for 72 h showed low toxicity even at concentration as high as 0.25 mg mL^{-1}. They undergo enhanced and fast endocytosis due to the specific interaction of galactose glycopolymer with HepG2 cells.

7.2. Hybrid Materials

The groups of Zhao and Chen have synthesized by RAFT process copolymers of MAG and methacrylic acid (MAA). They used these glycopolymers as templates to prepare fluorescent glycopolymer-functionalized silver nanoclusters (with particle size of ~5 nm) through microwave irradiation [76]. The bioactivity of these systems has been evaluated using GLUT-1 over-expressing cancer cells K562, showing great toxicity over them. Furthermore, they have obtained porphyrin-glycopolymer conjugates based on this MAG glycomonomer. Glycoparticles (~200 nm) showed enhanced binding ability toward Con A, due to the stronger multivalent effect displayed by the sufficient numbers of glucose groups on the surface of the nanoparticles. They were also able to efficiently bind cancer cells K562 and kill them under light irradiation. Therefore, these systems can be used as photosensitizer for cancer imaging and also in photodynamic therapy [77].

In addition, star-shaped porphyrin-cored block copolymers based on gluconamidoethyl methacrylate (GAMA) have been also synthesized by RAFT using as macroinitiator poly(L-Lactide) with anchored porphyrin (SPPLA) [78]. These glycopolymers showed specific recognition with Con A. Furthermore, SPPLA-*b*-PGAMA$_{26}$ block copolymer exhibited efficient singlet oxygen generation and high fluorescence quantum yields. Its cytotoxicity against COS-7 cells was very low and, when given a longer irradiation time, more BEL-7402 cancer cells died. Consequently, it can be useful for photodynamic therapy.

Glycopolymers have also been combined with boron-dipyrromethene (BODIPY) for direct tumor cell imaging [79]. Statistical glycopolymer showing high water solubility based on galactose-derivative of 2-hydroxyethyl methacrylate (GaEMA) glycomonomer and BODIPY fluorescent dye monomer was developed. It presented good photo-stability and no cytotoxicity against HepG2 and NIH3T3 cells. The viability of cells was higher than the corresponding parent copolymer based on 2-hydroxyethyl methacrylate (HEMA) and BODIPY, indicating that galactose units decrease the toxicity. Moreover, its fluorescence quantum yield was 1.7 times higher than that of a BODIPY monomer because of the incorporation of galactose. Besides, this glycopolymer leads to efficient internalization into HepG2 and clear visualizations in cytoplasm, distinguishing between HepG2 and NIH3T3 cells.

Very recently, glycopolymers of 1-(methacrylamido) glucopyranose (1-MAG) fluorescently labeled with fluorescein isothiocyanate (FITC) have been synthesized by RAFT process. Additionally, gold nanoparticles with poly(*N*-hydroxyethyl acrylamide) attached onto the surface by thiol group have also been prepared [80]. These nanoparticles were subsequently reacted with glucosamine hydrochloride to generate glycosylated nanoparticles. None of these systems are hemolytic and do not induce morphological changes in the cells. Therefore, they can be used in diagnostics or therapies.

One step further described in literature is the preparation of magnetic nanoparticles covered with FITC-labeled glycopolymer yielded fluorescent [81,82]. In these publications the visualization of the nanoparticles by fluorescence imaging is studied, but the possibility of application in magnetic resonance imaging (MRI) is also opened.

Magnetic nanoparticles represent one of the most important nanomaterials in the diagnosis and treatment of cancer. Nanosystems based on iron oxide nanoparticles are currently being applied as contrast agent in MRI [83]. The use of nanoparticles in cancer therapy is highly convenient because of so-called enhanced permeability and retention effect. The tumor vessels are altered in comparison with those in sane cells and the nanoparticles from the blood can be internalized better into tumor tissues, thus selectively accumulated. When this phenomenon is combined with other targeting strategies by incorporation of specific moieties the resulting nanosystems can exhibit enhanced diagnostic capacity for a more specific treatment and at the same time less aggressive to the human body as the required dose can be considerably reduced. The attachment of glycopolymers onto the iron oxide nanoparticles allows this selective localization of the nanoparticles concomitantly with an increase in the colloidal stability of the nanoparticles [84]. Several studies are found in literature concerning the functionalization of magnetic nanoparticles with glycopolymers for MRI applications. A series of diblock glycopolymers based on PEG and different carbohydrates, α-D-mannose, α-D-glucose and β-D-glucose, have been attached to iron oxide nanoparticle surfaces and the influence of the binding ability of the resulting nanosystems on the MRI signal was evaluated [85]. Remarkably, the α-D-mannose functionalized magnetic nanoparticles showed an improved cell uptake in a lung cancer cell line (A549) and also exhibited high r_2 transverse relaxivity when measured in a 9.4 T MRI. More important, the binding of the nanoparticles to the lectin Con A produces a significant change in T_2 relaxation, which is proportional to the lectin concentration, thus enhancing the diagnosis future perspectives.

In another study, iron oxide nanoparticles have been functionalized with poly (vinylbenzyl-O-β-D-galactopyranosyl-D-gluconamide) that contain galactose units, thus able to be recognized by asialoglycoprotein receptors on hepatocytes [86]. *In vivo* experiments in rats show that after injection the signal intensity of liver largely dropped on T_2-weighted MR image revealing that the nanoparticles are mainly accumulated in liver. Therefore, these nanoparticles can present a potential as contrast liver-targeting MRI contrast agent.

8. Future Developments and Conclusions

Up to here, promising applications of glycopolymers are presented; however, this type of materials can also be used in very different matters that could lead to a number of future applications. This is the case of glycopolymers used to optically observe the conformations of structurally well-defined polymers anchored to fluid lipid membranes [87]. Remarkable phase control on the crystallization of calcium carbonate by the stereochemistry of glycopolymers has been achieved. The selection of phase is based on the chelating character of the hydroxyl groups of the pyranoses (glucose or galactose) and their individual orientation, which allow to the control biomineralization processes [88]. Recently, ionotropic gelation has been reported on glycopolymers synthesized by copolymerization of

2-hydroxyethyl methacrylamide and acrylamide- and methacrylamide-type macromonomers obtained by modification of alginate-derived oligosaccharides [89]. These hydrogels formed under mild conditions have potential applications in cell encapsulation and *in vitro* 3D cell culture. Moreover, glycopolymers are highly hemocompatible and do not induce clot formation, red blood cell aggregation, and immune response, thereby these macromolecular structures could be very useful in many different *in vivo* applications [90]. In summary, these materials are promising systems for diverse biological, biomedical and other related activities. In this sense, the cost-effective and straightforward synthesis of adequate glycomonomers and glycopolymers is essential; therefore, chemistry as well as its combination with different areas, such as biology and biomedicine to understand their behavior, mathematics, and physics to model, such as behavior, analytical, and nanotechnology to implement their properties, and others, will allow enlarging the glycopolymeric material applications.

Acknowledgments

We thank the financial support of MINECO (Project MAT2013-47902-C2-1-R) and A. Muñoz-Bonilla thanks MINECO for her Ramon y Cajal contract.

Author Contributions

Both authors contributed equally to design and write the manuscript. All authors read and approved the final manuscript.

Conflicts of Interest

The authors declare no conflict of interest.

References

1. Lee, Y.C.; Lee, R.T. Carbohydrate-Protein Interactions: Basis of Glycobiology. *Acc. Chem. Res.* **1995**, *28*, 321–327.
2. Fasting, C.; Schalley, C.A.; Weber, M.; Seitz, O.; Hecht, S.; Koksch, B.; Dernedde, J.; Graf, C.; Knapp, E.-W.; Haag, R. Multivalency as a Chemical Organization and Action Principle. *Angew. Chem. Int. Ed.* **2012**, *51*, 10472–10498.
3. Ting, S.R.S.; Chen, G.; Stenzel, M.H. Synthesis of glycopolymers and their multivalent recognitions with lectins. *Polym. Chem.* **2010**, *1*, 1392–1412.
4. Slavin, S.; Burns, J.; Haddleton, D.M.; Becer, C.R. Synthesis of glycopolymers via click reactions. *Eur. Polym. J.* **2011**, *47*, 435–446.
5. Zhang, Q.; Haddleton, D. Synthetic glycopolymers: Some recent developments. In *Hierarchical Macromolecular Structures: 60 Years after the Staudinger Nobel Prize II*; Percec, V., Ed. Springer International Publishing: Berlin, Germany, 2013; pp. 39–59.
6. Kiessling, L.L.; Grim, J.C. Glycopolymer probes of signal transduction. *Chem. Soc. Rev.* **2013**, *42*, 4476–4491.

7. Xu, J.; Boyer, C.; Bulmus, V.; Davis, T.P. Synthesis of dendritic carbohydrate end-functional polymers via RAFT: Versatile multi-functional precursors for bioconjugations. *J. Polym. Sci. Part A Polym. Chem.* **2009**, *47*, 4302–4313.

8. Ladmiral, V.; Mantovani, G.; Clarkson, G.J.; Cauet, S.; Irwin, J.L.; Haddleton, D.M. Synthesis of Neoglycopolymers by a Combination of "Click Chemistry" and Living Radical Polymerization. *J. Am. Chem. Soc.* **2006**, *128*, 4823–4830.

9. Li, X.; Chen, G. Glycopolymer-based nanoparticles: synthesis and application. *Polym. Chem.* **2015**, *6*, 1417–1430.

10. Ruiz, C.; Sanchez-Chaves, M.; Cerrada, M.L.; Fernandez-Garcia, M. Glycopolymers Resulting from Ethylene-Vinyl Alcohol Copolymers: Synthetic Approach, Characterization, and Interactions with Lectins. *J. Polym. Sci. Part A Polym. Chem.* **2008**, *46*, 7238–7248.

11. Sanchez-Chaves, M.; Ruiz, C.; Cerrada, M.L.; Fernandez-Garcia, M. Novel glycopolymers containing aminosaccharide pendant groups by chemical modification of ethylene-vinyl alcohol copolymers. *Polymer* **2008**, *49*, 2801–2807.

12. Cerrada, M.L.; Sanchez-Chaves, M.; Ruiz, C.; Fernandez-Garcia, M. Recognition Abilities and Development of Heat-Induced Entangled Networks in Lactone-Derived Glycopolymers Obtained from Ethylene-vinyl Alcohol Copolymers. *Biomacromolecules* **2009**, *10*, 1828–1837.

13. Cerrada, M.L.; Ruiz, C.; Sanchez-Chaves, M.; Fernandez-Garcia, M. Molecular recognition capability and rheological behavior in solution of novel lactone-based glycopolymers. *Eur. Polym. J.* **2009**, *45*, 3176–3184.

14. Munoz-Bonilla, A.; Heuts, J.P.A.; Fernandez-Garcia, M. Glycoparticles and bioactive films prepared by emulsion polymerization using a well-defined block glycopolymer stabilizer. *Soft Matter* **2011**, *7*, 2493–2499.

15. Bordege, V.; Munoz-Bonilla, A.; Leon, O.; Cuervo-Rodriguez, R.; Sanchez-Chaves, M.; Fernandez-Garcia, M. Statistical Glycopolymers Based on 2-Hydroxyethyl Methacrylate: Copolymerization, Thermal Properties, and Lectin Interaction Studies. *Macromol. Chem. Phys.* **2011**, *212*, 1294–1304.

16. Leon, O.; Munoz-Bonilla, A.; Bordege, V.; Sanchez-Chaves, M.; Fernandez-Garcia, M. Amphiphilic Block Glycopolymers via Atom Transfer Radical Polymerization: Synthesis, Self-Assembly and Biomolecular Recognition. *J. Polym. Sci. Part A Polym. Chem.* **2011**, *49*, 2627–2635.

17. Bordegé, V.; Muñoz-Bonilla, A.; León, O.; Sánchez-Chaves, M.; Cuervo-Rodríguez, R.; Fernández-García, M. Glycopolymers with glucosamine pendant groups: Copolymerization, physico-chemical and interaction properties. *React. Funct. Polym.* **2011**, *71*, 1–10.

18. Muñoz-Bonilla, A.; Bordegé, V.; León, O.; Cuervo-Rodríguez, R.; Sánchez-Chaves, M.; Fernández-García, M. Influence of glycopolymers structure on the copolymerization reaction and on their binding behavior with lectins. *Eur. Polym. J.* **2012**, *48*, 963–973.

19. Muñoz-Bonilla, A.; León, O.; Cerrada, M.L.; Rodríguez-Hernández, J.; Sánchez-Chaves, M.; Fernández-García, M. Glycopolymers obtained by chemical modification of well-defined block copolymers. *J. Polym. Sci. Part A Polym. Chem.* **2012**, *50*, 2565–2577.

20. Munoz-Bonilla, A.; Leon, O.; Bordege, V.; Sanchez-Chaves, M.; Fernandez-Garcia, M. Controlled block glycopolymers able to bind specific proteins. *J. Polym. Sci. Part A Polym. Chem.* **2013**, *51*, 1337–1347.

21. León, O.; Bordegé, V.; Muñoz-Bonilla, A.; Sánchez-Chaves, M.; Fernández-García, M. Well-controlled amphiphilic block glycopolymers and their molecular recognition with lectins. *J. Polym. Sci. Part A Polym. Chem.* **2010**, *48*, 3623–3631.

22. Alvárez-Paino, M.; Juan-Rodríguez, R.; Cuervo-Rodríguez, R.; Muñoz-Bonilla, A.; Fernández-García, M. Preparation of amphiphilic glycopolymers with flexible long side chain and their use as stabilizer for emulsion polymerization. *J. Colloid Interface Sci.* **2014**, *417*, 336–345.

23. Alvarez-Paino, M.; Marcelo, G.; Munoz-Bonilla, A.; Rodriguez-Hernandez, J.; Fernandez-Garcia, M. Surface modification of magnetite hybrid particles with carbohydrates and gold nanoparticles via "click" chemistry. *Polym. Chem.* **2013**, *4*, 986–995.

24. Cerrada, M.L.; Bordege, V.; Munoz-Bonilla, A.; Leon, O.; Cuervo-Rodriguez, R.; Sanchez-Chaves, M.; Fernandez-Garcia, M. Amphiphilic polymers bearing gluconolactone moieties: Synthesis and long side-chain crystalline behavior. *Carbohydr. Polym.* **2013**, *94*, 755–764.

25. Bordege, V.; Munoz-Bonilla, A.; Leon, O.; Cuervo-Rodriguez, R.; Sanchez-Chaves, M.; Fernandez-Garcia, M. Gluconolactone-Derivated Polymers: Copolymerization, Thermal Properties, and Their Potential Use as Polymeric Surfactants. *J. Polym. Sci. Part A Polym. Chem.* **2011**, *49*, 526–536.

26. Godula, K.; Bertozzi, C.R. Synthesis of Glycopolymers for Microarray Applications via Ligation of Reducing Sugars to a Poly(acryloyl hydrazide) Scaffold. *J. Am. Chem. Soc.* **2010**, *132*, 9963–9965.

27. Lee, S.-G.; Brown, J.M.; Rogers, C.J.; Matson, J.B.; Krishnamurthy, C.; Rawat, M.; Hsieh-Wilson, L.C. End-functionalized glycopolymers as mimetics of chondroitin sulfate proteoglycans. *Chem. Sci.* **2010**, *1*, 322–325.

28. Tong, Q.; Wang, X.; Wang, H.; Kubo, T.; Yan, M. Fabrication of Glyconanoparticle Microarrays. *Anal. Chem.* **2012**, *84*, 3049–3052.

29. Michel, O.; Ravoo, B.J. Carbohydrate Microarrays by Microcontact "Click" Chemistry. *Langmuir* **2008**, *24*, 12116–12118.

30. Fukuda, T.; Onogi, S.; Miura, Y. Dendritic sugar-microarrays by click chemistry. *Thin Solid Films* **2009**, *518*, 880–888.

31. Godula, K.; Rabuka, D.; Nam, K.T.; Bertozzi, C.R. Synthesis and Microcontact Printing of Dual End-Functionalized Mucin-like Glycopolymers for Microarray Applications. *Angew. Chem. Int. Ed.* **2009**, *48*, 4973–4976.

32. Blohm, D.H.; Guiseppi-Elie, A. New developments in microarray technology. *Curr. Opin. Biotechnol.* **2001**, *12*, 41–47.

33. Park, S.; Gildersleeve, J.C.; Blixt, O.; Shin, I. Carbohydrate microarrays. *Chem. Soc. Rev.* **2013**, *42*, 4310–4326.

34. Piliarik, M.; Vaisocherová, H.; Homola, J. Surface Plasmon Resonance Biosensing. In *Biosensors and Biodetection*; Rasooly, A., Herold, K., Eds.; Humana Press: New York, NY, USA, 2009; pp. 65–88.

35. Wijaya, E.; Lenaerts, C.; Maricot, S.; Hastanin, J.; Habraken, S.; Vilcot, J.-P.; Boukherroub, R.; Szunerits, S. Surface plasmon resonance-based biosensors: From the development of different SPR structures to novel surface functionalization strategies. *Curr. Opin. Solid State Mater. Sci.* **2011**, *15*, 208–224.

36. Shi, H.; Liu, L.; Wang, X.; Li, J. Glycopolymer-peptide bioconjugates with antioxidant activity via RAFT polymerization. *Polym. Chem.* **2012**, *3*, 1182–1188.

37. Pasparakis, G.; Cockayne, A.; Alexander, C. Control of Bacterial Aggregation by Thermoresponsive Glycopolymers. *J. Am. Chem. Soc.* **2007**, *129*, 11014–11015.

38. Min, E.H.; Ting, S.R.S.; Billon, L.; Stenzel, M.H. Thermo-Responsive Glycopolymer Chains Grafted onto Honeycomb Structured Porous Films via RAFT Polymerization as a Thermo-Dependent Switcher for Lectin Concanavalin A Conjugation. *J. Polym. Sci. Part A Polym. Chem.* **2010**, *48*, 3440–3455.

39. Polizzotti, B.D.; Kiick, K.L. Effects of Polymer Structure on the Inhibition of Cholera Toxin by Linear Polypeptide-Based Glycopolymers. *Biomacromolecules* **2006**, *7*, 483–490.

40. Polizzotti, B.D.; Maheshwari, R.; Vinkenborg, J.; Kiick, K.L. Effects of Saccharide Spacing and Chain Extension on Toxin Inhibition by Glycopolypeptides of Well-Defined Architecture. *Macromolecules* **2007**, *40*, 7103–7110.

41. Miyagawa, A.; Kasuya, M.C.Z.; Hatanaka, K. Inhibitory effects of glycopolymers having globotriose and/or lactose on cytotoxicity of Shiga toxin 1. *Carbohydr. Polym.* **2007**, *67*, 260–264.

42. Remzi Becer, C.; Gibson, M.I.; Geng, J.; Ilyas, R.; Wallis, R.; Mitchell, D.A.; Haddleton, D.M. High-Affinity Glycopolymer Binding to Human DC-SIGN and Disruption of DC-SIGN Interactions with HIV Envelope Glycoprotein. *J. Am. Chem. Soc.* **2010**, *132*, 15130–15132.

43. Richards, S.-J.; Jones, M.W.; Hunaban, M.; Haddleton, D.M.; Gibson, M.I. Probing Bacterial-Toxin Inhibition with Synthetic Glycopolymers Prepared by Tandem Post-Polymerization Modification: Role of Linker Length and Carbohydrate Density. *Angew. Chem. Int. Ed.* **2012**, *51*, 7812–7816.

44. Jones, M.W.; Richards, S.-J.; Haddleton, D.M.; Gibson, M.I. Poly(azlactone)s: Versatile scaffolds for tandem post-polymerisation modification and glycopolymer synthesis. *Polym. Chem.* **2013**, *4*, 717–723.

45. Jones, M.W.; Otten, L.; Richards, S.J.; Lowery, R.; Phillips, D.J.; Haddleton, D.M.; Gibson, M.I. Glycopolymers with secondary binding motifs mimic glycan branching and display bacterial lectin selectivity in addition to affinity. *Chem. Sci.* **2014**, *5*, 1611–1616.

46. Menon, S.; Ongungal, R.M.; Das, S. Photoresponsive Glycopolymer Aggregates as Controlled Release Systems. *Macromol. Chem. Phys.* **2014**, *215*, 2365–2373.

47. Otsuka, I.; Osaka, M.; Sakai, Y.; Travelet, C.; Putaux, J.-L.; Borsali, R. Self-Assembly of Maltoheptaose-block-Polystyrene into Micellar Nanoparticles and Encapsulation of Gold Nanoparticles. *Langmuir* **2013**, *29*, 15224–15230.

48. Suriano, F.; Pratt, R.; Tan, J.P.K.; Wiradharma, N.; Nelson, A.; Yang, Y.-Y.; Dubois, P.; Hedrick, J.L. Synthesis of a family of amphiphilic glycopolymers via controlled ring-opening polymerization of functionalized cyclic carbonates and their application in drug delivery. *Biomaterials* **2010**, *31*, 2637–2645.

49. Sun, P.; He, Y.; Lin, M.; Zhao, Y.; Ding, Y.; Chen, G.; Jiang, M. Glyco-regioisomerism Effect on Lectin-Binding and Cell-Uptake Pathway of Glycopolymer-Containing Nanoparticles. *ACS Macro Lett.* **2014**, *3*, 96–101.

50. Jin, X.; Zhang, X.; Wu, Z.; Teng, D.; Zhang, X.; Wang, Y.; Wang, Z.; Li, C. Amphiphilic Random Glycopolymer Based on Phenylboronic Acid: Synthesis, Characterization, and Potential as Glucose-Sensitive Matrix. *Biomacromolecules* **2009**, *10*, 1337–1345.

51. Zheng, C.; Guo, Q.; Wu, Z.; Sun, L.; Zhang, Z.; Li, C.; Zhang, X. Amphiphilic glycopolymer nanoparticles as vehicles for nasal delivery of peptides and proteins. *Eur. J. Pharm. Sci.* **2013**, *49*, 474–482.

52. Pearson, S.; Scarano, W.; Stenzel, M.H. Micelles based on gold-glycopolymer complexes as new chemotherapy drug delivery agents. *Chem. Commun.* **2012**, *48*, 4695–4697.

53. Ahmed, M.; Mamba, S.; Yang, X.-H.; Darkwa, J.; Kumar, P.; Narain, R. Synthesis and Evaluation of Polymeric Gold Glyco-Conjugates as Anti-Cancer Agents. *Bioconjugate Chem.* **2013**, *24*, 979–986.

54. Adokoh, C.K.; Quan, S.; Hitt, M.; Darkwa, J.; Kumar, P.; Narain, R. Synthesis and Evaluation of Glycopolymeric Decorated Gold Nanoparticles Functionalized with Gold-Triphenyl Phosphine as Anti-Cancer Agents. *Biomacromolecules* **2014**, *15*, 3802–3810.

55. Fleming, C.; Maldjian, A.; Da Costa, D.; Rullay, A.K.; Haddleton, D.M.; St John, J.; Penny, P.; Noble, R.C.; Cameron, N.R.; Davis, B.G. A carbohydrate-antioxidant hybrid polymer reduces oxidative damage in spermatozoa and enhances fertility. *Nat. Chem. Biol.* **2005**, *1*, 270–274.

56. Wu, Y.; Wang, M.; Sprouse, D.; Smith, A.E.; Reineke, T.M. Glucose-Containing Diblock Polycations Exhibit Molecular Weight, Charge, and Cell-Type Dependence for pDNA Delivery. *Biomacromolecules* **2014**, *15*, 1716–1726.

57. Zhang, Q.; Su, L.; Collins, J.; Chen, G.; Wallis, R.; Mitchell, D.A.; Haddleton, D.M.; Becer, C.R. Dendritic Cell Lectin-Targeting Sentinel-like Unimolecular Glycoconjugates to Release an Anti-HIV Drug. *J. Am. Chem. Soc.* **2014**, *136*, 4325–4332.

58. Lin, K.; Kasko, A.M. Carbohydrate-Based Polymers for Immune Modulation. *ACS Macro Lett.* **2014**, *3*, 652–657.

59. Geng, J.; Mantovani, G.; Tao, L.; Nicolas, J.; Chen, G.; Wallis, R.; Mitchell, D.A.; Johnson, B.R.G.; Evans, S.D.; Haddleton, D.M. Site-Directed Conjugation of "Clicked" Glycopolymers to Form Glycoprotein Mimics: Binding to Mammalian Lectin and Induction of Immunological Function. *J. Am. Chem. Soc.* **2007**, *129*, 15156–15163.

60. Parry, A.L.; Clemson, N.A.; Ellis, J.; Bernhard, S.S.R.; Davis, B.G.; Cameron, N.R. 'Multicopy Multivalent' Glycopolymer-Stabilized Gold Nanoparticles as Potential Synthetic Cancer Vaccines. *J. Am. Chem. Soc.* **2013**, *135*, 9362–9365.

61. Qin, Q.; Yin, Z.; Bentley, P.; Huang, X. Carbohydrate antigen delivery by water soluble copolymers as potential anti-cancer vaccines. *MedChemComm* **2014**, *5*, 1126–1129.

62. Richards, S.-J.; Fullam, E.; Besra, G.S.; Gibson, M.I. Discrimination between bacterial phenotypes using glyco-nanoparticles and the impact of polymer coating on detection readouts. *J. Mater. Chem. B* **2014**, *2*, 1490–1498.

63. Tanaka, T.; Inoue, G.; Shoda, S.-I.; Kimura, Y. Protecting-group-free synthesis of glycopolymers bearing thioglycosides via one-pot monomer synthesis from free saccharides. *J. Polym. Sci. Part A Polym. Chem.* **2014**, *52*, 3513–3520.

64. Terada, Y.; Hashimoto, W.; Endo, T.; Seto, H.; Murakami, T.; Hisamoto, H.; Hoshino, Y.; Miura, Y. Signal amplified two-dimensional photonic crystal biosensor immobilized with glyco-nanoparticles. *J. Mater. Chem. B* **2014**, *2*, 3324–3332.

65. Otsuka, H.; Hagiwara, T.; Yamamoto, S. Carbohydrate-Based Amphiphilic Diblock Copolymers with Pyridine for the Sensitive Detection of Protein Binding. *J. Nanosci. Nanotechnol.* **2014**, *14*, 6764–6773.

66. Otsuka, H.; Muramatsu, Y.; Matsukuma, D. Gold Nanorods Functionalized with Self-assembled Glycopolymers for Ultrasensitive Detection of Proteins. *Chem. Lett.* **2015**, *44*, 132–134.

67. Wang, Y.; Narain, R.; Liu, Y. Study of Bacterial Adhesion on Different Glycopolymer Surfaces by Quartz Crystal Microbalance with Dissipation. *Langmuir* **2014**, *30*, 7377–7387.

68. Tanaka, T.; Ishitani, H.; Miura, Y.; Oishi, K.; Takahashi, T.; Suzuki, T.; Shoda, S.-I.; Kimura, Y. Protecting-Group-Free Synthesis of Glycopolymers Bearing Sialyloligosaccharide and Their High Binding with the Influenza Virus. *ACS Macro Lett.* **2014**, *3*, 1074–1078.

69. Ariga, K.; Yamauchi, Y.; Rydzek, G.; Ji, Q.; Yonamine, Y.; Wu, K.C.W.; Hill, J.P. Layer-by-layer Nanoarchitectonics: Invention, Innovation, and Evolution. *Chem. Lett.* **2014**, *43*, 36–68.

70. Uzawa, H.; Ito, H.; Neri, P.; Mori, H.; Nishida, Y. Glycochips from Polyanionic Glycopolymers as Tools for Detecting Shiga Toxins. *ChemBioChem* **2007**, *8*, 2117–2124.

71. Kottari, N.; Chabre, Y.; Sharma, R.; Roy, R. Applications of Glyconanoparticles as "Sweet" Glycobiological Therapeutics and Diagnostics. In *Multifaceted Development and Application of Biopolymers for Biology, Biomedicine and Nanotechnology*; Dutta, P.K., Dutta, J., Eds.; Springer: Berlin, Germany, 2013; pp. 297–341.

72. Wang, X.; Geng, X.; Ye, L.; Zhang, A.-Y.; Feng, Z.-G. Synthesis and characterization of novel glucose- and lactose-containing methacrylate-based radiopaque glycopolymers. *React. Funct. Polym.* **2009**, *69*, 857–863.

73. Obata, M.; Shimizu, M.; Ohta, T.; Matsushige, A.; Iwai, K.; Hirohara, S.; Tanihara, M. Synthesis, characterization and cellular internalization of poly(2-hydroxyethyl methacrylate) bearing α-d-mannopyranose. *Polym. Chem.* **2011**, *2*, 651–658.

74. Xu, L.Q.; Huang, C.; Wang, R.; Neoh, K.-G.; Kang, E.-T.; Fu, G.D. Synthesis and characterization of fluorescent perylene bisimide-containing glycopolymers for Escherichia coli conjugation and cell imaging. *Polymer* **2011**, *52*, 5764–5771.

75. Xing, T.; Yang, X.; Fu, L.; Yan, L. Near infrared fluorescence probe and galactose conjugated amphiphilic copolymer for bioimaging of HepG2 cells and endocytosis. *Polym. Chem.* **2013**, *4*, 4442–4449.

76. Lu, W.; Ma, W.; Lu, J.; Li, X.; Zhao, Y.; Chen, G. Microwave-Assisted Synthesis of Glycopolymer-Functionalized Silver Nanoclusters: Combining the Bioactivity of Sugar with the Fluorescence and Cytotoxicity of Silver. *Macromol. Rapid Commun.* **2014**, *35*, 827–833.

77. Lu, J.; Zhang, W.; Yuan, L.; Ma, W.; Li, X.; Lu, W.; Zhao, Y.; Chen, G. One-Pot Synthesis of Glycopolymer-Porphyrin Conjugate as Photosensitizer for Targeted Cancer Imaging and Photodynamic Therapy. *Macromol. Biosci.* **2014**, *14*, 340–346.

78. Dai, X.-H.; Wang, Z.-M.; Liu, W.; Dong, C.-M.; Pan, J.-M.; Yuan, S.-S.; Yan, Y.-S.; Liu, D.-M.; Sun, L. Biomimetic star-shaped porphyrin-cored poly(l-lactide)-b-glycopolymer block copolymers for targeted photodynamic therapy. *Colloid. Polym. Sci.* **2014**, *292*, 2111–2122.

79. Lu, Z.; Mei, L.; Zhang, X.; Wang, Y.; Zhao, Y.; Li, C. Water-soluble BODIPY-conjugated glycopolymers as fluorescent probes for live cell imaging. *Polym. Chem.* **2013**, *4*, 5743–5750.

80. Wilkins, L.E.; Phillips, D.J.; Deller, R.C.; Davies, G.-L.; Gibson, M.I. Synthesis and characterisation of glucose-functional glycopolymers and gold nanoparticles: study of their potential interactions with ovine red blood cells. *Carbohydr. Res.* **2015**, *405*, 47–54.

81. Babiuch, K.; Wyrwa, R.; Wagner, K.; Seemann, T.; Hoeppener, S.; Becer, C.R.; Linke, R.; Gottschaldt, M.; Weisser, J.; *et al.* Functionalized, biocompatible coating for superparamagnetic nanoparticles by controlled polymerization of a thioglycosidic monomer. *Biomacromolecules* **2011**, *12*, 681–691.

82. Pfaff, A.; Schallon, A.; Ruhland, T.M.; Majewski, A.P.; Schmalz, H.; Freitag, R.; Muller, A.H. Magnetic and fluorescent glycopolymer hybrid nanoparticles for intranuclear optical imaging. *Biomacromolecules* **2011**, *12*, 3805–3811.

83. Gallo, J.; Long, N.J.; Aboagye, E.O. Magnetic nanoparticles as contrast agents in the diagnosis and treatment of cancer. *Chem. Soc. Rev.* **2013**, *42*, 7816–7833.

84. Munoz-Bonilla, A.; Marcelo, G.; Casado, C.; Teran, F.J.; Fernandez-Garcia, M. Preparation of glycopolymer-coated magnetite nanoparticles for hyperthermia treatment. *J. Polym. Sci. Part A Polym. Chem.* **2012**, *50*, 5087–5096.

85. Basuki, J.S.; Esser, L.; Duong, H.T.T.; Zhang, Q.; Wilson, P.; Whittaker, M.R.; Haddleton, D.M.; Boyer, C.; Davis, T.P. Magnetic nanoparticles with diblock glycopolymer shells give lectin concentration-dependent MRI signals and selective cell uptake. *Chem. Sci.* **2014**, *5*, 715–726.

86. Yoo, M.K.; Kim, I.Y.; Kim, E.M.; Jeong, H.J.; Lee, C.M.; Jeong, Y.Y.; Akaike, T.; Cho, C.S. Superparamagnetic iron oxide nanoparticles coated with galactose-carrying polymer for hepatocyte targeting. *J. Biomed. Biotechnol.* **2007**, *2007*, 94740.

87. Godula, K.; Umbel, M.L.; Rabuka, D.; Botyanszki, Z.; Bertozzi, C.R.; Parthasarathy, R. Control of the Molecular Orientation of Membrane-Anchored Biomimetic Glycopolymers. *J. Am. Chem. Soc.* **2009**, *131*, 10263–10268.

88. Barz, M.; Götze, S.; Loges, N.; Schüler, T.; Theato, P.; Tremel, W.; Zentel, R. Well-defined carbohydrate-based polymers in calcium carbonate crystallization: Influence of stereochemistry in the polymer side chain on polymorphism and morphology. *Eur. Polym. J.* **2015**, doi:10.1016/j.eurpolymj.2015.02.010.

89. Ghadban, A.; Albertin, L.; Rinaudo, M.; Heyraud, A. Biohybrid Glycopolymer Capable of Ionotropic Gelation. *Biomacromolecules* **2012**, *13*, 3108–3119.

90. Ahmed, M.; Lai, B.F.L.; Kizhakkedathu, J.N.; Narain, R. Hyperbranched Glycopolymers for Blood Biocompatibility. *Bioconjugate Chem.* **2012**, *23*, 1050–1058.

3

Dispersion Process and Effect of Oleic Acid on Properties of Cellulose Sulfate- Oleic Acid Composite Film

Guo Chen *, Bin Zhang and Jun Zhao

Department of Biotechnology and Bioengineering, Huaqiao University, Xiamen 361021, China;
E-Mails: zhangbinzbgg@126.com (B.Z.); zhaojun@hqu.edu.cn (J.Z.)

* Author to whom correspondence should be addressed; E-Mail: chenguo@hqu.edu.cn;

Academic Editor: Carla Renata Arciola

Abstract: The cellulose sulfate (CS) is a newly developed cellulose derivative. The work aimed to investigate the effect of oleic acid (OA) content on properties of CS-OA film. The process of oleic acid dispersion into film was described to evaluate its effect on the properties of the film. Among the formulations evaluated, the OA addition decreased the solubility and water vapor permeability of the CS-OA film. The surface contact angle changed from 64.2° to 94.0° by increasing CS/OA ratio from 1:0 to 1:0.25 (w/w). The TS increased with OA content below 15% and decreased with OA over 15%, but the ε decreased with higher OA content. The micro-cracking matrices and micro pores in the film indicated the condense structure of the film destroyed by the incorporation of oleic acid. No chemical interaction between the OA and CS was observed in the XRD and FTIR spectrum. Film formulation containing 2% (w/w) CS, 0.3% (w/w) glycerol and 0.3% (w/w) OA, showed good properties of mechanic, barrier to moisture and homogeneity.

Keywords: packaging film; cellulose sulfate; oleic acid; composite films; hydrophobic properties

1. Introduction

Film and coating based on biomaterials as an alternative packaging draws lots of attention because of the consumers' demand for high quality foods and increased awareness of environmental issues [1].

The commonly used biomaterials are comprised of proteins, polysaccharides, lipids, and other degradable biomaterials [2].

Cellulose is D-glucopyranose unit of conformation chair bonded through β (1→4) glycosidic linkages. Cellulose ethers are a class of semi-synthetic polymers obtained by derivatization of the hydroxyl groups at positions 2, 3, and/or 6 of the anhydroglucose residues of cellulose. Hydroxypropylmethyl cellulose (HPMC), carboxymethyl cellulose (CMC) and methyl cellulose (MC) as derivatives with improved solubility have long been used in fiber, film and gel-based materials [3]. Films made from cellulose derivatives showed good tensile resistance and effective barrier against O_2/CO_2 [4,5]. Nonetheless, edible films prepared from cellulose derivatives do not act as an efficient water vapor barrier due to the hydrophilic nature of cellulose [6–8]. Cellulose nanocrystals (CNC) synthesized from microcrystalline cellulose by a sulfuric acid hydrolysis was added to PLA or PLA–PHB film to improve the thermal stability of the film and reduce water permeability of the film [9,10]. One method to improve the water vapor barrier of cellulose films is incorporation of hydrophobic substances (fatty acids, beeswax, lipids) into the hydrocolloid matrix either by emulsification of hydrophobic substances and hydrocolloid aqueous solution before drying to obtain film, or the formation of bilayer films with a hydrophobic layer over the hydrocolloid based film. Many authors have studied the influence of hydrophobic substances addition on the properties of edible films [11–15]. Some researchers applied edible film based on composites of cellulose derivatives and hydrophobic substances to the fruit coating [16].

Cellulose sulfate (CS), prepared by partial or complete substitution of the 6-hydroxyl groups (–OH) with sulfate group (–SO₃H), is a newly developed cellulose derivatives for several medical and biotechnological applications because of its biocompatibility and easy biodegradability [17–19]. According to our previous study [20,21], the film based on CS had poor water vapor barrier due to its excellent solubility. It is important to increase the water vapor barrier of the CS based packaging film for extending its application. Hydrocolloids films incorporation of lipid can result in better functionality than films of single component. Glycerol was also used in film as one of the most popular plasticizers used in film-making techniques, due to stability and compatibility with hydrophilic bio-polymer [22]. The objective of this work was to evaluate the influence of oleic acid incorporation into CS film on the mechanical, optical, structural, and water vapor barrier properties of CS films as compared with the pure CS films, using glycerol as plasticizer.

2. Results and Discussion

2.1. Rheological Behavior of the Film-Forming Emulsions

The viscosity of film-forming solution is important to avoid non-uniform in thin liquid film after coating. As shown in Figure 1, the viscosity of the CS/OA blends increased from 346 mPa·s⁻¹ to 423 mPa·s⁻¹ when the OA content varied from 0% to 25%. The emulsion viscosity is influenced by various factors, mainly by the continuous phase viscosity, interfacial film viscosity and droplet size [23]. In this work, the interfacial film viscosity was same to the continuous phase viscosity, because all of the emulsions were prepared with the same CS content under the same homogenization conditions. The variation of emulsion viscosity was mainly attributed to oil droplet size caused by variation of OA content in emulsion and the homogenization velocity. The O/W and W/O/W were possible oil-in-water structure, but O/W was more stable. In O/W structure, hydrogen bond formed between carboxyl (–COOH) of oleic acid and hydroxyl (–OH) of cellulose sulfate as shown in Graph 1. Generally, the size distributions of oil droplet in O/W emulsions were trimodal with peak maximum, respectively, at 0.2 μm (I), 10 μm (II) and 100 μm (III), in which the predominance was population II according to previous research [23,24]. The number of droplet population increased with OA addition, which lead to the viscosity increasing due to the enhanced contact probability between O/W interfacial area and coaxial cylinder of viscometer. When OA content reached a critical value, the number of droplet do not increase any more at the same homogenization conditions, whereas the size of droplet increased due to the coalescence of droplets. Hence, the OA content over critical value would decrease the homogeneity of the O/W emulsion; correspondingly, the structure of the film became uneven.

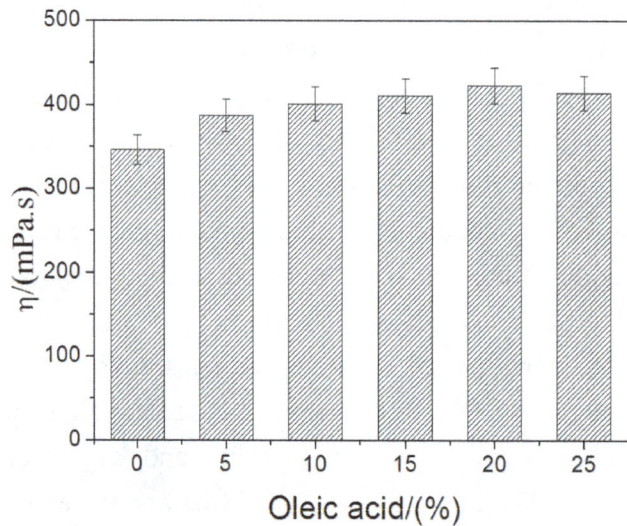

Figure 1. The effect of oleic acid on the viscosity of film forming emulsions.

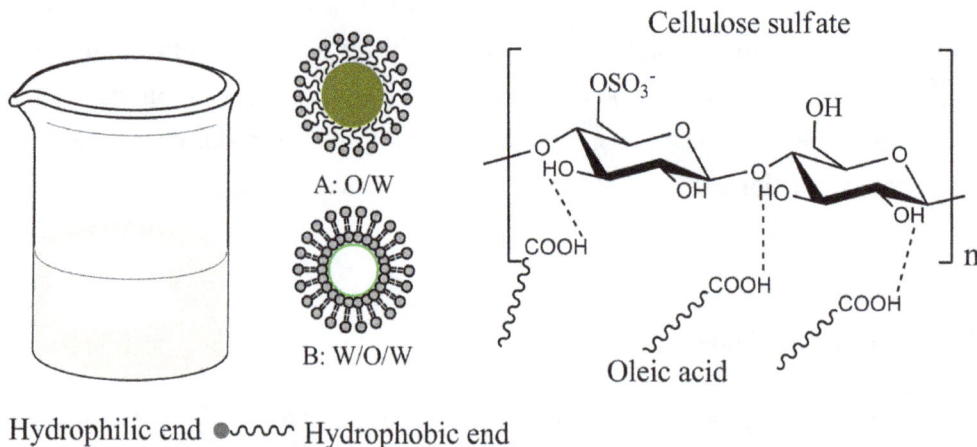

Graph 1. The schematic for emulsion of oleic acid, cellulose sulfate, glycerol and water.

2.2. Microstructure of CS-OA Films

The morphology of the CS film and composite film (CS + 25% OA) are compared in Figure 2 to investigate their microstructure. The surface of CS film and composite film were smooth. The circle splotch of different size on the film surface in Figure 2B and C was observed. It should be the function of the OA diffusion on the surface. The splotch formation was accompanied with the CS-OA film formation during water evaporating, as shown in Figure 2F. During water evaporating, the oil droplet in film emulsion will move down with the liquid level. The intermolecular distance of CS becomes closer with the decreasing of water content in emulsion; the hydrogen bonds between CS molecules were strong enough to prevent the oil droplet moving in film emulsion. Pressure at O/W interface increased dramatically with the water continuous evaporation, especially near the interface of the film, correspondingly spherical droplet became ellipsoidal and oil accelerated diffusion into the film matrix over time, like a droplet of oil on paper. Though the number of 100-μm (III) droplets was small, it definitely diffuses at film surface, because its size was much bigger than thickness of the film. Therefore, more large-size splotches were observed in film surface, as shown in Figure 2. Fabra *et al.* [25] reported that the lipid particle size in the film emulsion containing ι-carrageenan, glycerol and lipids mixtures of oleic acid (OA)/beeswax (BW) varied from 4.3 to 20.3 μm with the ratio variation between OA and BW lipid. Nevertheless, Ghasemlou *et al.* [26] reported that the $D_{3,2}$ (volume-surface mean diameter) of lipid particles in composite film emulsion based on kefiran and oleic acid (OA) varied from 1.13 μm to 2.06 μm. The results were quite different from each other. In our research, the lipid particle was not observed in the surface microstructure and the size of splotch on surface varied from ~10 to ~50 μm, according to Figure 2. Limpisophon *et al.* [27] also indicated that the films with oleic acid did not have crystalline particles like films with stearic acid, since oleic acid is in a liquid state at room temperature (m.p. = 13–14 °C) in edible films based on blue shark (*Prionace glauca*) skin gelatin. Cross-section images of CS films and CS-OA films with 25% OA are compared in Figure 2, in which the cross-sectional microstructure of CS-OA composite film was rougher than that of CS film. The CS-OA blend film also showed micro-cracking matrices and micro pores of different shapes and sizes, which meant the condense structure of film destroyed by the incorporation of oleic acid. The main pore size was 0.2 μm (I) according to Figure 2E, which was different from the droplet size in emulsion reported [23,24].

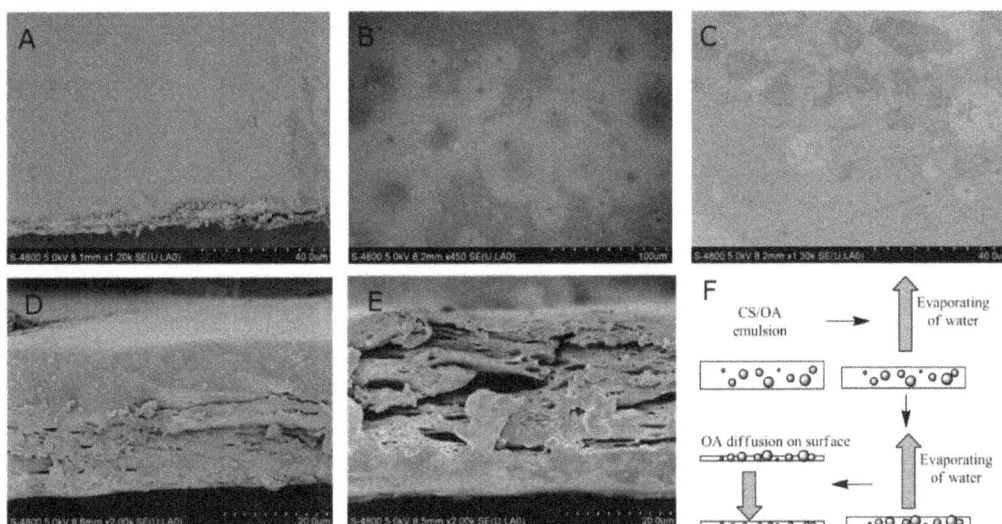

Figure 2. Surface appearance and cross-section appearance of films; surface of CS film (**A**); CS + 25% OA film (**B**); CS + 25% OA film (**C**) and cross section of CS film (**D**); CS + 25% OA film (**E**); and process of CS-OA film formation during water evaporating (**F**).

2.3. Surface Water Contact Angle of CS-OA Films

Contact angle is the angle (θ) between film surface and tangent at the droplet-film intersection. It can be used to indicate the hydrophobicity of the surface or the wettability of polymers [28]. The contact image between water droplet and CS-OA film at 0, 30, and 120 min is shown in Figure 3. The initial water contact angle for CS-OA film was 94°, whereas the water contact angle for pure CS film is 64.2°. Therefore the inclusion of OA in CS film increased the hydrophobicity. The droplet gradually permeated into film through the CS matrix. After 30 min the droplet left became small. After 120 min, some water unabsorbed was still observed on the surface of CS-OA films, compared with soluble hole formed on the film surface after 60 s for pure CS film [21]. This indicated that CS-OA film was less rapidly wetted, which explained the decreasing water solubility of CS-OA films as shown in Table 1.

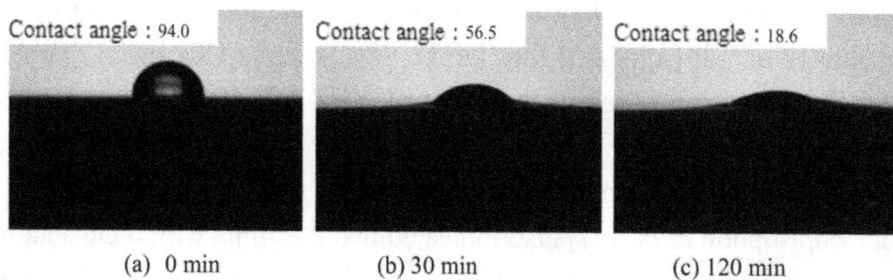

| Contact angle : 94.0 | Contact angle : 56.5 | Contact angle : 18.6 |
| (a) 0 min | (b) 30 min | (c) 120 min |

Figure 3. Surface water contact angle of oleic acid-cellulose sulfate films (CS + 25% OA film).

2.4. The Effect of Oleic Acid Content on Properties of CS-OA Films

Some properties of CS-OA films with different OA content, such as the thickness, flexibility, integrity, water solubility, oil permeability, transparency, mechanical properties and water vapor permeability are shown in Table 1. The integrity and folding property of all films are good, which was consistent with their easy peeling from the Teflon surface. The thickness, solubility and oil permeability of the CS-OA films did not change significantly under different OA content. The solubility and the transparency of composite films decreased gradually with addition of OA in film-forming emulsion, correspondingly the thickness of composite films and oil permeability increased with the OA content. The increasing of oil permeability can be attributed to the tunnel provided by the OA in the film.

Table 1. The effects of OA content on properties of oleic acid-cellulose sulfate films.

Sample	δ (μm)	Inte	Fold	S (h)	OP (%)	T (%)	TS (MPa)	ε (%)	WVP × 10⁻¹¹ (gm⁻¹s⁻¹Pa⁻¹)
2g CS + 0.3g Gly	25.00 ± 0.16 ᵃ	good	good	0.02 ± 0.01 ᵃ	0.2 ± 0.1% ᵃ	89.3 ± 3.2 ᵃ	14.5 ± 1.5 ᵃ	27.9 ± 0.8 ᵃ	3.92 ± 0.23 ᵃ
2g CS + 0.3g Gly + 0.1g OA	25.02 ± 0.10 ᵃ	good	good	1.50 ± 0.10 ᵇ	0.2 ± 0.2% ᵃ	85.0 ± 5.6 ᵃ	28.2 ± 2.1 ᵇ	18.6 ± 0.6 ᵇ	3.49 ± 0.21 ᵃ
2g CS + 0.3g Gly + 0.2g OA	25.85 ± 0.18 ᵇ	good	good	2.33 ± 0.08 ᶜ	1.9 ± 0.2% ᵇ	76.9 ± 2.8 ᵇ	37.1 ± 1.7 ᶜ	13.9 ± 0.3 ᵃ	2.41 ± 0.16 ᵇ
2g CS + 0.3g Gly + 0.3g OA	29.04 ± 0.12 ᶜ	good	good	4.45 ± 0.12 ᵈ	9.0 ± 0.3% ᶜ	73.5 ± 4.8 ᵇ	43.5 ± 1.3 ᵈ	9.2 ± 0.5 ᵈ	1.91 ± 0.12 ᵇᶜ
2g CS + 0.3g Gly + 0.4g OA	36.60 ± 0.26 ᵈ	good	good	5.13 ± 0.50 ᵈᵉ	14.6 ± 0.5% ᵈ	67.4 ± 2.3 ᶜ	36.8 ± 1.5 ᵉ	7.2 ± 0.3 ᵉ	1.57 ± 0.21 ᶜ
2g CS + 0.3g Gly + 0.5g OA	55.20 ± 0.32 ᵉ	good	good	5.55 ± 0.36 ᵉ	25.1 ± 0.8% ᵉ	68.2 ± 3.1 ᶜ	34.3 ± 2.2 ᵉ	6.4 ± 0.5 ᵉ	1.52 ± 0.14 ᶜ

δ: thickness; Inte: integrity; Fold: folding properties; S: solubility time; OP: oil permeability; T: transparency; TS: tension strength; ε: elongation at break; WVP: water vapor permeability; Mean ± standard deviation. Different letters represent significant differences ($p < 0.05$) according to the LSD test.

The ability of the coating to form a continuous layer over the product and the durability of the film are important, which can be reflected partly by mechanical properties listed in Table 1. The TS increased, but the ε decreased, when increasing OA content below 15%. The high TS and low ε indicated stronger and less extendible films. OA content over 15% decreased the values of all mechanical parameters, forming weaker and less extendible films. Elongation of CS-OA film significantly ($p < 0.05$) decreased when oleic acid was incorporated into the CS matrix, which has already been reported in other hydrocolloids films containing lipids [29–31]. It was explained by the fact that lipids were unable to form a cohesive and continuous matrix in the film. Oleic acid has also been reported to increase elongation of soy protein, corn zein, egg white and HPMC films [13,32,33]. They explained this as a plasticizing effect of unsaturated oleic acid. The effect of lipids on mechanical of hydrocolloids film may be dependent on the basic matrix properties of biomaterials, the interaction of the polymer molecular, the component of the film and the size distribution of the lipid droplet.

The WVP values of the CS-OA emulsified films are presented in Table 1. As expected, the water vapor permeability decreased when the OA content increased. Incorporation of fatty acids caused a significant difference ($p < 0.05$) between the WVP of the CS films containing different OA content. The WVP of hydrocolloids–fatty acid films decreased as the content of fatty acids increased. The WVP decreased from 3.92×10^{-11} to 1.52×10^{-11} gm^{-1}s^{-1}Pa^{-1} as OA content reached 25% of the CS in the CS-OA film. The OA dispersed in the CS film decreased the practical interfacial area exposed to water vapor. In general, the relative polarity of the support polymer and the type of lipid has the strongest influence on the water vapor barrier of emulsified films. Similar results were obtained by other researches [34–36]. However, as the amounts of fatty acids increased from 20% to 25%, no significant differences ($p > 0.05$) in the WVP values among the emulsified films were observed. Fabra et al. [37] showed that when the beeswax content added came to over 30% of the total lipid phase (70:30 OA:BW relationship) no further reduction in the WVP of sodium caseinate films was observed. The poor dispersion of lipid in the film system with the increasing lipid content may be account for the result. Size, distribution and physical state of the lipid, and polymorphism also seem to play a role in WVP, especially when the lipid content is over a critical value.

2.5. The Interactions among Components of Edible Film

When two or more substances are mixed, physical blends versus chemical interactions can be reflected by the changes in characteristic spectra peaks [38]. FTIR spectra of the CS film with different OA content are shown in Figure 4. The assignments proposed for the bands observed were annotated. The strong broad band observed in the 3500–3000 cm^{-1} range was attributable to hydrogen bond between different O–H groups from OH of glucosidic ring, OH carboxylic function of oleic acid acids and OH of glycerol. A strong broad band at 1207 cm^{-1} was attributed to the –O-SO$_3^-$ stretching mode. A strong and composite band with a maximum intensity at 1066 cm^{-1} was assigned to the stretching mode of the C–O bond. A medium weak band at 2949 cm^{-1} was assignable to C–H stretching mode, which was more intense for CS-OA film than that for pure CS film. Several medium bands in the region 1400–1200 cm^{-1} was assigned to the C-H bending and wagging modes, and O–H bending mode. A weak band at 807 cm^{-1} was assigned to the stretching mode of the glucosidic ring from CS. The new frequency at 2853 cm^{-1} for CS-OA film was attributed to the C–H stretching mode of the aliphatic chain of the fatty acids. The intense IR band at 1702 cm^{-1} and 1620 cm^{-1} corresponded to C=O and COO$^-$ stretching mode of oleic acid, respectively. These characteristic peaks of fatty acids only appeared in the CS-OA composite films. No additional chemical grouping was found on the FTIR spectra, which indicated no chemical bond formed between the lipid and CS.

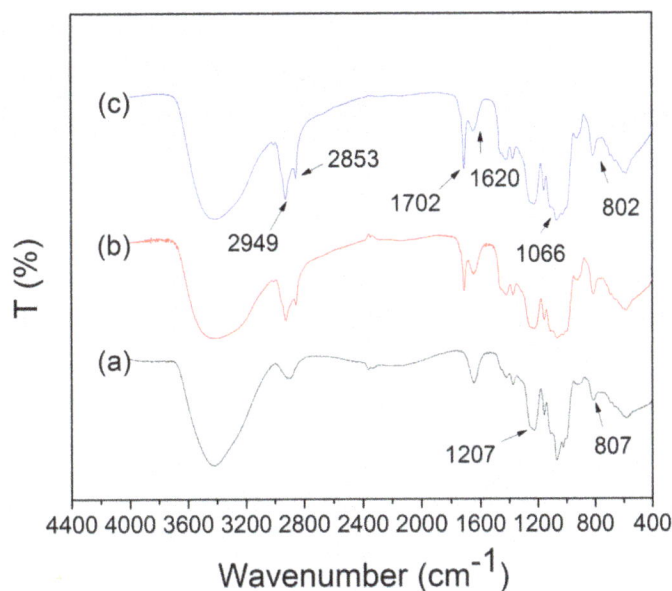

Figure 4. FTIR spectra of CS-OA films (a) 0%; (b) 10%; (c) 25% OA.

X-ray diffractogram (XRD) was used to investigate crystal structure, and assess the compatibility of CS and oleic acid. XRD of CS film with different contents of oleic acid blended were measured as shown in Figure 5. The XRD of CS film and CS-OA film showed two main crystalline reflections at 7.9° and 23.1°, presenting the characteristic of major amorphous structures for all films, which was similar with the fine structure of CMC [39]. No new diffraction peaks were observed in composite films, suggesting no intermolecular interactions between CS and OA.

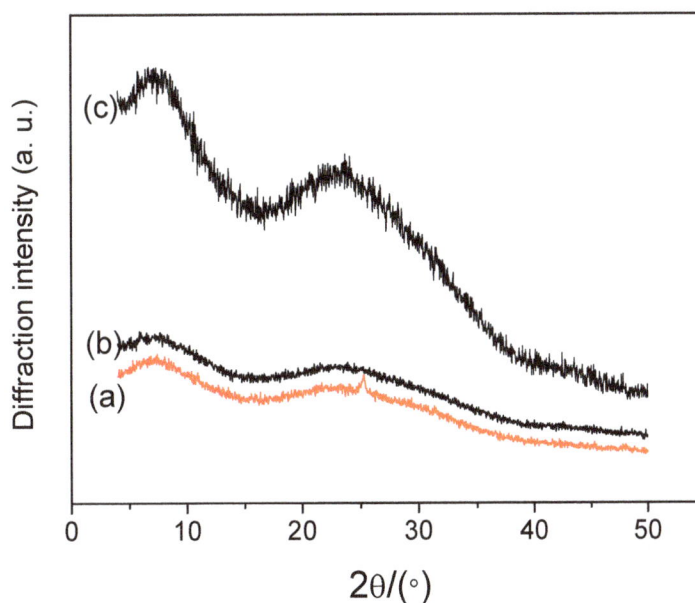

Figure 5. XRD patterns of CS-OA films (a) 0%; (b) 10%; (c) 25% OA.

3. Experimental Section

3.1. Materials

Cellulose sulfate (CS) was synthesized using heterogeneous method [17], with dynamic viscosity ($\eta_{2\%}$, 2 wt % solution) of 346 mPa·s^{-1} and average substitution degree of 0.4. Glycerol and oleic acid (OA, C18:1) were purchased from Sinopharm Chemical Reagent Co., Ltd. (China). Distilled water was used to form samples preparation.

3.2. Preparation of Cellulose Sulfate Films

Four different formulations based on cellulose sulfate (CS), glycerol (Gly) and oleic acid (OA) were prepared. CS (2 g) and Gly (0.3 g) were dispersed in 100 mL water in order to obtain polysaccharide dispersions. The OA fraction was incorporated in a CS–OA ratio of 1:0.05, 1:0.10, 1:0.15, 1:0.2, 1:0.25 and the mixture was homogenized at 13,000 rpm for 1 min, under vacuum, using a rotor-stator homogenizer (Ultraturrax T25, Janke and Kunkel, Germany). Then the film forming dispersions were gently spread over a leveled Teflon plate (150 mm diameter, Wei Xin Instrument Co., Ltd., Yixing, China) with 0.18 g cm^{-2} and dried for approximately 4 h at 60 °C and 45% RH. Afterwards, the films were peeled from the casting surface and stored in desiccators at 75% RH for further testing. All treatments were made in triplicate.

3.3. Rheological Behavior of the Film Forming Emulsions

The dynamic viscosity of the film-forming solution was measured at 30 ± 0.5 °C with a NDJ-5S rheometer (model NDJ-5S, Fangrui Instrument Co. Ltd., Shanghai, China). The range of shear rate, 100–300 s^{-1}, was used because it covered all the concentrations of the samples using the same coaxial cylinder device. Each sample was analyzed in triplicate.

3.4. Fourier Transform Infrared Spectroscopy (FTIR)

The FTIR transmission spectra of the film prepared was recorded on an instrument (Shimadzu FTIR 8400S, Kyoto, Japan) in the wavenumber range of 4000–400 cm^{-1}, using Attenuated Total Reflectance mode (ATR). Spectra were recorded at a resolution of 4 cm^{-1} and 400 scans were carried out to obtain a high signal-to-noise ratio spectrum.

3.5. X-ray Diffraction

X-ray diffraction (XRD) patterns of all samples were analyzed by a X'Pert PRO XRD system (PANalytical, Almelo, The Netherlands) at 25 °C and 75% RH, between $2\theta = 10°$ and $2\theta = 80°$ using Kα Cu radiation ($\lambda = 1.542$ Å), 40 kV and 40 mA with a step size of 4°. Samples were cut into 2 cm squares for analysis.

3.6. Film Thickness

The thickness (δ) of the sample was measured (exactness of ±0.001 mm) by a digital external micrometer (Mitutoyo Co., Tokyo, Japan) at ten different points of the film. Samples were conditioned at 25 °C and 75% RH (a saturated NaCl solution) for 24 h before they were measured.

3.7. Scanning Electron Microscopy

The samples were maintained in a desiccator with P_2O_5 for two weeks to ensure water in films completely removed. The cross-sections of the films were observed by cryofracture of films frozen in liquid N_2. The microstructure of the film was analyzed using a scanning electron microscope (Hitachi S4800, Tokyo, Japan). Samples were fixed on a copper stubs, gold coated, and observed using an accelerating voltage of 5 kV.

3.8. Mechanical Properties

Tensile strength (TS) and elongation at break (ε) of the film were measured using a Instron Universal Testing Machine (Instron Corp., model 5569, MA, USA) according to the standard method [40]. Test samples, 25 mm × 100 mm, were cut from each film and fixed with an initial grip separation of 30 mm. Five replicates of each film were then pulled apart at crosshead speed of 20 mm/s and preload of 2 N. The average thickness of film was 25 ± 2 µm. TS (MPa) was calculated by the Equation (1):

$$TS = F_{max}/A \tag{1}$$

where F_{max} is the maximum force (N) loaded on the specimen before pulling apart; A is the cross-sectional area (m^2) of the specimen. ε is defined as the Equation (2):

$$\varepsilon = \Delta L/L_0 \times 100\% \tag{2}$$

where ΔL is the film elongation at the moment of rupture (mm) and L_0 is the initial length between the grips.

3.9. Water Vapor Permeability (WVP)

WVP data were measured according to the standard ASTM method (1995) [41]. Test samples, 90 × 90 mm, were cut from each film and sealed on cups which was previously filled with fused anhydrous $CaCl_2$ (RH = 0%). And then the cups were placed into a humidity chamber at 25 °C and 75% RH (saturated NaCl solution) for 3 days. The sealed cups were weighed periodically (±0.0001 g) to calculate water vapor transported into the cup. The data, weight $vs.$ time, was linearly regressed to calculate the slope. The water vapor transmission rate (WVTR) through the film was calculated from the slope ($\Delta w/\Delta t$) of the fitted line divided by the test area (A) as Equation (3), (g $s^{-1}m^{-2}$).

$$WVTR = \Delta w/(\Delta t \cdot A) \tag{3}$$

where w is the weight of water transported into the cup (g), t is the time for weight change (s), A is the area exposed to water vapor transfer (m^2). The WVP ($gm^{-1}s^{-1}Pa^{-1}$) is calculated as Equation (4).

$$WVP = (WVTR \times \delta)/\Delta p \tag{4}$$

where δ is the film thickness (m) and Δp is the water vapor partial pressure difference across the two sides of the film (Δp = p(RH₂ − RH₁) = 2081.325 Pa, where p is the saturation vapor pressure of water at 25 °C, RH₂ = 75%, RH₁ = 0%) (Pa).

3.10. Contact Angle

Contact angle of the film was measured using a Video-Based Contact Angle Meter model OCA 20 (DataPhysics Instruments GmbH, Filderstadt, Germany). A droplet of 3 μL ultrapure water was dispensed on each film surface using a micro syringe. The contact angle was recorded by analyzing the shape of a sessile drop after it had been placed over the surface of each film at different time. Image analysis was carried out by SCA20 software. Each sample was tested with three drops and three measurements were conducted for each water drop.

3.11. Flexibility

The flexibility of the film was determined by a bending method [21]. Each sample was cut into the size of 20 × 40 mm and folded completely at the middle. The film was positively and negatively folded in turn until the film appeared rupture. The flexibility was rated as poor (folding number < 20), middle (20 ≤ folding number < 50), good (50 ≤ folding number < 100) and excellent (folding number ≥ 100).

3.12. Integrity

The CS film was deemed as good integrity of there was no breakage during peeling from the petri dish and *vice versa*.

3.13. Oil Permeability

Oil permeability of the film was determined following Hu's method [42]. The mouth of a glass test-tube filled with 3 g soybean oil (interior diameter: 25 mm, and outer diameter: 27 mm) was covered with CS film (50 × 50 mm) and sealed tightly. The tube was upside down on a piece of filter paper. The oil penetration was judged according to the infiltration of soybean oil on filter paper as time going. Each sample was observed in triple.

3.14. Water Solubility

The solubility of film in water was characterized by the resolve time of the film. The test sample was placed into deionized water at 25 °C until it was completely dissolved. Three samples of each film were tested.

3.15. Transparency

Film specimen was cut into a rectangle piece (20 × 10 mm) and attached on the wall of a test cell directly. The light transmission (T %) of sample was measured at 560 nm using a UV-722 spectroscope (Rayleigh Corp., model 722, Beijing, China) and air was used as the reference. Three samples of each film were tested.

3.16. Statistical Analysis

Data for each test were are presented as mean ± SD after statistical analysis. The significance in the difference between factors and levels was evaluated by the analysis of variance (ANOVA). Comparison of the means was done employing a Tukey test to identify which groups were significantly different from others ($p < 0.05$).

4. Conclusions

The above findings indicated that the hydrophobicity of CS films can be regulated by incorporation of OA. The OA incorporation into CS film decreased the cohesive matrix of the film, which was severely affected by the size distribution of the OA droplet in film-forming solution. The solubility and transparence of the CS-OA decreased, while the thickness and oil permeability dramatically increased with the OA content increasing. Films with the addition of OA presented better water vapor barrier properties as compared to pure CS films. The WVP decreased from 3.92×10^{-11} to 1.45×10^{-11} $gm^{-1}s^{-1}Pa^{-1}$ when OA content varied from 0 to 30%. The TS and E% of the CS-OA film decreased, which indicated the film was more fragile than pure CS film. Films with a 1:0.15:0.15 CS:glycerol:OA ratio showed the most adequate functional properties when considering both tensile and water transport properties (TS: 43.5 ± 1.3 MPa, E%: 9.2 ± 0.5, WVP: $1.91 \pm 0.12 \times 10^{-11}$ g $m^{-1}s^{-1}Pa^{-1}$).

Acknowledgments

The authors express their thanks for the support from the Nature Science Foundation of China (20906035), the Promotion Program for Young and Middle-aged Teacher in Science and Technology Research of Huaqiao University (ZQN-PY109) and fok ying-tong education foundation for young teachers (133037).

Author Contributions

Guo Chen and Jun Zhao conceived and designed the experiments; Bin Zhang performed the experiments; Bin Zhang and Guo Chen analyzed the data; Guo Chen contributed reagents/materials/analysis tools; Guo Chen and Bin Zhang wrote the paper.

Conflicts of Interest

The authors declare no conflict of interest.

References

1. Debeaufort, F.; Quezada-Gallo, J.A.; Voilley, A. Edible films and coatings: Tomorrow's packagings: A review. *Crit. Rev. Food Sci. Nutr.* **1998**, *38*, 299–313.
2. Jimenez, A.; Fabra, M.J.; Talens, P.; Chiralt, A. Edible and biodegradable starch films: A review. *Food Bioprocess Technol.* **2012**, *5*, 2058–2076.
3. Clasen, C.; Kulicke, W.M. Determination of viscoelastic and rheo-optical material functions of water-soluble cellulose derivatives. *Progr. Polym. Sci.* **2001**, *26*, 1839–1919.

4. Ortega-Toro, R.; Jiménez, A.; Talens, P.; Chiralt, A. Properties of starch-hydroxypropyl methylcellulose based films obtained by compression molding. *Carbohydr. Polym.* **2014**, *109*, 155–165.

5. Slavutsky, A.M.; Bertuzzi, M.A. Water barrier properties of starch films reinforced with cellulose nanocrystals obtained from sugarcane bagasse. *Carbohydr. Polym.* **2014**, *110*, 53–61.

6. De Moura, M.R.; Lorevice, M.V.; Mattoso, L.H.C.; Zucolotto, V. Highly stable, edible cellulose films incorporating chitosan nanoparticles. *J. Food Sci.* **2011**, *76*, S25–S29.

7. Malmiri, H.J.; Osman, A.; Tan, C.P.; Rahman, R.A. Effects of edible surface coatings (sodium carboxymethyl cellulose, sodium caseinate and glycerol) on storage quality of berangan banana (*musa sapientum cv. Berangan*) using response surface methodology. *J. Food Process. Preserv.* **2012**, *36*, 252–261.

8. Wang, S.; Kuang, X.; Li, B.; Wu, X.; Huang, T. Physical properties and antimicrobial activity of chilled meat pads containing sodium carboxymethyl cellulose. *J. Appl. Polym. Sci.* **2013**, *127*, 612–619.

9. Arrieta, M.P.; Fortunati, E.; Dominici, F.; López, J.; Kenny, J.M. Bionanocomposite films based on plasticized pla–phb/cellulose nanocrystal blends. *Carbohydr. Polym.* **2015**, *121*, 265–275.

10. Fortunati, E.; Peltzer, M.; Armentano, I.; Torre, L.; Jiménez, A.; Kenny, J.M. Effects of modified cellulose nanocrystals on the barrier and migration properties of pla nano-biocomposites. *Carbohydr. Polym.* **2012**, *90*, 948–956.

11. Ayranci, E.; Tunc, S. A method for the measurement of the oxygen permeability and the development of edible films to reduce the rate of oxidative reactions in fresh foods. *Food Chem.* **2003**, *80*, 423–431.

12. Bravin, B.; Peressini, D.; Sensidoni, A. Development and application of polysaccharide-lipid edible coating to extend shelf-life of dry bakery products. *J. Food Eng.* **2006**, *76*, 280–290.

13. Navarro-Tarazaga, M.L.; Del Rio, M.A.; Krochta, J.M.; Perez-Gago, M.B. Fatty acid effect on hydroxypropyl methylcellulose-beeswax edible film properties and postharvest quality of coated 'ortanique' mandarins. *J. Agric. Food Chem.* **2008**, *56*, 10689–10696.

14. Navarro-Tarazaga, M.L.; Massa, A.; Perez-Gago, M.B. Effect of beeswax content on hydroxypropyl methylcellulose-based edible film properties and postharvest quality of coated plums (cv. Angeleno). *Lwt-Food Sci. Technol.* **2011**, *44*, 2328–2334.

15. Soazo, M.; Rubiolo, A.C.; Verdini, R.A. Effect of drying temperature and beeswax content on moisture isotherms of whey protein emulsion film. *Procedia Food Sci.* **2011**, *1*, 210–215.

16. Contreras-Oliva, A.; Rojas-Argudo, C.; Perez-Gago, M.B. Effect of solid content and composition of hydroxypropyl methylcellulose-lipid edible coatings on physico-chemical and nutritional quality of 'oronules' mandarins. *J. Sci. Food Agric.* **2012**, *92*, 794–802.

17. Chen, G.; Zhang, B.; Zhao, J.; Chen, H.W. Improved process for the production of cellulose sulfate using sulfuric acid/ethanol solution. *Carbohydr. Polym.* **2013**, *95*, 332–337.

18. Sanz-Nogues, C.; Horan, J.; Ryan, G.; Kassem, M.; O'Brien, T. Encapsulation of human mesenchymal stem cells in sodium cellulose sulfate-based microcapsules requires immortalization. *Cytotherapy* **2014**, *16*, S94.

19. Zhu, L.Y.; Yan, X.Q.; Zhang, H.M.; Lin, D.Q.; Yao, S.J.; Jiang, L. Determination of apparent drug permeability coefficients through chitosan-sodium cellulose sulfate polyelectrolyte complex films. *Acta Phys.-Chim. Sin.* **2014**, *30*, 365–370.

20. Chen, G.; Liu, B.; Zhang, B. Characterization of composite hydrocolloid film based on sodium cellulose sulfate and cassava starch. *J. Food Eng.* **2014**, *125*, 105–111.

21. Chen, G.; Zhang, B.; Zhao, J.; Chen, H.W. Development and characterization of food packaging film from cellulose sulfate. *Food Hydrocoll.* **2014**, *35*, 476–483.

22. Chillo, S.; Flores, S.; Mastromatteo, M.; Conte, A.; Gerschenson, L.; Del Nobile, M.A. Influence of glycerol and chitosan on tapioca starch-based edible film properties. *J. Food Eng.* **2008**, *88*, 159–168.

23. Camino, N.A.; Pilosof, A.M.R. Hydroxypropylmethylcellulose at the oil-water interface. Part ii. Submicron-emulsions as affected by ph. *Food Hydrocoll.* **2011**, *25*, 1051–1062.

24. Mitidieri, F.; Wagner, J. Coalescence of o/w emulsiones stabilized by whey and isolate soybean proteins. Influence of thermal denaturation, salt adittion and competitive interfacial adsorption. *Food Res. Int.* **2002**, *35*, 547–557.

25. Fabra, M.J.; Hambleton, A.; Talens, P.; Debeaufort, F.; Chiralt, A.; Voilley, A. Influence of interactions on water and aroma permeabilities of ι-carrageenan-oleic acid-beeswax films used for flavour encapsulation. *Carbohydr. Polym.* **2009**, *76*, 325–332.

26. Ghasemlou, M.; Khodaiyan, F.; Oromiehie, A.; Yarmand, M.S. Characterization of edible emulsified films with low affinity to water based on kefiran and oleic acid. *Int. J. Biol. Macromol.* **2011**, *49*, 378–384.

27. Limpisophon, K.; Tanaka, M.; Osako, K. Characterisation of gelatin-fatty acid emulsion films based on blue shark (prionace glauca) skin gelatin. *Food Chem.* **2010**, *122*, 1095–1101.

28. Hong, S.D.; Ha, M.Y.; Balachandar, S. Static and dynamic contact angles of water droplet on a solid surface using molecular dynamics simulation. *J. Colloid Interface Sci.* **2009**, *339*, 187–195.

29. Zahedi, Y.; Ghanbarzadeh, B.; Sedaghat, N. Physical properties of edible emulsified films based on pistachio globulin protein and fatty acids. *J. Food Eng.* **2010**, *100*, 102–108.

30. Bertan, L.C.; Tanada-Palmu, P.S.; Siani, A.C.; Grosso, C.R.F. Effect of fatty acids and 'brazilian elemi' on composite films based on gelatin. *Food Hydrocoll.* **2005**, *19*, 73–82.

31. Yang, L.; Paulson, A.T. Effects of lipids on mechanical and moisture barrier properties of edible gellan film. *Food Res. Int.* **2000**, *33*, 571–578.

32. Handa, A.; Gennadios, A.; Hanna, M.A.; Weller, C.L.; Kuroda, N. Physical and molecular properties of egg-white lipid films. *J. Food Sci.* **1999**, *64*, 860–864.

33. Shellhammer, T.H.; Krochta, J.M. Whey protein emulsion film performance as affected by lipid type and amount. *J. Food Sci.* **1997**, *62*, 390–394.

34. De la Caba, K.; Pena, C.; Ciannamea, E.M.; Stefani, P.M.; Mondragon, I.; Ruseckaite, R.A. Characterization of soybean protein concentrate-stearic acid/palmitic acid blend edible films. *J. Appl. Polym. Sci.* **2012**, *124*, 1796–1807.

35. Nobrega, M.M.; Olivato, J.B.; Muller, C.M.O.; Yamashita, F. Addition of saturated fatty acids to biodegradable films: Effect on the crystallinity and viscoelastic characteristics. *J. Polym. Environ.* **2013**, *21*, 166–171.

36. Rezvani, E.; Schleining, G.; Sumen, G.; Taherian, A.R. Assessment of physical and mechanical properties of sodium caseinate and stearic acid based film-forming emulsions and edible films. *J. Food Eng.* **2013**, *116*, 598–605.

37. Fabra, M.J.; Talens, P.; Chiralt, A. Tensile properties and water vapor permeability of sodium caseinate films containing oleic acid-beeswax mixtures. *J. Food Eng.* **2008**, *85*, 393–400.

38. Wang, Z.; Zhou, J.; Wang, X.-X.; Zhang, N.; Sun, X.-X.; Ma, Z.-S. The effects of ultrasonic/microwave assisted treatment on the water vapor barrier properties of soybean protein isolate-based oleic acid/stearic acid blend edible films. *Food Hydrocoll.* **2014**, *35*, 51–58.

39. Martins, S.; Fernandes, J.B. A simple method to prepare high surface area activated carbon from carboxyl methyl cellulose by low temperature physical activation. *J. Therm. Anal. Calorim.* **2013**, *112*, 1007–1011.

40. American Society for Testing and Materials (ASTM). Standard test method for tensile properties of thin plastic sheeting. In *Standard D882. Annual Book of American Standard Testing Methods*; ASTM: Philadelphia, PA, USA, 2001.

41. American Society for Testing and Materials (ASTM). Standard test methods for water vapour transmission of materials. In *Standards Designations: E96–95. In Annual Book of ASTM Standards*; ASTM: Philadelphia, PA, USA, 1995; pp. 406–413.

42. Hu, G.; Chen, J.; Gao, J. Preparation and characteristics of oxidized potato starch films. *Carbohydr. Polym.* **2009**, *76*, 291–298.

4

Effect of Solids-To-Liquids, Na_2SiO_3-To-NaOH and Curing Temperature on the Palm Oil Boiler Ash (Si + Ca) Geopolymerisation System

Zarina Yahya [1,*], Mohd Mustafa Al Bakri Abdullah [2,3,*], Kamarudin Hussin [1,2], Khairul Nizar Ismail [4], Rafiza Abd Razak [1] and Andrei Victor Sandu [5]

[1] Center of Excellence Geopolymer and Green Technology, School of Materials Engineering, Universiti Malaysia Perlis (UniMAP), P.O. Box 77, D/A Pejabat Pos Besar, Kangar 01000, Perlis, Malaysia; E-Mails: vc@unimap.edu.my (K.H.); rafizarazak@unimap.edu.my (R.A.R.)

[2] Faculty of Engineering Technology, Uniciti Alam Campus, Universiti Malaysia Perlis, Sungai Chuchuh 02100, Padang Besar, Perlis, Malaysia

[3] Faculty of Technology, Universitas Ubudiyah Indonesia, Jl. Alue Naga, Kec. Syiah Kuala Desa Tibang 23536, Banda Aceh, Indonesia

[4] School of Environmental Engineering, Universiti Malaysia Perlis (UniMAP), P.O. Box 77, D/A Pejabat Pos Besar, Kangar 01000, Perlis, Malaysia; E-Mail: nizar@unimap.edu.my

[5] Faculty of Materials Science and Engineering, Gheorghe Asachi Technical University of Iasi, Blvd. D. Mangeron 41, Iasi 700050, Romania; E-Mail: sav@tuiasi.ro

* Authors to whom correspondence should be addressed; E-Mails: zarinayahya@unimap.edu.my (Z.Y.); mustafa_albakri@unimap.edu.my (M.M.A.B.A.)

Academic Editor: Jérôme Chevalier

Abstract: This paper investigates the effect of the solids-to-liquids (S/L) and Na_2SiO_3/NaOH ratios on the production of palm oil boiler ash (POBA) based geopolymer. Sodium silicate and sodium hydroxide (NaOH) solution were used as alkaline activator with a NaOH concentration of 14 M. The geopolymer samples were prepared with different S/L ratios (0.5, 1.0, 1.25, 1.5, and 1.75) and Na_2SiO_3/NaOH ratios (0.5, 1.0, 1.5, 2.0, 2.5, and 3.0). The main evaluation techniques in this study were compressive strength, X-Ray Diffraction (XRD), Fourier Transform Infrared Spectroscopy (FTIR), and Scanning Electron Microscope (SEM). The results showed that the maximum compressive strength (11.9 MPa) was obtained at a S/L ratio and Na_2SiO_3/NaOH ratio of 1.5 and 2.5 at seven days of testing.

Keywords: POBA; geopolymer; alkaline activator; NaOH; sodium silicate

---

1. Introduction

The study of alkali-activated binder was started by Purdon in the 1940s in which blast furnace slag was activated with sodium hydroxide (NaOH) solution [1]. After that, in the late 1950s and 1960s Glukhovsky invented an alkali activated system which contained calcium silicate hydrate (CSH) and aluminosilicate phases [2]. Besides that, the term geopolymer was introduced by Davidovits in 1972 where it was described as tri-dimensional alumina-silicate which formed at low temperature and short time from naturally occurring alumina-silicates such as kaolin [3]. Geopolymerisation is a geosynthesis where the reaction integrates minerals from alumino-silicate sources [4,5] and an exothermic process is involved. Any raw material rich in silica and alumina or pozzolanic materials which can be dissolved in an alkaline activator solution can hence undergo a geopolymerisation process.

Geopolymer systems can be divided into two types of binding system which are silica-aluminum (Si + Al) with medium to high alkaline solution and silica-calcium (Si + Ca) with a mild alkaline solution [6]. For the (Si + Al) binding system, the source materials included in this system are class F fly ash and metakaolin due to having silica and alumina content as the main composition. Meanwhile, for the (Si + Ca) system, ground granulated blast furnace slag (GGBS) was included in this system due to its main composition which is silica and calcium. The hydration products of these two systems are also different where for the (Si + Ca) system, calcium silicate hydrate (CSH) is the main product and zeolite like polymers are the main products for the (Si + Al) system.

For the alkaline activator solution, the combination of NaOH and sodium silicate solution (Na_2SiO_3) leads to higher geopolymerisation rates compared to hydroxide alone [6]. Moreover, it was proved by Xu and Van Deventer [7] where different source materials of alumina-silicate mineral are used to produce geopolymer, it required additional silica (Si) for the geopolymerisation process. Alkali hydroxide is required for the dissolution process of aluminosilicate sources, while Na_2SiO_3 solution acts as binder [2].

The mix design for geopolymers can be divided into solids/liquid (S/L) and sodium silicate/NaOH (Na_2SiO_3/NaOH) ratio which is important in developing the mechanical strength of the geopolymer [6,8]. Fresh geopolymer paste with a high S/L ratio had a low viscosity while geopolymer paste with a low S/L ratio resulted in high viscosity [9]. Besides that, geopolymer with a low S/L ratio could accelerate the dissolution rate, but it was not applicable to the polycondensation process when high concentrations of NaOH were used [10]. In addition, the low S/L ratio contributed to low strength due to insufficient formation of binder [11].

The influence of the Na_2SiO_3/NaOH ratio (0.4, 1.5, 5.0, 10.0, and 15.0) in natural zeolite based geopolymers showed that increasing the Na_2SiO_3/NaOH ratio up to 1.5 increased the compressive strength, but beyond that the strength was decreased [12]. This may be due to excessive sodium silicate that retarded the geopolymerisation process by the precipitation of Al-Si phase, which prevented contact between the reacting material and activating solution and decreased the activator content [13]. Researchers have suggested that the optimum Na_2SiO_3/NaOH ratio to produce high strength geopolymer is in the range 0.67–1.00 [14]. Meanwhile, Hardjito *et al.* [15] investigated the effect of two different Na_2SiO_3/NaOH ratios (0.4 and 2.5) on the performance of fly ash based geopolymer. They found that when the ratio increased the strength of geopolymer also increased.

Malaysia is one of the world's largest producers of palm oil products and as such the waste material from this industry is also widely available and estimated to be about 44.9 million tonnes. The solid wastes are burned in the boiler to generate electricity at the palm oil mill and the palm oil boiler ash (POBA) or bottom ash is obtained at the lower compartment of the boiler. Generally, palm oil fuel ash (POFA) is used as a cement replacement in concrete as POBA contains coarse particles. As such this study investigates the utilization of POBA in geopolymers and the effect of solids-to-liquids ratio (S/L), alkaline activator ratio (Na_2SiO_3/NaOH) and curing temperature on geopolymer paste. The results of the geopolymer paste are evaluated in terms of compressive strength, X-Ray Diffraction (XRD), Fourier Transform Infrared Spectroscopy (FTIR), and Scanning Electron Microscope (SEM).

2. Results and Discussion

2.1. Compressive Strength

The strength of the geopolymer paste with different solid/liquid (S/L) and Na_2SiO_3/NaOH ratios is shown in Figure 1. When the S/L and Na_2SiO_3/NaOH ratio increased, the compressive strength also increased. The maximum compressive strength (11.9 MPa) was achieved at S/L and Na_2SiO_3/NaOH ratio of 1.5 and 2.5. Furthermore, at S/L ratio 1.0, 1.25 and 1.5 the maximum compressive strength was obtained at a Na_2SiO_3/NaOH ratio of 2.5. The compressive strength of geopolymer paste increased when the amount of Na_2SiO_3 increased. Moreover, the use of Na_2SiO_3 helps to improve the geopolymerisation process by accelerating the dissolution of source material [7]. It was observed by Hardjito and Rangan [16] that increasing the Na_2SiO_3/NaOH ratio increased the geopolymerisation rate. Conversely, when the Na_2SiO_3/NaOH ratio was more than 3.0 the compressive strength tended to decrease for all S/L ratios. This may be due to excessive alkali content which retards the geopolymerisation process. It occurs when Al-Si phase precipitation prevents interaction between reacting material and alkaline activator thus reducing the activator concentration [12].

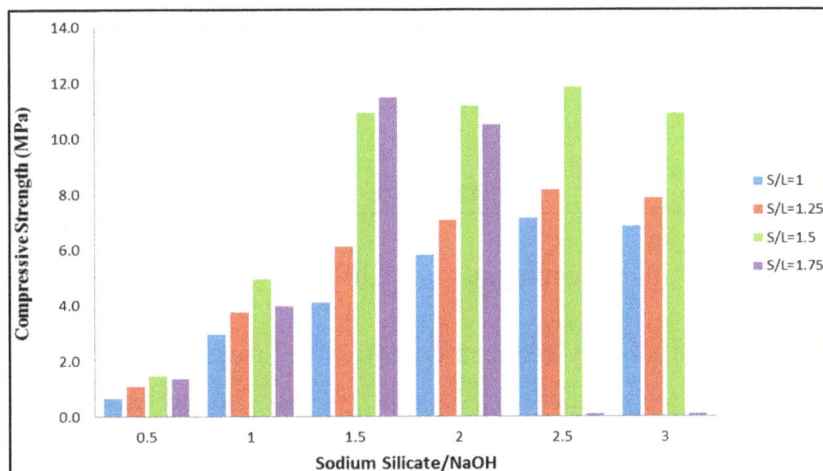

Figure 1. Compressive strength of geopolymer samples with different solid/liquid (S/L) and Na_2SiO_3/NaOH ratios at 7 days.

However, at S/L ratio 1.75 with Na_2SiO_3/NaOH ratios 2.5 and 3.0 the geopolymer paste was unable to mix due to lower workability. As such, geopolymer strength was unable to be acquired. This condition showed that the Na_2SiO_3/NaOH ratio was correlated to the workability and compressive strength of the geopolymer paste. For this S/L ratio, the maximum strength (11.5 MPa) was contributed by a geopolymer sample with a Na_2SiO_3/NaOH ratio 1.5.

Meanwhile, for S/L ratio 1.0 and 1.25 contributed a low strength of geopolymer paste and can be due to a high content of alkaline activator. The high content of alkaline activator produced excessive OH^- that was left in the system, thus weakening the geopolymer structure [17]. From this study it was shown that S/L and Na_2SiO_3/NaOH ratios did influence the compressive strength and the workability of geopolymers. The optimum mix design for POBA geopolymer paste was obtained at S/L and Na_2SiO_3/NaOH ratios of 1.5 and 2.5, respectively.

Figure 2 shows the effect of curing temperature on the compressive strength of geopolymer paste where at room temperature (RT) the lowest compressive strength (0.4 MPa) was obtained. Meanwhile, the maximum compressive strength (11.5 MPa) was obtained at a curing temperature 80 °C for 24 h. When the curing temperature increased, the compressive strength also resulted in an increment. The geopolymer samples displayed increment in strength in the range 3%–96% when cured at different temperatures. Hence, from this result it is indicated that geopolymer paste produced using POBA, required heat curing in order to increase the compressive strength. Palomo et al. [6] also mentioned that heat curing acts as an accelerator in the geopolymer production. Heat curing in geopolymers leads to a quicker geopolymerisation process, thus producing adequate strength within a very short period [16,18–20].

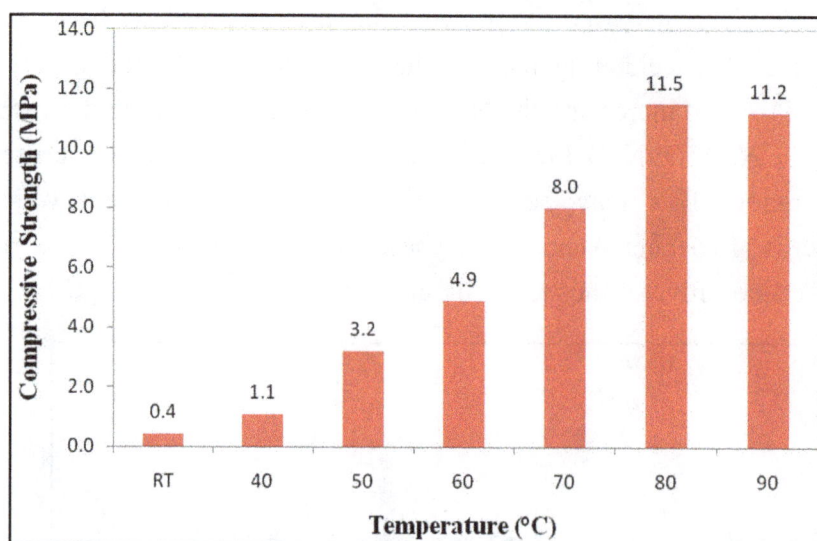

Figure 2. Compressive strength of geopolymer with different curing temperatures at 7 days.

Geopolymer samples exhibited slightly reduced compressive strength after curing at temperature 90 °C. Bakharev [21] found that geopolymer samples cured at 90 °C experienced a significant loss of moisture. As such, Hardjito et al. [17] concluded that curing samples at higher temperature does not essentially produce higher strength geopolymer products. Moreover, since the geopolymer samples that were produced using a 50mm mould had a high surface-to-volume ratio, this is more vulnerable to heat curing and also loss of moisture, and thus could lead to strength reduction when cured at high temperature [14].

Additionally, Chindaprasirt *et al.* [14] stated that to produce geopolymer with good strength requires the presence of moisture.

2.2. X-Ray Diffraction (XRD) Analysis

The X-Ray Diffraction (XRD) analysis for S/L of 1.0, 1.25, 1.5, and 1.75 with maximum compressive strength is presented in Figure 3. From the figure it is shown that the highest peak for all mix design was contributed by quartz. At ratio S/L 1.0, only quartz and cristobalite peaks were detected compared to S/L 1.25, 1.5, and 1.75. However, at ratio S/L 1.25, 1.5 and 1.75 peaks of albite were found at $2\theta = 13°$ and $2\theta = 28°$, approximately. The highest peak of albite was contributed by S/L 1.5 which is consistent with optimum compressive strength. The peak of albite was attributed to the strength of the geopolymer paste by forming a crystalline phase of the N-A-S-H (aluminosilicate gel) system [22].

The peaks of cristobalite and quartz at $2\theta = 21°$ and $2\theta = 22°$ in POBA still exists in geopolymer paste. Besides that, the geopolymer paste with the different mix design also demonstrated an amorphous to semi crystalline phase which is the same as with POBA. Since the crystalline peaks were detected more with a S/L ratio of 1.5, it can be concluded that the existence of these peaks helps to increase the strength of the geopolymer paste [23].

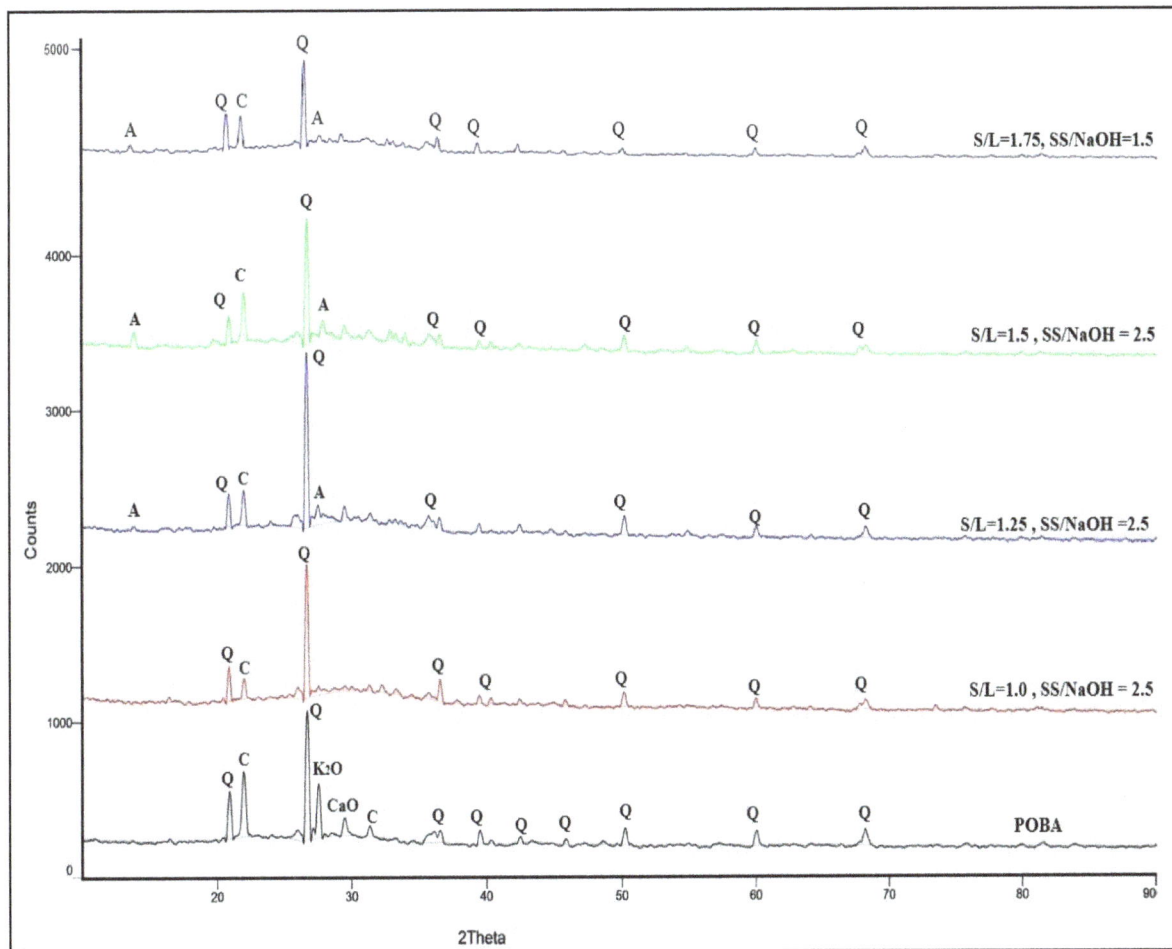

Figure 3. X-Ray Diffraction (XRD) analysis on geopolymer paste with different mix design (Q = quartz, C = cristobalite, A = albite, K₂O = potassium oxide, CaO = calcium oxide, C = calcium).

Figure 4 demonstrates the XRD analysis on geopolymer samples cured with different curing temperatures for 24 h. Quartz and cristobalite were spotted in all samples with maximum peaks of quartz. All the geopolymer samples still displayed the amorphous phase despite being cured at different temperatures. The remaining quartz and cristobalite in geopolymer samples showed that the originally used quantity had not fully reacted during the geopolymerisation process for all temperatures.

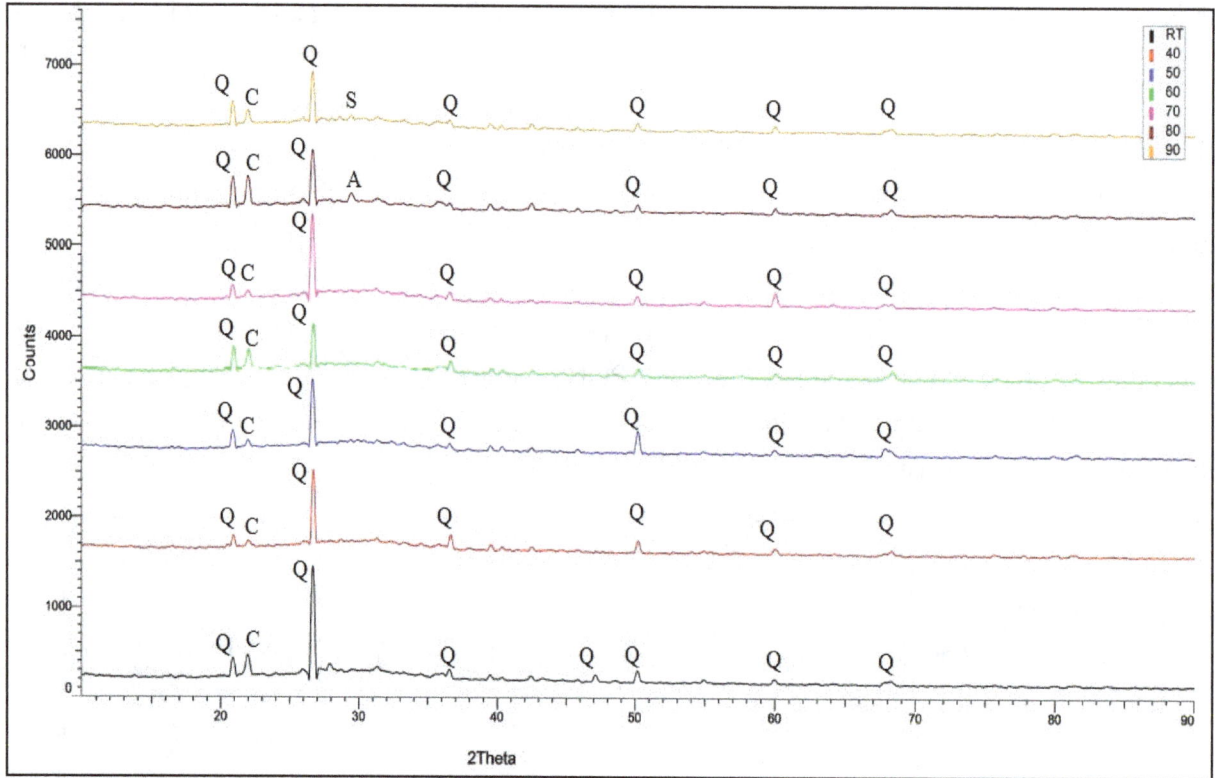

Figure 4. X-Ray Diffraction (XRD) analysis on geopolymer paste with different curing temperatures (Q = quartz, C = cristobalite, A = albite, S = sodium silicate).

2.3. Fourier Transform Infrared Spectroscopy (FTIR) Analysis

Figure 5 displays the IR spectra of geopolymer samples with the different mix design of S/L ratio. The geopolymer samples with optimum strength at each S/L ratio were analyzed in this study. The broad bands appearing at 2316–3351 cm^{-1} were due to stretching vibrations OH and HOH. In addition, the bending vibration of HOH was detected at 1651–1655 cm^{-1}. The existence of these bondings was related to entrapped water molecules in the geopolymeric network.

The stretching vibration of O-C-O was still detected in each of the geopolymer samples at 1411–1412 cm^{-1} [23] which was attributed to the carbonation reaction. The carbonation process occurred because an excessive amount of Na was available from the alkaline activator solution where it reacted with CO_2 from the atmosphere [24].

The band attributed to asymmetric stretching vibration of Si-O-Si and Al-O-Si around area 1011–1027 cm^{-1} indicated the formation of aluminosilicate gel [25]. Besides that, symmetric stretching vibrations Si-O-Si were located at 778–795 cm^{-1}. At 583–721 cm^{-1}, symmetric stretching vibrations of Si-O-Si and Al-O-Si were identified. In the meantime, the bending vibrations of Si-O-Si and O-Si-O were found at area 467–479 cm^{-1}.

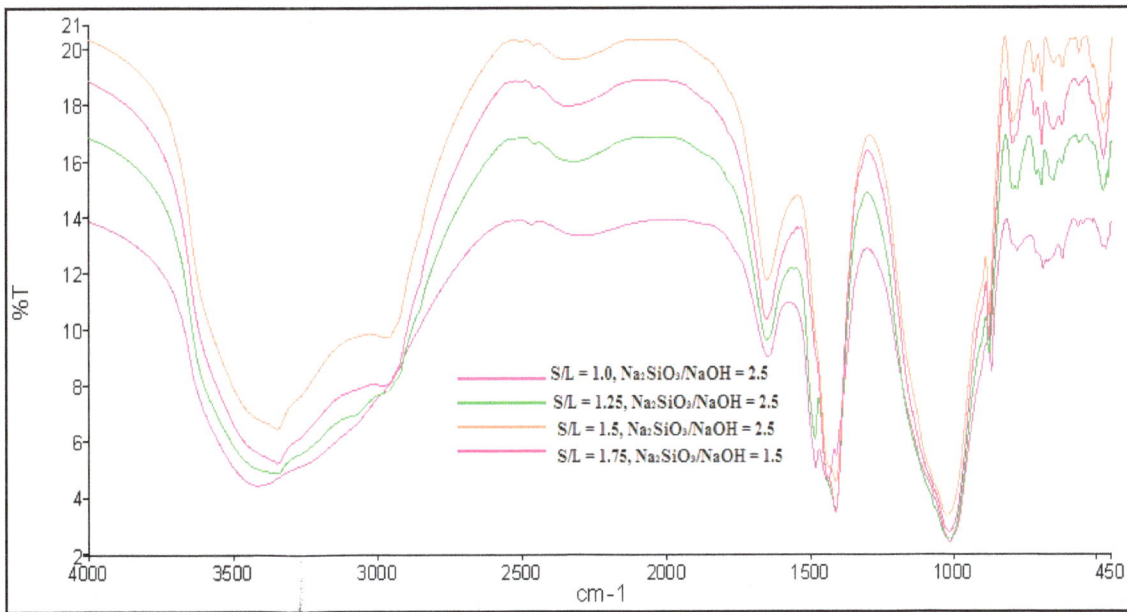

Figure 5. IR spectra of geopolymer paste with different mix design.

The broad band appeared in all IR spectra in the region 2287–3435 cm^{-1} indicating the presence of stretching vibrations OH and HOH as in Figure 6. Meanwhile bending vibration HOH was detected at 1648–1658 cm^{-1} where all these bondings represent water molecule. The band at 1412–1432 cm^{-1} represents the stretching vibration of O-C-O. The aluminosilicate gel (asymmetric stretching vibrations Si-O-Si and Al-O-Si) were detected at 1011–1023 cm^{-1}. When the curing temperature increased, the asymmetric stretching vibrations Si-O-Si and Al-O-Si also shift to lower frequency. Besides that, the symmetric stretching vibration Si-O-Si was found at 776–794 cm^{-1}.

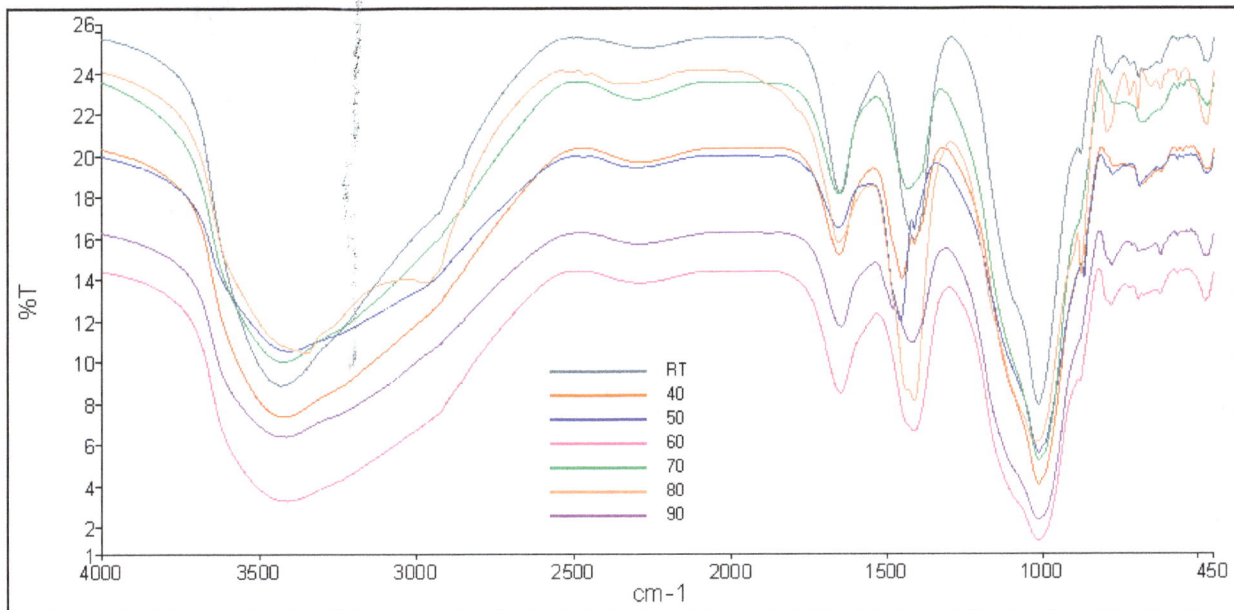

Figure 6. IR spectra of geopolymer paste with different curing temperatures.

In addition, the symmetric stretching vibrations Si-O-Si and Al-O-Si were located at 621–721 cm^{-1}. However, at 467–476 cm^{-1}, bending vibrations Si-O-Si and O-Si-O were identified which represents quartz, thus indicating that quartz which originally exists in POBA did not fully react with the alkaline activator solution which correlates with the finding in XRD analysis.

2.4. Scanning Electron Microscope (SEM) Analysis

For each S/L ratio (1.0, 1.25, 1.5, and 1.75) the samples that contributed the maximum compressive strength are displayed in Figure 7a–d. Geopolymer samples with ratio S/L 1.0 and Na_2SiO_3/NaOH 2.5 are shown in Figure 7a where it demonstrates incomplete geopolymerisation. This leads to a less dense geopolymer matrix and lower compressive strength (7.2 MPa). Since the quantity of POBA and alkaline activator was equal in this sample, it takes time for the geopolymerisation process to complete.

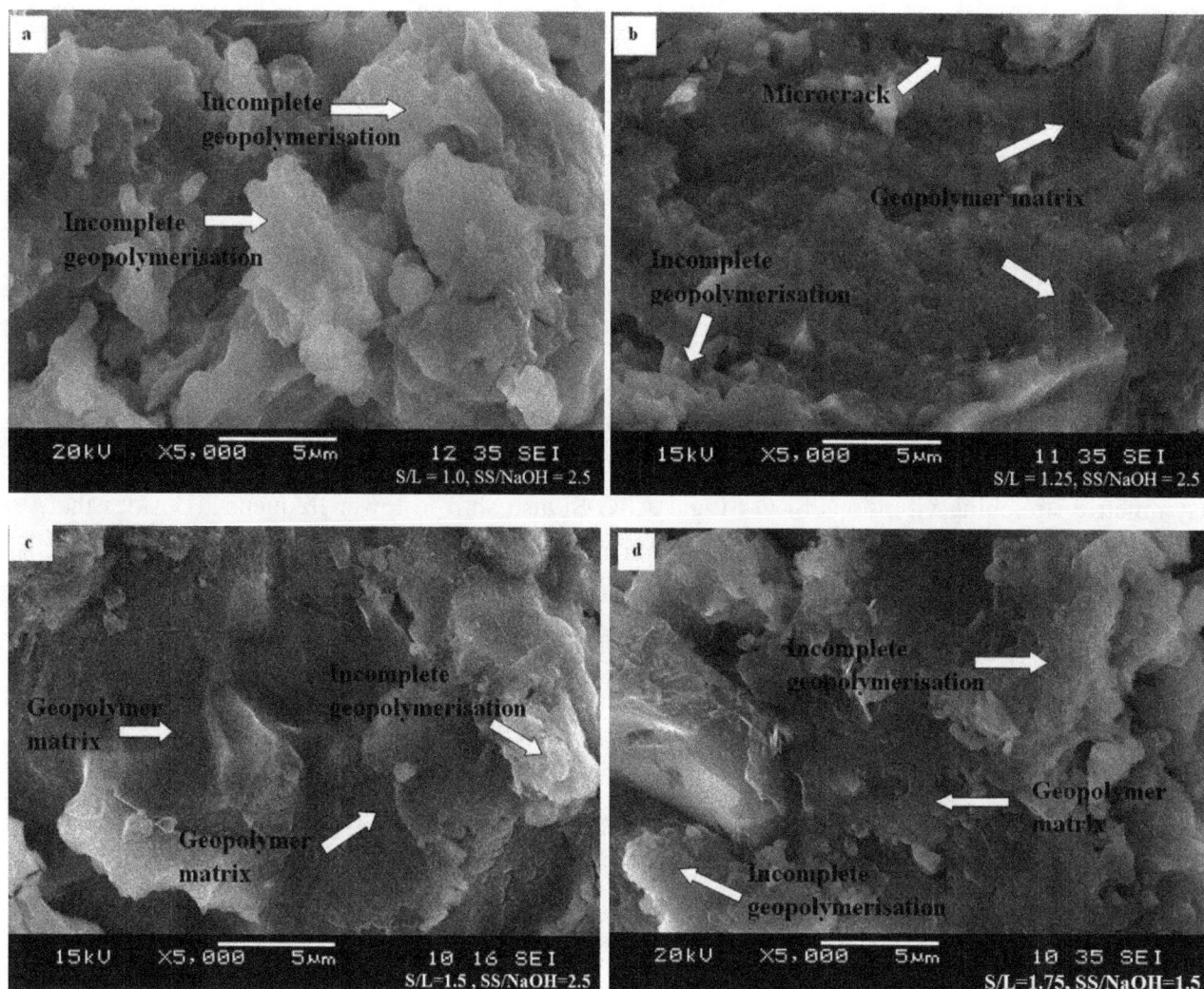

Figure 7. Geopolymer with different mix design. (**a**)S/L = 1.0, Na_2SiO_3/NaOH=2.5; (**b**) S/L = 1.25, Na_2SiO_3/NaOH=2.5; (**c**) S/L = 1.5, Na_2SiO_3/NaOH=2.5; (**d**) S/L = 1.75, Na_2SiO_3/NaOH=1.5.

Besides that, Figure 7b with ratio S/L 1.25 and Na_2SiO_3/NaOH 2.5 demonstrated a denser geopolymer matrix compared to previous figures. Therefore, the quantity of alkaline activator affects the saturation rate of the geopolymerisation process and the strength of geopolymer. In this sample, microcrack was detected and may be due to the sample preparation for the SEM analysis.

Geopolymer samples with ratio S/L 1.25 and Na_2SiO_3/NaOH 2.5 as in Figure 7c show a denser geopolymer matrix compared to others. It shows POBA reacts homogeneously with alkaline activator thus leading to maximum compressive strength (11.9 MPa). Nevertheless, incomplete geopolymerisation is still observed in this sample.

Figure 7d illustrates a geopolymer sample with ratio S/L 1.75 and Na_2SiO_3/NaOH 1.5 and the maximum compressive strength (11.5 MPa) was obtained. For this mix design with Na_2SiO_3/NaOH more than 2.0, the geopolymer samples were unable to be prepared due to low workability. As such, from the figure it shows incomplete geopolymerisation due to less alkaline activator available to react with POBA.

The morphology of geopolymer samples with different curing temperature is displayed in Figure 8a–g. Geopolymer cured at room temperature (Figure 8a) displayed more incomplete geopolymerisation compared to other samples. The reaction rate between POBA and alkaline activator occurred very slowly where no solid geopolymer matrix is formed. Subsequently, due to a slow geopolymerisation process, the strength was also low.

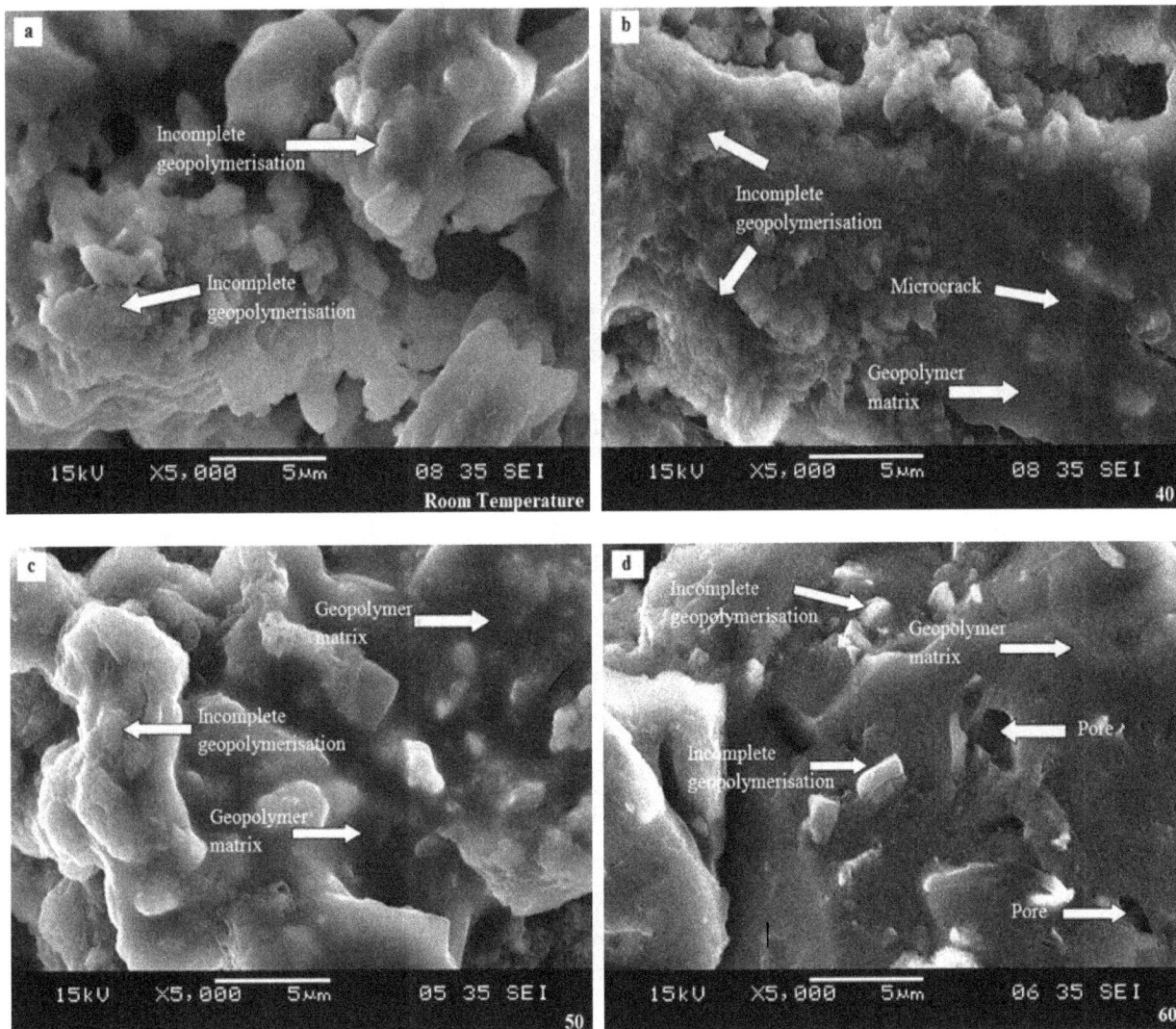

Figure 8. *Cont.*

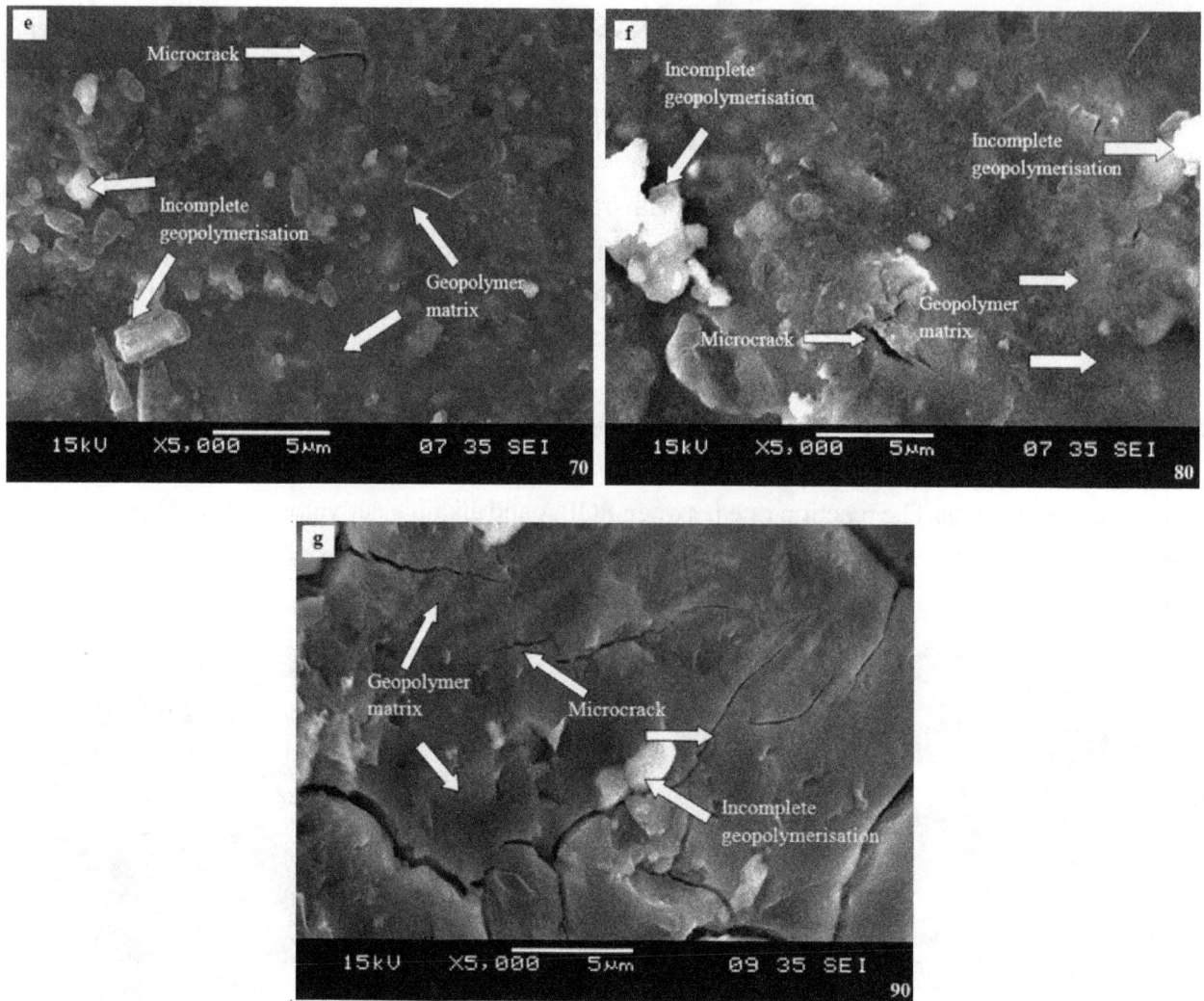

Figure 8. Geopolymer samples with different curing temperatures. (**a**) Room Temperature; (**b**) 40 °C; (**c**) 50 °C; (**d**) 60 °C; (**e**) 70 °C; (**f**) 80 °C; (**g**) 90 °C.

In the meantime, a geopolymer sample cured at 40 °C showed less incomplete geopolymerisation and a dense matrix was formed. At this temperature, the geopolymerisation process starts only slowly to form the geopolymer matrix. A dense gel-like matrix imbedded with POBA particles is seen in Figure 8b. Besides that, the microcrack detected was due to sample preparation for morphological analysis. In Figure 8c, the hardening process in the geopolymerisation covered POBA particles with a dense gel-like matrix. The unreacted POBA particles were seen on the surface of the dense geopolymer matrix. Jaarsveld *et al.* [26] mentioned that the dissolution of source material is not complete where in many cases the surface reaction is responsible for the formation of the geopolymer final structure in bonding the undissolved particles.

At a curing temperature of 60 °C (Figure 8d), a dense geopolymer matrix was observed with some unreacted POBA particles. The existence of pores was also detected in this sample. Since a more dense geopolymer structure was produced, the strength also increased. For geopolymer samples cured at 70 °C, 80 °C, and 90 °C, dense geopolymer matrix was produced as well as maximum compressive strength at 80 °C. The geopolymerisation process completely occurred when the geopolymer sample was cured at 70 °C. However, many microcracks were observed in the geopolymer sample cured at

90 °C. This may be due to the high curing temperature that causes a quick hardening process thus leading to microcracks. The strength also reduces when cured at high temperature.

3. Experimental Section

3.1. Material

The POBA was obtained from United Palm Oil Mill in Penang, Malaysia where it contained large particles which included unburned nutshells, fibers, and kernels as in Figure 9a. Then the POBA was ground using a heavy duty grinder in order to obtain finer particles. After that, the ground POBA was sieved using 100 μm sieves. The POBA that passed through the 100 μm sieve (Figure 9b) was used to produce geopolymer paste and the chemical composition is as in Table 1 below [27]. The POBA was classified as a silica-calcium (Si + Ca) geopolymerisation system due to the high content of silica (Si) and calcium (Ca).

Figure 9. **(a)** Palm oil boiler ash (POBA); **(b)** fine POBA.

Table 1. Chemical composition of fine palm oil boiler ash (POBA).

No.	Compositions	POBA (wt%)
1	SiO_2	40.60
2	Al_2O_3	3.71
3	Fe_2O_3	15.74
4	CaO	19.60
5	MgO	1.30
6	P_2O_5	2.73
7	K_2O	13.80
8	SO_3	0.44
9	TiO_2	0.35
10	MnO	0.28

NaOH solution and sodium silicate (Na_2SiO_3) solution were used as alkaline activator to synthesize POBA. NaOH pellets with 99% purity brand name of Formosoda-P, from the Formosa Plastic Corporation, Taiwan were used to produce the NaOH solution. The NaOH solution with 14 M concentration was prepared by diluting NaOH pellets with distilled water. Sodium silicate (Na_2SiO_3) solution was obtained from South Pacific Chemical Industries Sdn. Bhd. (SPCI), Malaysia.

3.2. Mix Design

3.2.1. Solids-to-Liquids (S/L) and Na$_2$SiO$_3$/NaOH

The fine POBA was mixed with an alkaline activator with four different S/L (POBA/alkaline activator) ratios such as 1.0, 1.25, 1.5, and 1.75. Meanwhile, for alkaline activator solution (Na$_2$SiO$_3$/NaOH ratio), six solutions were prepared according to the ratio 0.5, 1.0, 1.5, 2.0, 2.5, and 3.0 [28–31]. Table 2 shows the experimental details for solid/liquid ratios and Na$_2$SiO$_3$/ NaOH ratios. All the samples were cured at 80 °C for 24 h and left at room temperature for 7 days for compressive strength testing.

3.2.2. Various Curing Temperatures

After the optimum S/L ratio and Na$_2$SiO$_3$/NaOH was obtained, further study was conducted to investigate the effect of curing temperature. The samples were cured at room temperature (RT), 40 °C, 50 °C, 60 °C, 70 °C, 80 °C, and 90 °C for 24 h [32]. Then, all the samples were cured at room temperature for 7 days for compressive strength testing.

Table 2. Mix design details for geopolymer pastes.

S/L Ratio	Na$_2$SiO$_3$/NaOH Ratio	Mass of Solid (g)	Na$_2$SiO$_3$ Solution (g)	NaOH Solution (g)
1.0	0.5	480	160.0	320.0
	1.0		240.0	240.0
	1.5		288.0	192.0
	2.0		320.0	160.0
	2.5		342.9	137.1
	3.0		360.0	120.0
1.25	0.5	480	128.0	256.0
	1.0		192.0	192.0
	1.5		230.4	153.6
	2.0		256.0	128.0
	2.5		274.3	109.7
	3.0		288.0	96.0
1.5	0.5	480	106.7	213.3
	1.0		160.0	160.0
	1.5		192.0	128.0
	2.0		213.3	106.7
	2.5		228.6	91.4
	3.0		240.0	80.0
1.75	0.5	480	91.4	182.9
	1.0		137.1	137.1
	1.5		164.6	109.7
	2.0		182.9	91.4
	2.5		195.9	78.4
	3.0		205.7	68.6

3.3. Mixing Process

The alkaline activator solution was prepared by mixing NaOH solution with Na_2SiO_3 solution until a homogeneous solution was achieved. Then, the alkaline activator was mixed with POBA in the mechanical mixer for about 5 min approximately. The geopolymer paste was placed in a mould ($50 \times 50 \times 50$ mm) and then placed in a vibrating table for 10 s to remove entrapped air. The geopolymer samples that underwent heat curing were covered with a plastic sheet to avoid moisture loss.

3.4. Testing

3.4.1. Compressive Strength

The strength of geopolymer pastes was measured using compressive strength testing based on American Society for Testing and Materials (ASTM C109). The testing was carried out using an Instron machine series 5569 Mechanical Tester (Instron, Singapore) with maximum loading 50 KN and speed rate 50 mm/min. Three samples were used for each mix design to determine the optimum compressive strength.

3.4.2. X-Ray Diffraction (XRD)

The phase of geopolymer paste that leads to maximum compressive strength was determined using XRD. The geopolymer paste was crushed into powder and tested using a XRD-6000, Shimadzu X-ray diffractometer using Cu-Kα radiation generated at 30 Ma and 40 kV. The samples were tested in powder form starting from 10° to 90° (2θ) at 0.04° steps with step time 1.0 s.

3.4.3. Fourier Transform Infrared Spectroscopy (FTIR)

The FTIR analysis was conducted using a Perkin Elmer FTIR Spectrum RX1 Spectrometer. The samples of geopolymer paste were prepared in powder form where they were mixed with potassium bromide (KBr), then a cold press machine was used with a 4 ton loading for 2 min. All the samples used wavelengths from 450 cm^{-1} to 4000 cm^{-1}.

3.4.4. Microstructure Analysis

The microstructure of POBA geopolymer paste was observed using Scanning Electron Microscope (SEM). The geopolymer paste samples were cut into small pieces and coated with platinum by using an Auto Fine Coater. A JSM-6460LA model Scanning Electron Microscope (JEOL, Pleasanton, CA, USA) was used in this analysis.

4. Conclusions

From this study the results led to the conclusions below:
(a) The optimum mix design for geopolymer paste using POBA is S/L = 1.5 and Na_2SiO_3/NaOH = 2.5 with maximum compressive strength 11.9 MPa. During XRD analysis, the existence of albite which is due to the formation of an aluminosilicate gel was detected in the

optimum mix design. The ratio of $Na_2SiO_3/NaOH$ also plays an important role in the mix design of the geopolymer paste. When the ratio of $Na_2SiO_3/NaOH$ is more than 2.5, the compressive strength for S/L (1.0, 1.25, and 1.5) tends to decrease due to excessive alkali content that retards the geopolymerisation process.

(b) The optimum curing temperature for POBA in this study was 80 °C which led to maximum compressive strength (11.5 MPa) at 24 h curing period. Thus, heat curing for geopolymer is needed in order to obtain sufficient strength and with heat curing, the geopolymerisation becomes more rapid. The presence of albite during XRD analysis was also detected in the geopolymer sample cured at 80 °C. The morphology of the geopolymer samples showed changes in the matrix when the curing temperature was increased.

Acknowledgments

We would like to extend our appreciation to the United Palm Oil, Center of Excellence Geopolymer and Green Technology in the School of Materials Engineering at the Universiti Malaysia Perlis (UniMAP).

Author Contributions

ZarinaYahya designed the overallanalysis and interpretation of data. Mohd Mustafa Al Bakri Abdullah responsible for research publication. Kamarudin Hussin and Khairul Nizar Ismail contributed to the design of the study. Rafiza Abd Razak involves in the production of samples. Andrei Victor Sandu involves in X-Ray Diffraction (XRD) analysis.

Conflicts of Interest

The authors declare no conflict of interest.

References

1. Pacheco-Torgal, F.; Castro-Gomes, J.; Jalali, S. Alkali-activated binders: A review. Part 1. Historical background, terminology, reaction mechanism and hydration products. *Constr. Build. Mater.* **2008**, *22*, 1305–1314.

2. Komnitsas, K.; Zaharaki, D. Geopolymerisation: A review and prospect for the mineral industry. *Miner. Eng.* **2007**, *20*, 1261–1277.

3. Davidovits, J. Soft mineralurgy and geopolymer. In Proceedings of the 1st International Conference on Geopolymer '88, Compiegne, France, 1–3 June 1988; Davidovits, J., Orlinski, J., Eds.; Volume 1, pp. 19–23.

4. Hermann, E.; Kunze, C.; Gatzweiler, R.; Davidovits, J. Solidification of various radioactive residues by geopolymer with special emphasis on long term stability. In Proceedings of the Geopolymers Conference, Saint-Quentin, France, 30 June 1999.

5. Rafiza, A.R.; Mustafa Al Bakri, M.A.; Kamarudin, H.; Khairul Nizar, I.; Ioan, G.S.; Hardjito, D.; Zarina, Y.; Andrei, V.S. Assessment on the potential of volcano ash as artificial lightweight aggregates using geopolymerisation method. *Revista de Chimie* **2014**, *65*, 828–834.

6. Palomo, A.; Grutzeck, M.W.; Blanco, M.T. Alkali-activated fly ashes, a cement for the future. *Cement Concrete Res.* **1999**, *29*, 1323–1329.

7. Xu, H.; van Deventer J.S.J. The geopolymerisation of alumino-silicate minerals. *Int. J. Miner. Process.* **2000**, *59*, 247–266.

8. Van jaarsveld, J.G.S.; van Deventer, J.S.J.; Lukey, G.C. The effect of composition and temperature on the properties of fly ash and kaolinite-based geopolymers. *Chem. Eng. J.* **2002**, *89*, 63–73.

9. Yao, X.; Zhang, Z.; Zhu, H.; Chen, Y. Geopolymerization Process of alkali-metakaolinite characterized by isothermal calorimetry. *Thermochim. Acta* **2009**, *493*, 49–54.

10. Zuhua, Z.; Xiao, Y.; Yue, C. Role of water in the synthesis of calcined kaolin-based geopolymer. *Appl. Clay Sci.* **2009**, *43*, 218–223.

11. Provis, J.L.; Yong, C.Z.; Duxson, P.; van Deventer, J.S.J. Correlating mechanical and thermal properties of sodium silicate-fly ash geopolymers. *Colloids Surf. A: Physicochem. Eng. Asp.* **2009**, *336*, 57–63.

12. Villa, C.; Pecina, E.T.; Torres, R.; Gomez, L. Geopolymer synthesis using alkaline activation of natural zeolite. *Constr. Build. Mater.* **2010**, *24*, 2084–2090.

13. Lee, W.K.W.; van Deventer, J.S.J. The effects of inorganic salt contamination on the strength and durability of geopolymers. *Colloids Surf. A: Physicochem. Eng. Asp.* **2002**, *211*, 115–126.

14. Chindaprasirt, P.; Chareerat, T.; Sirivivatnanon, V. Workability and strength of coarse high calcium fly ash geopolymer. *Cement Concrete Compos.* **2007**, *29*, 224–229.

15. Hardjito, D.; Wallah, S.E.; Sumajouw, D.M.J.; Rangan, B.V. Development fly ash-based geopolymer concrete. *ACI Mater. J.* **2004**, *101*, 467–472.

16. Hardjito, D.; Wallah, S.E.; Sumajouw, D.M.J.; Rangan, B.V. Fly ash-based geopolymer concrete. *Austr. J. Struct. Eng.* **2005**, *6*, 1–9.

17. Hardjito, D.; Cheak, C.C.; Ing, C.H.L. Strength and setting times of low calcium fly ash-based geopolymer mortar. *Mod. Appl. Sci.* **2008**, *2*, 3–11.

18. Hardjito, D.; Rangan, B.V. Development and properties of low-calcium fly ash based geopolymer concrete. In *Research Report GC1*; Faculty of Engineering, Curtin University of Technology: Perth, Australia, 2005.

19. Matthew, R.; Brian, O.C. Chemical optimisation of the compressive strength of luminosilicate geopolymers synthesised by sodium silicate activation of metakaolinite. *J. Mater. Chem.* **2003**, *13*, 1161–1165.

20. Wan Mastura, W.I.; Mustafa Al Bakri, A.M.; Andrei, V.S.; Kamarudin, H.; Ioan G.S.; Khairul Nizar, I.; Aeslina, A.K.; Binhussain, M. Processing and characterization of fly ash-based geopolymer bricks. *Revista de Chimie* **2014**, *65*, 1340–1345.

21. Bakharev, T. Geopolymeric materials prepared using class F fly ash and elevated temperature curing. *Cement Concrete Res.* **2005**, *25*, 1224–1232.

22. Garcia-Lodeiro, I.; Fernandez-Jimenez, A.; Palomo, A.; Macphee, D.E. Effect of Fresh C-S-H gels of the simultaneous addition of alkali and aluminium. *Cement Concrete Res.* **2010**, *40*, 27–32.

23. Alvarez-Ayuso, E.; Querol, X.; Plana, F.; Alastuey, A.; Moreno, N.; Izquierdo, M.; Font, O.; Moreno, T.; Diez, S.; Vázquez, E.; Barra, M. Environmental, physical and structural characterisation of geopolymer matrixes synthesised from coal (co-) combustion fly ashes. *J. Hazard. Mater.* **2008**, *154*, 175–183.

24. Panias, D.; Giannopoulou, I.P.; Perraki, T. Effect of synthesis parameters on the mechanical properties of fly ash-based geopolymers. *Colloids Surf. A* **2007**, *301*, 246–254.

25. Nath, S.K.; Kumar, S. Influence of iron making slags on strength and microstructure of fly ash geopolymer. *Constr. Build. Mater.* **2013**, *38*, 924–930.

26. Jaarsveld, J.G.S.; Deventer, J.S.J.; Lukey, G.C. The characterization of source materials in fly ash-based geopolymers. *Mater. Lett.* **2003**, *57*, 1272–1280.

27. Zarina, Y.; Mustafa Al Bakri, A.M.; Kamarudin, H.; Khairul Nizar, I.; Andrei, V.S.; Petrica, V.; Rafiza, A.R. Chemical and physical characterization of boiler ash from palm oil industry waste for geopolymer composite. *Revista de Chimie* **2013**, *64*, 1408–1412.

28. Chub-uppakarn, T.; Thaenlek, N.; Thaisiam, R. Palm ash-based geopolymer mortar incorporating metakaolin. In Proceedings of the Pure and Applied Chemistry International Conference, Bangkok, Thailand, 5–7 January 2011; pp. 347–350.

29. Ariffin, M.A.M.; Hussin, M.; Warid, M.; Rafique Bhutta, M.A. Mix design and compressive strength of geopolymer concrete containing blended ash from agro-industrial wastes. *Adv. Mater. Res.* **2011**, *339*, 452–457.

30. Abdullah, M.M.A.; Kamarudin, H.; Mohammed, H.; Khairul Nizar, I.; Rafiza, A.R.; Zarina, Y. The relationship of NaOH molarity, Na_2SiO_3/NaOH ratio, fly ash/alkaline activator ratio and curing temperature to the strength of fly ash-based geopolymer. *Adv. Mater. Res.* **2011**, *328–330*, 1475–1482.

31. Mustafa Al Bakri, A.M.; Kamarudin, H.; Binhussain, M.; Rafiza, A.R.; Zarina, Y. Effect of Na_2SiO_3/NaOH ratios and NaOH molarities on compressive strength of fly-ash-based geopolymer. *ACI Mater. J.* **2012**, *109*, 503–508.

32. Mustafa Al Bakri, A.M.; Kamarudin, H.; BinHussain, M.; Khairul Nizar, I.; Zarina, Y.; Rafiza, A.R. The effect of curing temperature on physical and chemical properties of geopolymers. *Phys. Procedia* **2011**, *22*, 286–291.

5

Ionic Liquid-Doped Gel Polymer Electrolyte for Flexible Lithium-Ion Polymer Batteries

Ruisi Zhang [1], Yuanfen Chen [1] and Reza Montazami [1,2,*]

[1] Department of Mechanical Engineering, Iowa State University, Ames, IA 50011, USA;
E-Mails: ruisizhang@yahoo.com (R.Z.); yuanfenc@iastate.edu (Y.C.)

[2] Center for Advanced Host Defense Immunobiotics and Translational Comparative Medicine,
Iowa State University, Ames, IA 50011, USA

* Author to whom correspondence should be addressed; E-Mail: reza@iastate.edu;

Academic Editor: Christof Schneider

Abstract: Application of gel polymer electrolytes (GPE) in lithium-ion polymer batteries can address many shortcomings associated with liquid electrolyte lithium-ion batteries. Due to their physical structure, GPEs exhibit lower ion conductivity compared to their liquid counterparts. In this work, we have investigated and report improved ion conductivity in GPEs doped with ionic liquid. Samples containing ionic liquid at a variety of volume percentages (vol %) were characterized for their electrochemical and ionic properties. It is concluded that excess ionic liquid can damage internal structure of the batteries and result in unwanted electrochemical reactions; however, samples containing 40–50 vol % ionic liquid exhibit superior ionic properties and lower internal resistance compared to those containing less or more ionic liquids.

Keywords: ionic liquid; gel polymer electrolyte; flexible electronics; lithium-ion polymer battery

1. Introduction

Li-ion batteries are among the most promising, efficient and common high-energy-density systems for electrochemical energy storage. In recent years application of Li-ion batteries in common electronic devices, and thus demand for more efficient and safer batteries, has increased significantly [1–4]. Batteries with higher efficiency, superior mechanical properties, and smaller size [5] are needed for handheld electronics to keep up with the rapidly increasing computing power, larger screens and thinner and lighter designs of such devices. In addition, there is increasing need for polymer-based batteries to be incorporated with flexible, soft and micro electronics [6–11]. There has also been a significant increase with concerns regarding the issues associated with such batteries. Use of flammable organic solvents as electrolyte, formation of lithium dendrites, and large volume change due to poor structural stability are among the main concerns associated with Li-ion batteries. Use of gel polymer electrolytes (GPEs) has addressed some concerns regarding leakage of liquid electrolytes and the resultant fire hazards; however, charge transfer through GPE doped with organic solvents is not as efficient as that in liquid electrolytes. Also, doping GPE with organic solvents poses some limiting difficulties.

Generally, gel polymer electrolytes' synthesis is achieved by incorporating an organic electrolyte solution into a polymer matrix with a trapping structure enhanced by carbonate esters [12,13]. Polymer matrices with high chemical stability and strong electron-withdrawing functional groups to induce a net dipole moment are desirable as the polymer host [14]. One polymer commonly used in gel polymer electrolytes is polyvinylidene fluoride (PVdF), containing -C-F functional groups. The PVdF base gel polymer electrolyte membranes attract ions in the organic electrolyte solution due to the electric field at the surface of the PVdF membrane. Because of the semi-crystalline structure of PVdF, parts of the attracted ions are drafted into the membrane when the rest of the ions stay at the surface [15–20]. Thus, a gel polymer electrolyte membrane with fully interconnected open micropores, *i.e.*, higher interfacial surface area, enhances ion storage and mobility [17,21–28].

Compared to liquid electrolytes, gel polymer electrolytes have several advantages, such as faster charging/discharging, and potentially higher power density [29–31]. However, ion permeability of gel polymer electrolytes is orders of magnitude lower than that of liquid electrolytes, mainly because of the polymeric structure which limits the ion mobility [1,14].

Room temperature ionic liquids have been used as substitutes for organic electrolytes to increase ion mobility throughout the electrolyte and also to eliminate hazards associated with organic electrolytes. Application of ionic liquids in lithium-ion batteries has been the focus of several studies in the recent years. Fernicaola *et al.* incorporated ionic liquids in an organic electrolyte solution to increase the ionic conductivity and stabilize the lithium ions carried on the surface of PVdF based membrane [32]. In one study Egashira *et al.* have shown than the ion mobility through the gel electrolyte containing ionic liquids depends on the miscibility of polymer component in the ionic liquid. It was shown that, for example, gel electrolytes containing hexyltrimethylammonium bis(trifluoromethane sulfone)imide ionic liquid exhibits high lithium ion permeability, whereas no obvious lithium ion mobility was detected through a gel electrolyte containing 1-ethylk-3-methyl imidazolium bis(trifluoromethane sulfone)imide ionic liquid [33]. In other studies it was demonstrated that the ion permeability of the gel electrolyte could be improved by the addition of carbonate esters. Carbonate esters play the role of ion dissociation enhancer and improve ion mobility because of their relatively high dielectric constants (ε). Ethylene carbonate ($\varepsilon = 89.78$ at 40 °C) and propylene carbonate ($\varepsilon = 64.93$ at 25 °C) are among the most common carbonate esters used in lithium-ion polymer batteries. They both have excellent thermal stability and a boiling point of above 240 °C [34]. Ye *et al.* doped the gel electrolyte by a small amount of ethylene carbonate and observed a significant increase in lithium ion transport through the gel electrolyte [35].

Sirisopanaporn *et al.* demonstrated higher ion permeability and interfacial stability by the addition of small amounts of ethylene carbonate and propylene carbonate to the gel electrolyte [36]. A vapor-free lithium-ion polymer battery with high discharge performance based on lithium salt dissolved in ionic liquid and ultra-high molecular weight ionic liquid polymer was reported by Sato *et al.*; it was demonstrated that the discharge performance is higher than that of a conventional lithium polymer batteries [37].

In this work we attempt to fill the gap between efficient systems based on organic solvents and safe and reliable systems based on ionic liquids. We have investigated GPEs doped with a mixture of organic electrolyte and ionic liquid at different ratios, in the presence of carbonate esters, to enhance ion permeability and electrochemical properties of the GPEs. We have characterized GPEs for their electronic properties and have also investigated their performance in lithium-ion polymer batteries as a function of ionic liquid content.

2. Materials and Methods

2.1. Materials

Copper foil single-side coated by 0.1 mm of composite graphite anode and aluminum foil single-side coated by 0.1 mm of lithium manganese oxide ($LiMn_2O_4$) cathode were purchased from MTI corporation (Richmond, CA, USA) and used as received. *N*-Methyl-2-pyrrolidone (NMP), ethylene carbonate (EC), propylene carbonate (PC), lithium hexafluorophosphate ($LiPF_6$), polyvinylidene fluoride (PVdF), and 1-Ethyl-3-methylimidazolium triluoromethanesufonate (EMI-Tf) were purchased from Sigma-Aldrich (St. Louis, MO, USA) and used as received. $LiPF_6$ was selected as the lithium salt due to its high conductivity in carbonates solvent mixtures, also due to its ability to prevent aluminum corrosion at the cathode aluminum current collector by forming a passivation layer. EMI-Tf was selected because of its high ionic conductivity (10^{-2} S/cm) and wide electrochemical window.

2.2. Electrolyte

Electrolyte solutions with different compositions were prepared by dissolving 1 M of $LiPF_6$ in a solvent consisting of EC, PC and EMI-Tf at different ratios. EC, PC and EMI-Tf were mixed at desired ratios (see Table 1) and stirred for at least 2 h; lithium hexafluorophosphate was then added to the solvent under an inert gas environment to achieve 1 M concentration, and stirred for 24 h.

Table 1. Gel polymer electrolyte (GPE) composition. Volume percentage (vol%) of EMI-Tf is also used as samples' names.

Compound	GPE Composition (vol%)							
EMI-Tf	0%	25%	30%	40%	50%	60%	75%	100%
EC	50%	37.5%	35%	30%	25%	20%	12.5%	0%
PC	50%	37.5%	35%	30%	25%	20%	12.5%	0%

2.3. Membrane Synthesis and Activation

The membrane was synthesized by first preparing a carbonate ester mixture. A 1:1 weight ratio mixture of EC and PC was heated to 80 °C to achieve complete dissolution. The resultant clear carbonate ester solution was mixed with PVdF and 1-Methyl-2-pyrrolidone at 10:4:11 weight ratio. The mixture was then heated to 110 °C and stirred on a magnetic stirrer until a clear solution was obtained with a relatively high viscosity. The solution was then casted on a glass template and dried under vacuum at 80 °C for 2 h to form membranes. The membranes were then soaked in a 10% ethanol aqueous solution overnight. Pale yellow membranes with 50 μm thickness were obtained and cut into 22 mm × 22 mm squares and stored under ambient conditions. Membranes were then activated by full exposure (immersion) to electrolyte solution for 24 h. Electrolyte content was measured as the weight percentage (wt %) of dry weight of the membrane, and calculated from Equation (1).

$$W_e(\%) = \frac{W_f - W_d}{W_f} \times 100 \tag{1}$$

where W_e (%) is the weight-percent of the electrolyte; and, W_d and W_f are the weights of dry and doped membranes, respectably. All membranes reached an approximate 35 wt % electrolyte content which is the typical upper limit saturation.

Schematic of the fabrication and activation processes are presented in Figure 1.

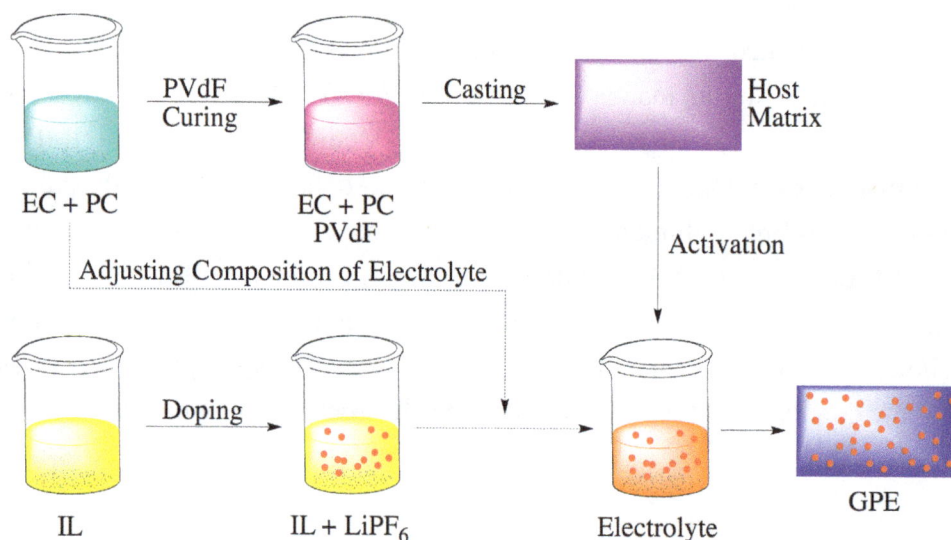

Figure 1. Schematic of the process for fabrication and activation of GPEs.

2.4. Full Cell Assembly

The anode material of Graphite was casted on the surface of copper foil as the current collector; the cathode material of $LiMn_2O_4$ was casted on the surface of aluminum foil as the current collector. Schematic of the test cell is presented in Figure 2, with GPE located in between the cathode and anode. Cathode and anode are exactly 20 mm × 20 mm; but GPE film is larger than cathode and anode so the cell would not be shorted. The surface of protection cover that faces inside of the cell is adhesive, which helped airtight enclosure of the whole system. The actual model is displayed in Figure 3a; the flexibility of the cell is shown in Figure 3b.

Figure 2. Schematic of test cell structure.

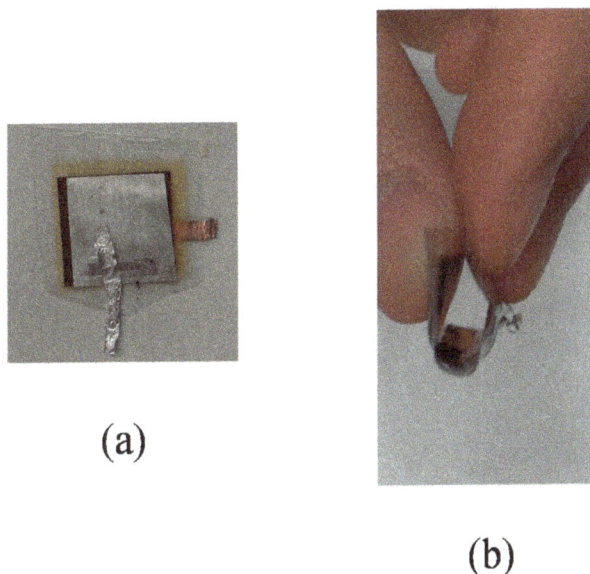

(a)

(b)

Figure 3. (**a**) Photograph of a sample battery structure; (**b**) bending of the battery under mechanical stress.

2.5. Measurement

A VersaSTAT-4 potentiostat (Princeton Applied Research, Oak Ridge, TN, USA) was used for electrochemical and impedance spectroscopy studies of the gel polymer electrolyte membranes. For these studies GPE membranes were secured between two steel-disks electrodes of 200 μm thickness and 15.5 mm diameter, and two pieces of adhesive plastic were used as a pouch to seal and hold each sample. The impedance spectroscopy of full cells studies were carried at a frequency between of 1.0×10^5 and 0.1 Hz and a potential difference (ΔV) of 10 mV, after completion of 10 charging/discharging cycles. In each sample, the membrane was cut slightly larger than the electrodes to prevent short-circuit. Charging-discharging tests were carried out with a computer controlled BST8-MA battery analyzer (MTI corporation), between 0.5 and 5 V with a constant current of 5 mA.

3. Results and Discussion

3.1. Ionic Conductivity

Ion conductivity of GPEs was studied by AC impedance spectroscopy. GPEs doped with electrolytes of different composition were secured between two steel disks and studied at a high frequency range

(10–100 kHz). As presented in Figure 4, the Nyquist plot exhibited approximately vertical lines, suggesting nearly pure resistive behavior at high frequencies, for all GPE samples. Here, the effect of the imaginary part of the impendence can be neglected and the system can be considered as a pure resistor with minimum dependence on frequency. The internal resistance values of the bulk electrolytes can be induced from the intercept of the extended impedance plots with the x-axis (Z_{re}) (see Figure 5). Ionic conductivity of GPEs can be calculated using internal resistance, thickness and surface area of the GPEs. The ionic conductivity σ for each GPE sample was calculated using Equation (2):

$$\sigma = \frac{t}{RA} \tag{2}$$

where t is the thickness of each sample, R is the internal resistance and A is the surface area (1.89 cm^2). Ionic conductivities of the GPE samples are presented in Table 2.

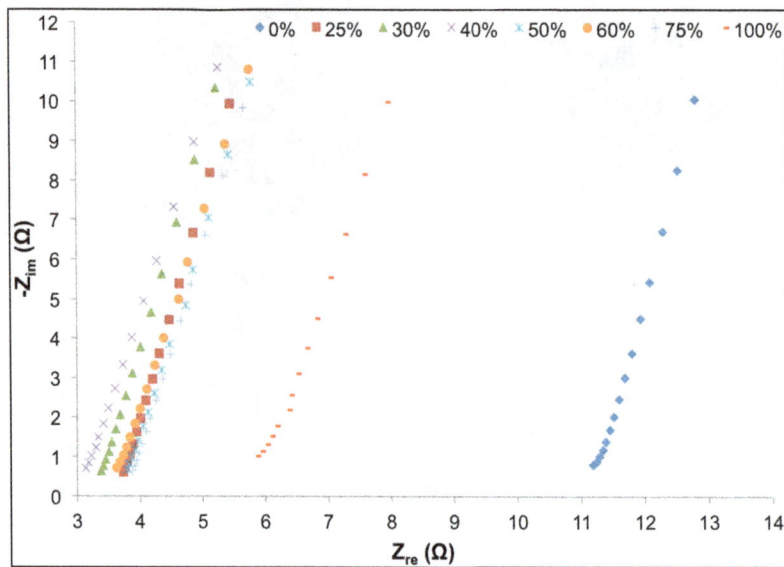

Figure 4. Nyquist plots of systems consisting of GPEs with different volume percent ionic liquid at high frequency.

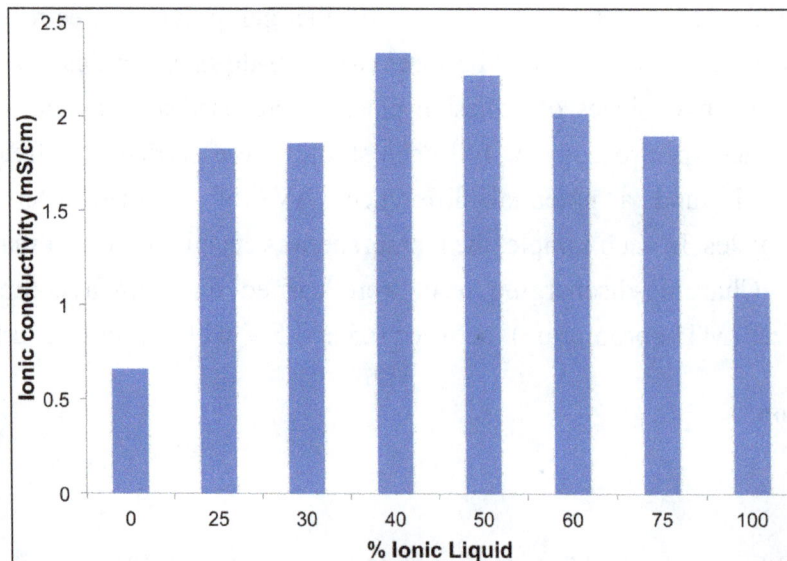

Figure 5. Ionic conductivity of GPEs as a function of IL content.

Table 2. Characteristic properties of GPEs as a function of IL content.

A (cm^2)	IL (vol %)	R (Ω)	t (cm)	σ (mS·cm^{-1})
1.89	0	11.2379	0.014	0.66
1.89	25	3.7595	0.013	1.83
1.89	30	3.4118	0.012	1.86
1.89	40	3.1770	0.014	2.34
1.89	50	3.8264	0.016	2.22
1.89	60	3.6806	0.014	2.02
1.89	75	3.9111	0.014	1.90
1.89	100	5.9292	0.012	1.07

Samples containing EMI-Tf ionic liquid exhibited ~65% to ~270% improvement on their ion permeability compared to a sample without EMI-TF (100% EC-PC solvent), and samples with only EMI-TF. Specifically, samples with 40 and 50 vol % ionic liquid had the highest ionic conductivity. Interestingly, ion permeability of the GPEs showed an increasing trend as the ionic liquid content increased to ~40% and decreased thereafter, suggesting the significant contribution of EC and PC toward ion permeability of the GPEs; and potentially, formation of ionic double layers at electrodes which retard ion mobility, as observed and reported previously [9]. It is important to note that in this study the ion permeability of PVdF-based host matrix is deduced from the movement of all salt and ionic liquid ions (Li$^+$, EMI$^+$, Tf$^-$, and PF$_6^-$, EMI$^+$ and Tf$^-$ only when ionic liquid was used) in the GPE. Previous studies [24,38–40] have shown that among cations, the EMI$^+$ always diffuses faster than smaller Li$^+$, and that Li$^+$ and anions are more likely to form ion complexes and diffuse together at a slower rate. Also, high ionic conductivity of EMI-Tf (6.4 mS/cm) at room temperature contributes significantly to the ion permeability of the GPE membrane [41,42]. Yet, GPE's containing EMI-Tf as the only solvent exhibited relatively low ionic conductivity, comparable to that of samples only containing EC-PC solvent.

3.2. Interfacial Properties

The interfacial properties of GPEs were determined by impedance spectroscopy (1.0 × 10^5 to 0.1 Hz, ΔV = 10 mV) in thin-film cell pack configuration, after completing 10 cycles of charging and discharging. As presented in Figure 6, the curves of Nyquist plots for all cells are semicircular, followed by a linear segment, at approximately 45° slope, indicating Warburg impedance.

At high frequencies, close to the origin of the x-axis, the electrochemical systems exhibited pure resistance behavior. Intersection of the semicircular plots with the x-axis, at high frequency regions, manifests the solution resistance (R_s), as presented in the Randles equivalent electrical circuit (Figure 7). Solution resistance depended on the ion content of the entire system and transportation of ions between anode and cathode; and it is slightly different than ionic conductivity, discussed in the previous section. Solution resistance of samples, listed in Table 3, suggests that the addition of ionic liquids results in reduction of solution resistance, an observation that is in agreement with expected effect of any ion-rich electrolyte, such as ionic liquids. A slight increase in solution resistance was observed for 100% ionic liquid electrolyte that is anticipated to be a result of high concentration of ions in the ionic liquid, limiting mobility of charge.

Table 3. Electrical properties of samples as a function of ionic liquid content, in accordance with Randles equivalent electrical circuit.

ILs (vol %)	R_S (Ω)	R_{CT} (Ω)	C_{DL} (F)	σ_S
0	9.983	98.940	3.99×10^{-4}	78.814
25	7.475	216.578	3.03×10^{-4}	95.976
30	5.517	177.381	4.25×10^{-4}	99.088
40	3.785	136.122	4.43×10^{-4}	113.52
50	2.849	190.677	4.80×10^{-4}	97.602
60	3.968	189.382	4.43×10^{-4}	95.877
75	3.065	155.311	5.36×10^{-4}	84.32
100	2.931	154.235	6.54×10^{-4}	98.015

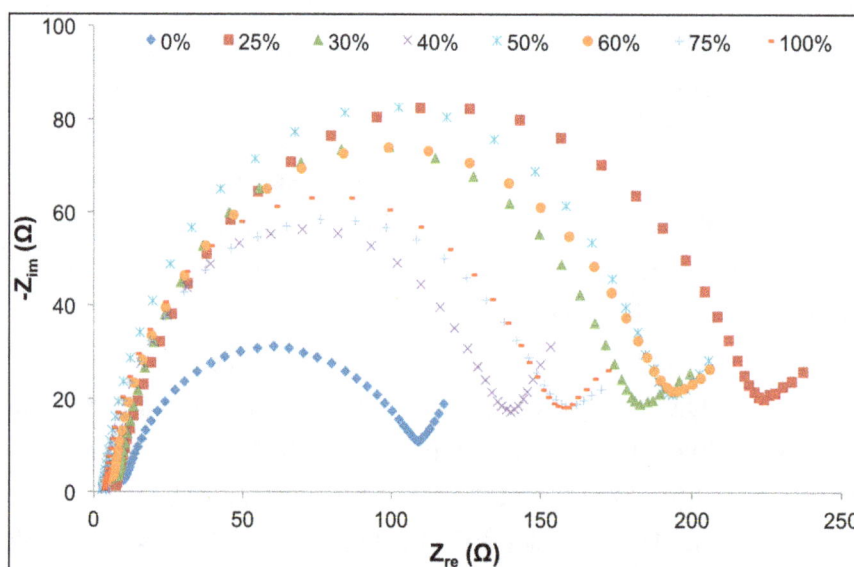

Figure 6. Nyquist plots of Li-ion polymer batteries consisting of GPEs with different volume percent ionic liquid.

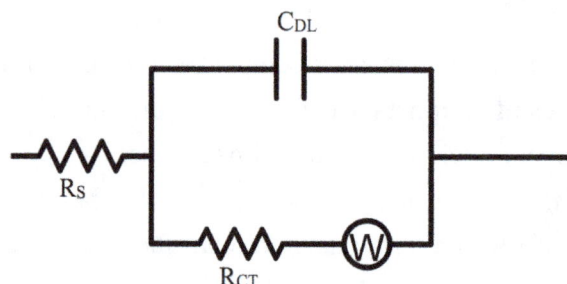

Figure 7. Randles equivalent electrical circuit is used to describe Lithium-ion polymer battery cells.

The diameters of the semicircles represent charge transfer resistance (R_{CT}); by assuming the semicircles are associated with parallel combination of charge transfer resistance (R_{CT}), and double layer capacitance (C_{DL}) and Warburg impedance (W) in series, the system could be described as the equivalent circuit presented in Figure 7 [43,44]; R_S and R_{CT} data are presented in Figure 8. Double layer capacitance (C_{DL}) was calculated using Z_{im} from the Nyquist plots at pure capacitance points (phase angle = $\pi/2$) and the corresponding frequency (Equation (3)) [45,46]. Generally, C_{DL} showed an increasing trend with an increase of ionic liquid content in the GPE, which translates to higher storage capacity, due to increased concentration of ions and resultant decreased solution resistance.

$$C_{DL} = \frac{2\pi}{f \times Z_{im}} \tag{3}$$

Warburg coefficient (σ_S) is deduced from the slope of the plot of square root of radial frequency at low frequency domain *vs.* the real impedance values. The corresponding plot is presented in Figure 9. All electrical properties of the samples, in accordance with Randles equivalent electrical circuit, are summarized in Table 3.

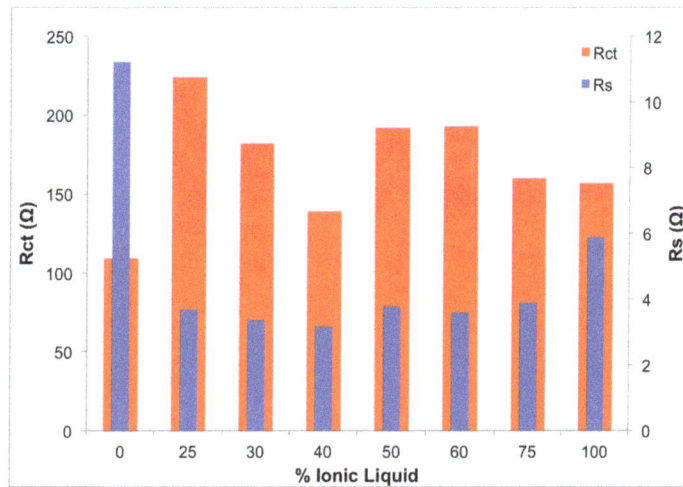

Figure 8. Solution and charge transfer resisstance as a function of ionic liquid content.

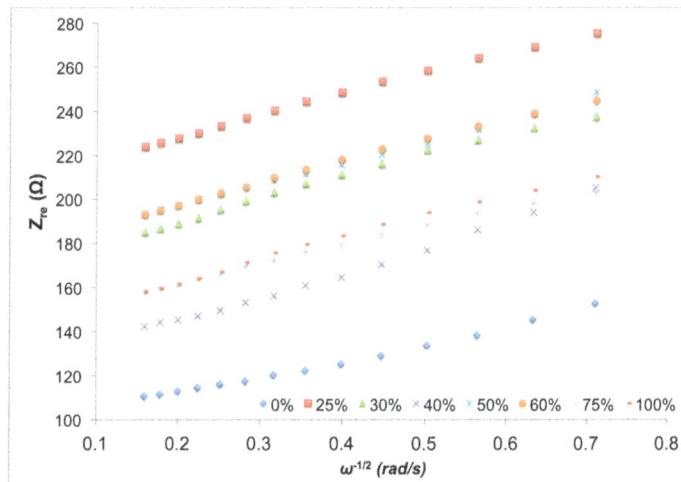

Figure 9. Real impedance verses the inverse of the square root of radial frequency at low frequency domain (63–3 Hz).

3.3. Battery Performance

Galvan static charging and discharging were used to evaluate to performance of lithium-ion polymer batteries *containing GPE of different ionic liquid content.* Each cycle included constant current (0.5 A) discharging followed by an 8 min rest period, then constant current and constant voltage charging.

As presented in Figure 10, samples with ionic liquid exhibited an overall higher discharge capacity (~90 mAh/g) compared to the system with no ionic liquid. However, more specifically samples with 0 and >50 vol % ionic liquid exhibited poor stability and discharge capacity when samples with ionic liquid content between 25 and 50 vol % showed higher discharge capacity and stability. Low discharge capacity of 0 vol % samples is mainly due to the lack of ion conductivity. EC forms an ultra-thin film that acts as a protective layer, preventing damage to active materials on the electrodes. At higher concentrations (>50 vol %) of ionic liquid, this protective film is damaged over cycling and compromises the stability of the system, thus the average capacity of the system drops. At lower ionic liquid concentrations, this film remains intact, which results in maintenance of the system's stability and performance. Average discharge capacities of the samples over 50 charging/discharging cycles are presented in Figure 10. Samples with 40–50 vol % ionic liquid show superior performance, which is in agreement with results from other reported experiments. Average rest voltages and discharge capacities are summarized in Table 4.

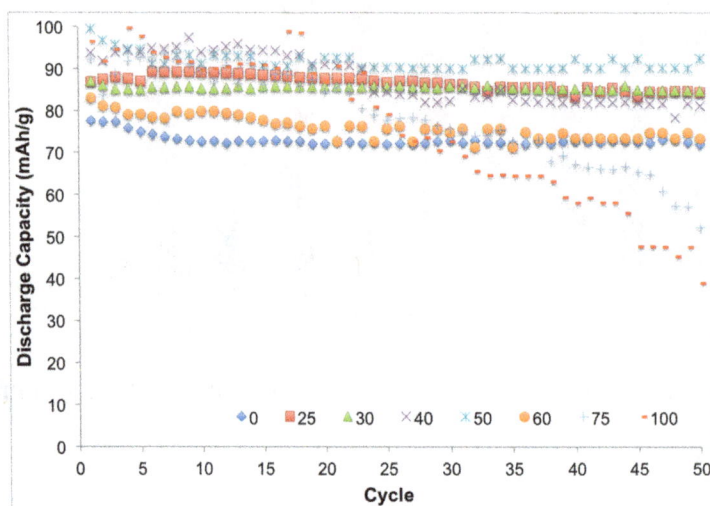

Figure 10. Consecutive cycling behavior of batteries containing GPE of different ionic liquid content. Samples containing 40–50 vol % ionic liquid manifested superior stability and performance.

Table 4. The average rest voltages and discharge capacity of Li-ion polymer batteries as function of GPE composition.

Volume Percent of ILs (vol %)	Avg. Rest Voltage (V)	Avg. Discharge Capacity (mAh/g)
0	3.342	73.04
25	3.800	86.74
30	3.853	85.47
40	3.807	87.61
50	3.942	91.79
60	3.630	76.03
75	3.202	78.48
100	3.073	75.71

4. Conclusions

GPEs were characterized for their electrochemical properties. Chemical composition of GPEs was altered by the addition of ionic liquid at different volume ratios, and an optimum system was identified. GPEs consisting of 40–50 vol % EMI-Tf ionic liquid, in a 1:1 EC-PC solution exhibited highest performance, with considerably lower solution and charge transfer resistances, as well as highest rest voltage and discharge capacity.

Acknowledgments

This material is based upon work supported in part by Iowa State University. A portion of this work was supported by a funding from Health Research Initiative and Presidential Initiative for Interdisciplinary Research at Iowa State University.

Author Contributions

Ruisi Zhang prepared samples and conducted experimental work and some data analysis and wrote most of the sections in the manuscript. Yuanfen Chen contributed toward sample preparation and experimental work. Prof. Montazami oversaw the project and contributed toward writing the manuscript.

Conflicts of Interest

The authors declare no conflict of interest.

References

1. Song, J.Y.; Wang, Y.Y.; Wan, C.C. Review of gel-type polymer electrolytes for lithium-ion batteries. *J. Power Sources* **1999**, *77*, 183–197.
2. Yoo, E.; Kim, J.; Hosono, E.; Zhou, H.-S.; Kudo, T.; Honma, I. Large reversible Li storage of graphene nanosheet families for use in rechargeable lithium ion batteries. *Nano Lett.* **2008**, *8*, 2277–2282.
3. Tarascon, J.M.; Armand, M. Issues and challenges facing rechargeable lithium batteries. *Nature* **2001**, *414*, 359–367.
4. Wakihara, M. Recent developments in lithium ion batteries. *Mater. Sci. Eng. R Rep.* **2001**, *33*, 109–134.
5. Patil, A.; Patil, V.; Shin, D.W.; Choi, J.-W.; Paik, D.-S.; Yoon, S.-J. Issue and challenges facing rechargeable thin film lithium batteries. *Mater. Res. Bull.* **2008**, *43*, 1913–1942.
6. Acar, H.; Çınar, S.; Thunga, M.; Kessler, M.R.; Hashemi, N.; Montazami, R. Study of physically transient insulating materials as a potential platform for transient electronics and bioelectronics. *Adv. Funct. Mater.* **2014**, *24*, 4135–4143.
7. Meis, C.; Hashemi, N.; Montazami, R. Investigation of spray-coated silver-microparticle electrodes for ionic electroactive polymer actuators. *J. Appl. Phys.* **2014**, *115*, 134302.

8.   Hong, W.; Meis, C.; Heflin, J.R.; Montazami, R. Evidence of counterion migration in ionic polymer actuators via investigation of electromechanical performance. *Sens. Actuators B Chem.* **2014**, *205*, 371–376.

9.   Hong, W.; Almomani, A.; Montazami, R. Influence of ionic liquid concentration on the electromechanical performance of ionic electroactive polymer actuators. *Org. Electron.* **2014**, *15*, 2982–2987.

10. Amiri Moghadam, A.A.; Hong, W.; Kouzani, A.; Kaynak, A.; Zamani, R.; Montazami, R. Nonlinear dynamic modeling of ionic polymer conductive network composite actuators using rigid finite element method. *Sens. Actuators A Phys.* **2014**, *217*, 168–182.

11. Alfeeli, B.; Ali, S.; Jain, V.; Montazami, R.; Heflin, J.; Agah, M. MEMS-based gas chromatography columns with nano-structured stationary phases. In Proceedings of the 2008 IEEE on Sensors, Lecce, Italy, 26–29 October 2008; pp. 728–731.

12. Appetecchi, G.B.; Romagnoli, P.; Scrosati, B. Composite gel membranes: A new class of improved polymer electrolytes for lithium batteries. *Electrochem. Commun.* **2001**, *3*, 281–284.

13. Wachtler, M.; Ostrovskii, D.; Jacobsson, P.; Scrosati, B. A study on PVdF-based SiO_2-containing composite gel-type polymer electrolytes for lithium batteries. *Electrochim. Acta* **2004**, *50*, 357–361.

14. Esterly, D.M. Manufacturing of Poly (vinylidene fluoride) and Evaluation of its Mechanical Properties. Master's Thesis, Virginia Polytechnic Institute and State University, Blacksburg, Virginia, 9 August 2002.

15. Gentili, V.; Panero, S.; Reale, P.; Scrosati, B. Composite gel-type polymer electrolytes for advanced, rechargeable lithium batteries. *J. Power Sources* **2007**, *170*, 185–190.

16. Salimi, A.; Yousefi, A.A. Analysis Method: FTIR studies of β-phase crystal formation in stretched PVDF films. *Polym. Test.* **2003**, *22*, 699–704.

17. Kim, J.R.; Choi, S.W.; Jo, S.M.; Lee, W.S.; Kim, B.C. Electrospun PVdF-based fibrous polymer electrolytes for lithium ion polymer batteries. *Electrochim. Acta* **2004**, *50*, 69–75.

18. Zhang, H.P.; Zhang, P.; Li, Z.H.; Sun, M.; Wu, Y.P.; Wu, H.Q. A novel sandwiched membrane as polymer electrolyte for lithium ion battery. *Electrochem. Commun.* **2007**, *9*, 1700–1703.

19. Ji, G.-L.; Zhu, B.-K.; Cui, Z.-Y.; Zhang, C.-F.; Xu, Y.-Y. PVDF porous matrix with controlled microstructure prepared by TIPS process as polymer electrolyte for lithium ion battery. *Polymer* **2007**, *48*, 6415–6425.

20. Nakagawa, H.; Izuchi, S.; Kuwana, K.; Nukuda, T.; Aihara, Y. Liquid and polymer gel electrolytes for lithium batteries composed of room-temperature molten salt doped by lithium salt. *J. Electrochem. Soc.* **2003**, *150*, A695–A700.

21. Wang, Y.; Travas-Sejdic, J.; Steiner, R. Polymer gel electrolyte supported with microporous polyolefin membranes for lithium ion polymer battery. *Solid State Ion.* **2002**, *148*, 443–449.

22. Boudin, F.; Andrieu, X.; Jehoulet, C.; Olsen, I.I. Microporous PVdF gel for lithium-ion batteries. *J. Power Sources* **1999**, *81–82*, 804–807.

23. Choi, S.W.; Jo, S.M.; Lee, W.S.; Kim, Y.R. An electrospun poly(vinylidene fluoride) nanofibrous membrane and its battery applications. *Adv. Mater.* **2003**, *15*, 2027–2032.

24. Montazami, R.; Liu, S.; Liu, Y.; Wang, D.; Zhang, Q.; Heflin, J.R. Thickness dependence of curvature, strain, and response time in ionic electroactive polymer actuators fabricated via layer-by-layer assembly. *J. Appl. Phys.* **2011**, *109*, 104301.

25. Montazami, R.; Wang, D.; Heflin, J.R. Influence of conductive network composite structure on the electromechanical performance of ionic electroactive polymer actuators. *Int. J. Smart Nano Mater.* **2012**, *3*, 204–213.

26. Liu, Y.; Zhao, R.; Ghaffari, M.; Lin, J.; Liu, S.; Cebeci, H.; de Villoria, R.G.; Montazami, R.; Wang, D.; Wardle, B.L.; *et al.* Equivalent circuit modeling of ionomer and ionic polymer conductive network composite actuators containing ionic liquids. *Sens. Actuators A Phys.* **2012**, *181*, 70–76.

27. Liu, Y.; Liu, S.; Lin, J.; Wang, D.; Jain, V.; Montazami, R.; Heflin, J.R.; Li, J.; Madsen, L.; Zhang, Q.M. Ion transport and storage of ionic liquids in ionic polymer conductor network composites. *Appl. Phys. Lett.* **2010**, *96*, 223503.

28. Liu, S.; Montazami, R.; Liu, Y.; Jain, V.; Lin, M.; Zhou, X.; Heflin, J.R.; Zhang, Q.M. Influence of the conductor network composites on the electromechanical performance of ionic polymer conductor network composite actuators. *Sens. Actuators A Phys.* **2010**, *157*, 267–275.

29. Manuel Stephan, A. Review on gel polymer electrolytes for lithium batteries. *Eur. Polym. J.* **2006**, *42*, 21–42.

30. Scrosati, B. Recent advances in lithium ion battery materials. *Electrochim. Acta* **2000**, *45*, 2461–2466.

31. Scrosati, B.; Croce, F.; Panero, S. Progress in lithium polymer battery R&D. *J. Power Sources* **2001**, *100*, 93–100.

32. Fernicola, A.; Scrosati, B.; Ohno, H. Potentialities of ionic liquids as new electrolyte media in advanced electrochemical devices. *Ionics* **2006**, *12*, 95–102.

33. Egashira, M.; Todo, H.; Yoshimoto, N.; Morita, M. Lithium ion conduction in ionic liquid-based gel polymer electrolyte. *J. Power Sources* **2008**, *178*, 729–735.

34. Xu, K. Nonaqueous liquid electrolytes for lithium-based rechargeable batteries. *Chem. Rev.* **2004**, *104*, 4303–4418.

35. Ye, H.; Huang, J.; Xu, J.J.; Khalfan, A.; Greenbaum, S.G. Li ion conducting polymer gel electrolytes based on ionic liquid/PVDF-HFP blends. *J. Electrochem. Soc.* **2007**, *154*, A1048–A1057.

36. Sirisopanaporn, C.; Fernicola, A.; Scrosati, B. New, ionic liquid-based membranes for lithium battery application. *J. Power Sources* **2009**, *186*, 490–495.

37. Sato, T.; Marukane, S.; Narutomi, T.; Akao, T. High rate performance of a lithium polymer battery using a novel ionic liquid polymer composite. *J. Power Sources* **2007**, *164*, 390–396.

38. Hayamizu, K.; Aihara, Y.; Nakagawa, H.; Nukuda, T.; Price, W.S. Ionic conduction and ion diffusion in binary room-temperature ionic liquids composed of [emim][BF$_4$] and LiBF$_4$. *J. Phys. Chem. B* **2004**, *108*, 19527–19532.

39. Li, J.; Wilmsmeyer, K.; Hou, J.; Madsen, L. The role of water in transport of ionic liquids in polymeric artificial muscle actuators. *Soft Matter* **2009**, *5*, 2596–2602.

40. Hou, J.; Zhang, Z.; Madsen, L.A. Cation/anion associations in ionic liquids modulated by hydration and ionic medium. *J. Phys. Chem. B* **2011**, *15*, 4576–4582.

41. Li, T.; Balbuena, P.B. Theoretical studies of lithium perchlorate in ethylene carbonate, propylene carbonate, and their mixtures. *J. Electrochem. Soc.* **1999**, *146*, 3613–3622.

42. Pandey, G.; Hashmi, S. Experimental investigations of an ionic-liquid-based, magnesium ion conducting, polymer gel electrolyte. *J. Power Sources* **2009**, *187*, 627–634.

43. Rodrigues, S.; Munichandraiah, N.; Shukla, A. A review of state-of-charge indication of batteries by means of a.c. impedance measurements. *J. Power Sources* **2000**, *87*, 12–20.

44. Wang, X.; Hao, H.; Liu, J.; Huang, T.; Yu, A. A novel method for preparation of macroposous lithium nickel manganese oxygen as cathode material for lithium ion batteries. *Electrochim. Acta* **2011**, *56*, 4065–4069.

45. Appetecchi, G.; Croce, F.; de Paolis, A.; Scrosati, B. A poly (vinylidene fluoride)-based gel electrolyte membrane for lithium batteries. *J. Electroanal. Chem.* **1999**, *463*, 248–252.

46. Reale, P.; Panero, S.; Scrosati, B. Sustainable high-voltage lithium ion polymer batteries. *J. Electrochem. Soc.* **2005**, *152*, A1949–A1954.

6

Mechanical Properties and Cytocompatibility Improvement of Vertebroplasty PMMA Bone Cements by Incorporating Mineralized Collagen

Hong-Jiang Jiang [1], Jin Xu [2], Zhi-Ye Qiu [3,4], Xin-Long Ma [5], Zi-Qiang Zhang [4], Xun-Xiang Tan [1], Yun Cui [4] and Fu-Zhai Cui [3,*]

[1] Wendeng Orthopaedic Hospital, No. 1 Fengshan Road, Wendeng 264400, Shandong, China;
E-Mails: boneman@163.com (H.-J.J.); jboneman@sina.com (X.-X.T.)

[2] Kangda College of Nanjing Medical University, No. 8 Chunhui Road, Xinhai District,
Lianyungang 222000, Jiangsu, China; E-Mail: xujin33@hotmail.com

[3] School of Materials Science and Engineering, Tsinghua University, Haidian District,
Beijing 100084, China; E-Mail: ye841215@gmail.com

[4] Beijing Allgens Medical Science and Technology Co., Ltd., No. 1 Disheng East Road,
Yizhuang Economic and Technological Development Zone, Beijing 100176, China;
E-Mails: zhangzq@allgensmed.com (Z.-Q.Z.); cuiyun@allgensmed.com (Y.C.)

[5] Tianjin Hospital, No. 406 Jiefang South Road, Tianjin 300211, China;
E-Mail: maxinlong8686@sina.com

* Author to whom correspondence should be addressed; E-Mail: cuifz@mail.tsinghua.edu.cn;

Academic Editor: Amir A. Zadpoor

Abstract: Polymethyl methacrylate (PMMA) bone cement is a commonly used bone adhesive and filling material in percutaneous vertebroplasty and percutaneous kyphoplasty surgeries. However, PMMA bone cements have been reported to cause some severe complications, such as secondary fracture of adjacent vertebral bodies, and loosening or even dislodgement of the set PMMA bone cement, due to the over-high elastic modulus and poor osteointegration ability of the PMMA. In this study, mineralized collagen (MC) with biomimetic microstructure and good osteogenic activity was added to commercially available PMMA bone cement products, in order to improve both the mechanical properties and the cytocompatibility. As the compressive strength of the modified bone cements remained well, the compressive elastic modulus could be significantly down-regulated by the MC, so as to reduce the pressure on the adjacent vertebral bodies. Meanwhile, the adhesion and proliferation of pre-osteoblasts on the modified bone cements were improved compared with cells on those unmodified, such result is beneficial for a good osteointegration formation between the bone cement and the host bone tissue in clinical applications. Moreover, the modification of the PMMA bone cements by adding MC did not significantly influence the injectability and processing times of the cement.

Keywords: mineralized collagen; polymethyl methacrylate bone cement; vertebroplasty; compressive elastic modulus; cytocompatibility

1. Introduction

Vertebral compression fractures (VCF) are one of the most common fractures for the elders with osteoporosis. In the United States, it was reported that about 25% of postmenopausal women suffered from VCF, and such morbidity rate was estimated to be 40% for those women over 80 years old [1]. With the current accelerated trend of the aging of the world population, the occurrence of VCF will continue increasing. Besides osteoporosis, VCF can also be induced by other disease, such as osteogenesis imperfecta [2], spinal tumors [3], and so on.

Percutaneous vertebroplasty (PVP) and percutaneous kyphoplasty (PKP) are the major applications of the polymethyl methacrylate (PMMA) bone cement in the treatment of VCF. In either PVP or PKP, the bone cement is injected into the vertebral body for the augmentation of the fractured vertebral body. The immediate effect and safety of the PMMA bone cements used in PVP and PKP have been deeply investigated and verified by long-term clinical practices. However, existing commercially available PMMA bone cement products for PVP and PKP have been reported to cause some complications, mainly includes secondary fractures of the adjacent vertebral bodies, and loosening or even dislodgement of the set PMMA bone cement, due to the high elastic modulus and bioinert of the PMMA.

The compressive elastic modulus of normal human vertebral body is 50–800 MPa [4–6], while the PMMA bone cements form hard solid body with an elastic modulus of 2000–3700 MPa [7,8], which is much higher than that of normal human vertebral body. The vertebral body filled with PMMA bone cement has a significantly higher stiffness than the adjacent segments, and the resulting stress concentration will easily cause secondary fracture on adjacent vertebral bodies and endplate near the surgical segment [9,10]. The incidence of the secondary fracture of the adjacent bodies after PVP and PKP was reported as high as 7%–20% [11], which is 4.62 times than those occurred on other segments [12].

On the other hand, PMMA is a bioinert material that neither form chemical bonding, nor form osteointegration with the bone tissue at the implant site [13], resulting in obvious interface and weak combination strength between the bone cement and the host bone. Micro motion cannot be avoided under such weak combination in daily activities, and small wear debris produced by the micro motion would cause osteolysis and further aseptic loosening or even dislodgement of the bone cement implant [14,15].

A new PVP or PKP surgery, or even more are necessary for the treatment of the secondary fracture on the adjacent vertebral body, which increase pain and economic burden of the patient. For serious loosening or dislodgement of the bone cement, further revision surgery is inevitable. Therefore, the modification of

PMMA bone cement for the treatment of VCF is important and extremely urgent for clinical applications. Many approaches were tried to improve mechanical properties and/or biocompatibility of the PMMA bone cement by, for example, adding biocompatible hydroxyapatite (HA) powder, or partially modifying methyl methacrylate (MMA) monomer. However, ideal results were not achieved by previous reported modification studies, since the compressive strength decreased too much to meet the requirement of corresponding standard (ISO 5833-2002), or the compressive elastic modulus increased rather than decreased, or the injectability was limited and is not available in the use of PVP or PKP.

Mineralized collagen (MC) is a biomimetic biomaterial with the same chemical composition and hierarchical structures to natural bone tissue. The MC is usually prepared by an *in vitro* biomimetic mineralization process that is similar to the formation of natural bone tissue [16,17]. Within the MC, the organic type-I collagen is orderly arranged with the inorganic nano-sized HA [16]. Many laboratory studies and clinical practices have demonstrated that the MC could be used to fill bone defects and is able to promote new bone formation at the bone defect sites [18,19].

In this study, MC particles were added to commercially available PMMA bone cement products to improve both the mechanical properties and the cytocompatibility. The modification parameters, including MC particle size range and additive percentage were investigated for each PMMA bone cement. Injectability, mechanical properties, maximum temperature and setting time were tested to determine the modification availability and effectiveness. Cell experiments were performed to evaluate cytocompatibility improvement of the modification by observing adhesion and quantifying proliferation of pre-osteoblasts on the modified bone cements.

2. Materials and Methods

2.1. PMMA Bone Cement Products

Three commercially available PMMA bone cement products for PVP and PKP were purchased. The three products were Osteopal® V (Heraeus Medical GmbH, Hanau, Germany), Mendec® Spine (Tecres S. P. A., Verona, Italy) and Spineplex™ (Stryker Instruments, Kalamazoo, MI, USA). All these three bone cements were certified by medical administration of many countries and regions, and have been used in clinics for many years.

2.2. Preparation of MC Particles

MC particles used for the modification of the PMMA bone cements were made from a commercially available artificial bone graft "BonGold" produced by Beijing Allgens Medical Science and Technology Co., Ltd. (Beijing, China). The MC bone grafts were prepared by following main steps described in [20]. Briefly, water-soluble calcium salt solution and phosphate salt solution were added into acidic collagen solution to form MC deposition by adjusting pH value and temperature of the reaction system. This step is a biomineralization process, which was similar to the mineralization process of the natural bone tissue that the HA crystal nucleation and growth were directed by collagen molecular templates. The deposition was then collected by centrifugation and freeze-dried to obtain MC bone graft product.

The MC bone graft was ground into small particles and screened out 4 groups with different particle sizes by sieving. The particle size range for each group was: <200 μm, 200–300 μm, 300–400 μm, and

400–500 μm, respectively. Since the inner diameter of bone filler device for delivering bone cement in PVP and PKP are usually 2.5–4.0 mm, MC particles less than 500 μm were used in this modification study.

2.3. Addition Methods of the MC

MC particles with different addition amounts and size ranges were added into the bone cements for the modification. In our preliminary experiments, too much MC addition (>20wt % of the powder part of the bone cement) would lead to hard stirring of the bone cement and losing injectability. Therefore, 4 addition amount groups, 5 wt%, 10 wt%, 15 wt%, and 20 wt% of the powder part of the bone cement, were studied for each particle size range.

In the modification process, powder and liquid parts of the bone cement were firstly mixed for 30 s to form a uniform flowing phase, and MC particles were then added into with rapid stirring for 30 s to ensure homogeneous distribution within the bone cement. There were two adding methods for the MC particles. One is direct addition of a certain amount of MC particles, the other is partial replacement of the powder part of the bone cement by equivalent amount of MC particles. Specifically, in the replacement method, a portion of the powder part of the bone cement was firstly removed, and then the MC particles equivalent to the removed bone cement powder in weight would be added into the mixed bone cement. The direct addition is preferred since such operation is convenient for clinical use.

2.4. Injectability of the Modified Bone Cements

A bone filler device with an inner diameter of 2.8 mm (Shanghai Kinetic Co., Ltd., Shanghai, China) was used to investigate the injectability of the MC modified PMMA bone cements. The uniformly mixed bone cement was extracted into a 20 mL syringe, injected into the bone filler device, and then pushed out to determine whether the modified bone cement was injectable or not.

2.5. Mechanical Property Tests

Mechanical properties of the MC modified PMMA bone cements were tested by using a universal materials testing machine (Instron-5880, Instron, Norwood, MA, USA) according to annex E and F of ISO 5833-2002. Cylindrical specimens with 6 diameter and 12 mm height were prepared for compressive strength and compressive modulus tests, and flat plate specimens with 75 mm length, 10 mm width and 3.3 mm depth were prepared for four-point bending strength and bending modulus tests.

The compressive strength, bending strength and bending modulus for each specimen were calculated according to related expressions provided by ISO 5833-2002. The compressive modulus for each specimen was calculated as the slope of the linear region of the stress-strain curve, which was derived from the displacement-load curve recorded by the testing machine, the height and the diameter of the specimen.

2.6. Maximum Temperature and Setting Time Tests

Maximum temperature and setting time of the MC modified PMMA bone cement were tested and recorded as described by annex C of ISO 5833-2002. Briefly, approximately 25 g immediately mixed bone cement was filled into a polytetrafluoroethylene (PTFE) mold, and the temperature was measured via a thermocouple and an electronic converting device having an accuracy of ±0.1 °C. The maximum temperature would be directly recorded by the electronic converting device, and the setting time was determined as the time corresponding to the average value of the maximum and the ambient temperature [21]. The best modification solution screened by above mechanical property tests was tested for each PMMA bone cement product, and the two parameters of each unmodified original product were also tested as the control. The tested were performed at 23 °C and relative humidity of 50%.

2.7. Processing Times Tests

Processing times are of importance for clinical operation of the bone cements by a surgeon. The processing times consisted of four phases, including mixing, waiting, application, and setting. In this study, processing times were tested for each bone cement before and after the modification to investigate the influence of the MC addition on the operation properties of the bone cements.

The measurement principles for the four phases were as follows:

Mixing time: time for completely mixing of the powder part and liquid part of the bone cement, as well as the MC particles;

Waiting time: time from the bone cement being extracted in to the syringe to being suitable for the injection;

Application time: time from the bone cement being applicable to being hard to inject;

Setting time: time from the injection of the bone cement to it become hardened.

2.8. In Vitro Cytocompatibility Evaluation

Cytocompatibility improvement of the MC modified bone cement were evaluated by culturing pre-osteoblasts on modified and unmodified Osteopal® V and Mendec® Spine bone cements. The use of these two bone cements was because that they contained different contrast agents, Osteopal® V contained ZrO_2 and Mendec® Spine contained $BaSO_4$ in their respect powder part. A clonal osteogenic cell line derived from newborn mouse calvarias, MC3T3-E1 (purchased from Cell Bank of Chinese Academy of Sciences, Shanghai, China), was used in this cytocompatibility evaluation. The cells were cultured in Dulbecco's Modified Eagle Medium (DMEM) with 10% fetal bovine serum (FBS), 100 U/mL penicillin and 0.1 mg/mL streptomycin at 37 °C in an incubator with 5% CO_2.

To prepare bone cement samples for cell culturing, the modified and unmodified bone cements were injected into respective 5 wells in a 96-well plate with 0.1 mL per well, immediately after all the components were fully mixed together. After setting for 24 h, cells were seeded on the set bone cements by adding 100 µL cell suspension into each well at a concentration of 1×10^5 cells/mL. Wells without bone cement were seeded with cells as the control group. Four such 96-well plates were maintained at 37 °C in an incubator with 5% CO_2, and the culture medium was replaced by fresh medium 1, 3, 5 and 7 days after the cell seeding.

Cell proliferation on both modified and unmodified PMMA bone cements were tested by cell counting kit-8 (CCK-8, Dojindo, Japan) at the 1st, 3rd, 5th and 7th day after cell seeding. At each time point, one 96-well plate was randomly selected after refreshing culturing medium, and 10 µL of CCK-8 solution was added into each well. After 2 h incubation at 37 °C, 100 µL solution of each well was transferred to another 96-well plate. Optical density (OD) values at 450 nm of all the wells were measured by a microplate reader (Bio-Rad, Model 680, Hercules, CA, USA).

Cytocompatibility improvement of the modified bone cement was also studied by observing cell attachment on the bone cements before and after the modification. The bone cement samples used for SEM observation of cell attachment were discs with 10 mm diameter and 2 mm thickness. The cell attachment was observed by scanning electron microscopy (SEM; FEI Quanta 200, Hillsboro, OR, USA) 48 h after cell seeding. Samples for the SEM observation were prepared as follows: bone cement samples with the cells were washed with phosphate buffer saline (PBS) to remove any non-adherent cells, and fixed in 2.5% glutaraldehyde in PBS for 24 h; the samples were then dehydrated in ascending series of ethanol solution from 50% to 100% and stored in frozen tert-butyl alcohol (TBA); followed by thoroughly freeze-drying, cell samples were sputter-coated with nano gold particles and observed by SEM.

2.9. Statistical Methods

The results were compared using standard analysis of Student's *t*-test and expressed as means ± SD. $p < 0.05$ was considered statistically significant.

3. Results

3.1. Injectability of the Modified Bone Cements

Tables 1–3 list the injectability of the MC modified bone cements. The symbol "○" refers to injectable, and "×" refers to uninjectable. The expression "100/x" means the direct addition method, and "(100 − x)/x" means the replacement method.

Table 1. The injectability of the MC modified Osteopal® V bone cement.

Particle size (μm)	Powder part of the bone cement/MC particle (w/w)								
	100/0	100/5	100/10	100/15	100/20	95/5	90/10	85/15	80/20
<200	○	○	×	×	×	○	×	×	×
200–300	○	○	×	×	×	○	○	○	×
300–400	○	○	○	×	×	○	○	○	○
400–500	○	○	○	×	×	○	○	○	×

Table 2. The injectability of the MC modified Mendec® Spine bone cement.

Particle size (μm)	Powder part of the bone cement/MC particle (w/w)								
	100/0	100/5	100/10	100/15	100/20	95/5	90/10	85/15	80/20
<200	○	○	×	×	×	○	×	×	×
200–300	○	○	○	○	×	○	○	○	○
300–400	○	○	○	○	×	○	○	○	○
400–500	○	○	○	○	×	○	○	○	○

Table 3. The injectability of the MC modified Spineplex™ bone cement.

Particle size (µm)	Powder part of the bone cement/MC particle (w/w)								
	100/0	100/5	100/10	100/15	100/20	95/5	90/10	85/15	80/20
<200	○	○	×	×	×	○	×	×	×
200–300	○	○	○	×	×	○	○	○	×
300–400	○	○	○	○	×	○	○	○	×
400–500	○	○	○	○	×	○	○	○	×

The results show that either small particles or high MC addition amount largely affected the injectability of the bone cement. For Osteopal® V bone cement, the equivalent placement method less influenced the injectability than the direct addition method. Once a MC modified bone cement was extracted into the syringe, it can easily be injected into the bone filler device and then be pushed out.

3.2. The Appearance of the Modified Bone Cements

Figure 1 shows the appearance of the unmodified and MC modified Spineplex™ bone cement. MC particles were homogeneously dispersed in the polymerized PMMA without obvious aggregation or vacancy, indicating that the MC particles were mixed well within the bone cement during flowing phase and MC was compatible with the PMMA material. The homogeneity of the MC modified bone cement ensures uniform mechanical properties throughout the bone cement, thus avoiding stress concentration in clinical applications.

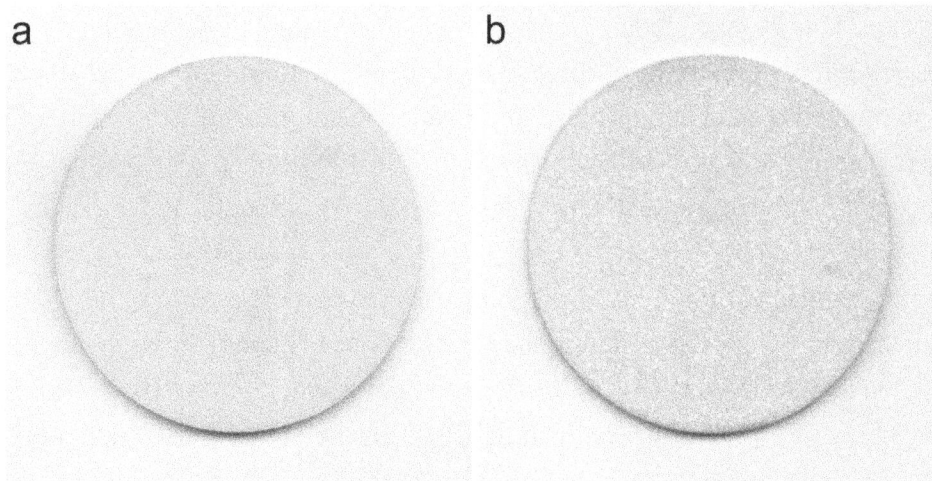

Figure 1. Appearance of the (**a**) unmodified PMMA bone cement and (**b**) MC modified PMMA bone cement.

3.3. Mechanical Properties of the Modified Bone Cements

3.3.1. Mechanical Properties of the Modified Osteopal® V Bone Cement

For Osteopal® V bone cement, partial replacement of the powder part of the bone cement by equivalent 400–500 µm MC particles kept injectable. Therefore, mechanical properties of different amount of 400–500 µm MC modified bone cement were tested, so as to screen out the best modification resolution. The compressive strength and modulus are shown in Figure 2.

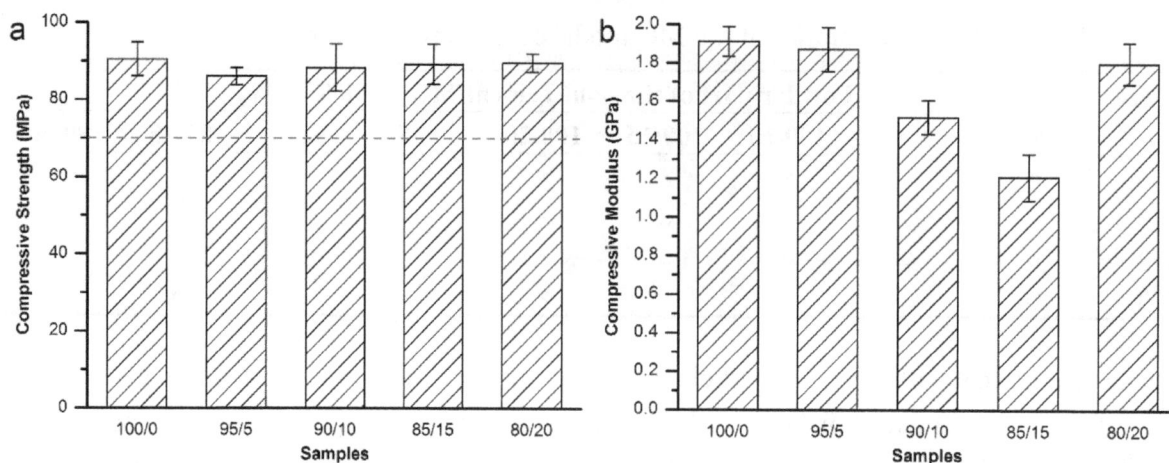

Figure 2. (**a**) Compressive strength and (**b**) compressive modulus of the MC modified Osteopal® V bone cement.

As shown in Figure 2a, the addition of the MC particles did not affected the compressive strength of the Osteopal® V bone cement. There were no significant differences between the control and each experimental group, or among experimental groups. The compressive strength for each group was higher than the 70 MPa specified by ISO 5833-2002 (red dash line in Figure 2a), thus meeting the requirement of clinical applications. Figure 2b demonstrates that replacement of the bone cement powder part by 10 wt% or 15 wt% could obtain significant down-regulation effects. 90/10 group down-regulated 20.6% and 85/15 group down-regulated 36.8%. There were statistical differences between 85/15 group and each of the other groups. Other experimental groups achieved very small down-regulation effects.

In order to investigate the effects of the particle size range on the compressive mechanical properties of the modified bone cements, and obtain the best modification result, nearby factors and levels of above experimental groups were further tested. MC particles with 200–300 μm and 300–400 μm were used to prepared 90/10 and 85/15 groups, respectively. The compressive strength and modulus are shown in Figure 3.

As shown in Figure 3a, the addition of MC particles with the particle size of either 200–300 μm or 300–400 μm did not affect the compressive strength of the set bone cements (red dash line in Figure 3a). Figure 3b demonstrated that the replacement of the bone cement powder part by 10wt % of 400–500 μm MC particles could obtain a 16.4% down-regulation effect on the compressive modulus, which was statistically different from the control (100/0) group or (85/15, 400–500) group. However, the modification results were much inferior to the 90/10 and 85/15 groups shown in Figure 2b. Therefore, equivalent replacement of Osteopal® V bone cement powder part by 15 wt% MC particles with 300–400 μm particle size achieved the best modification result for the compressive modulus of the bone cement.

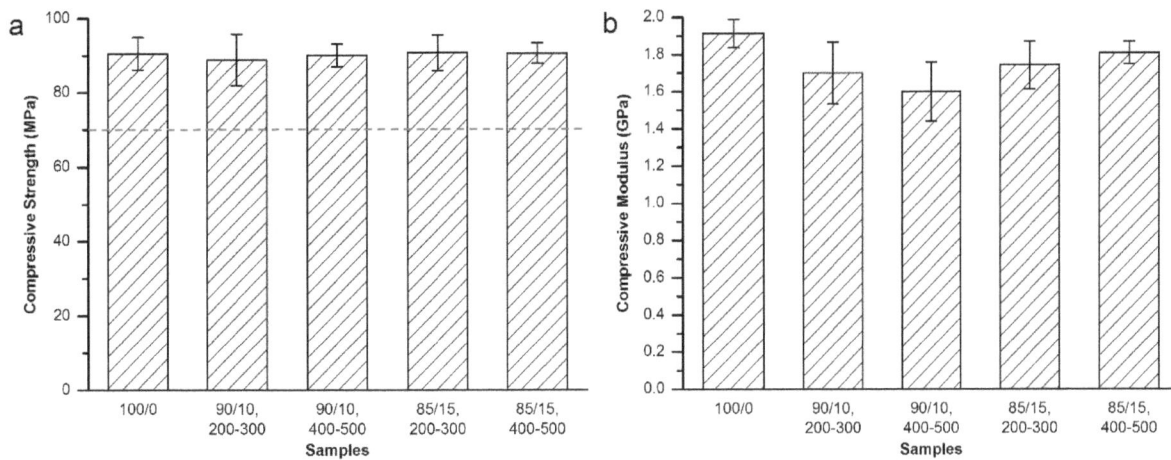

Figure 3. (**a**) Compressive strength and (**b**) compressive modulus of 200–300 μm and 300–400 μm MC particles modified Osteopal® V.

Then, bending strength and modulus were tested for the Osteopal® V bone cement specimens modified by equivalent replacement of the bone cement powder part by 10 wt% and 15 wt% MC particles with 300–400 μm particle size. As shown in Figure 4, it can be seen from the Figure 4 that both of the bending strength and bonding modulus decreased with the partial replacement of the powder part by MC particles. However, both of the bending strength and bending modulus were in conformity with related requirements in ISO 5833-2002 (red dash lines in Figure 4).

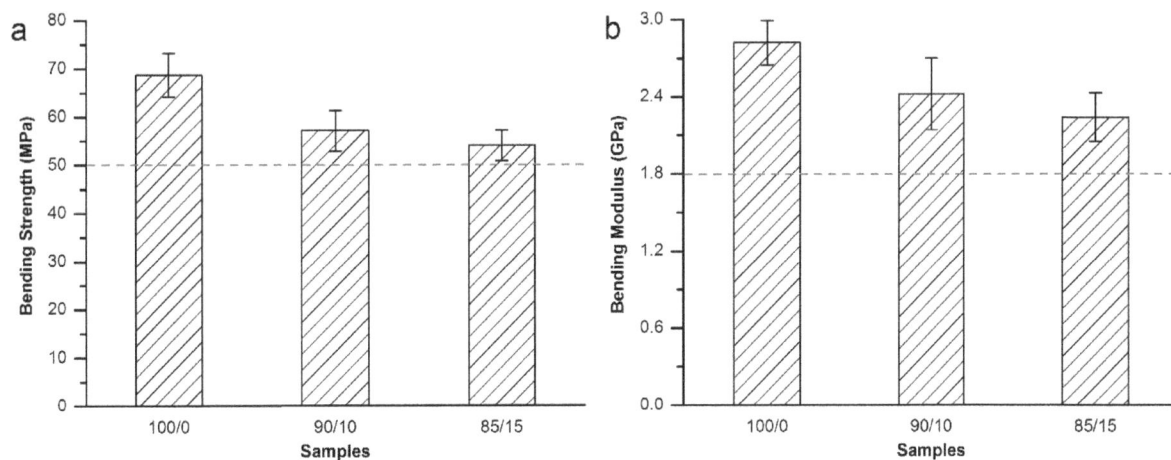

Figure 4. (**a**) Bending strength and (**b**) bending modulus of the MC modified Osteopal® V bone cement.

As a result, equivalent replacement of the bone cement powder part by 15 wt% MC particles with 300–400 μm particle size was considered to be the best resolution for the modification of the Osteopal® V bone cement. The mechanical properties met the requirement of the standard and the clinical applications after the modification.

In light of the modification study on Osteopal® V bone cement, 10 wt%–15 wt% were found to obtain better modification effects than other addition amounts. Moreover, the more MC contained within the bone cement, the better modification effects achieve for cytocompatibility improvement. Therefore,

15 wt% addition amount of MC particles was considered prior to other amounts, and different particle size ranges were investigated on the premise of injectability.

3.3.2. Mechanical Properties of the Modified Mendec® Spine Bone Cement

MC particles with the size ranges of 200–300 μm, 300–400 μm and 400–500 μm were used for the modification of Mendec® Spine bone cement. The addition amount was 15 wt% for each group, and both direct addition and equivalent replacement methods were investigated. The compressive strength and modulus are shown in Figure 5.

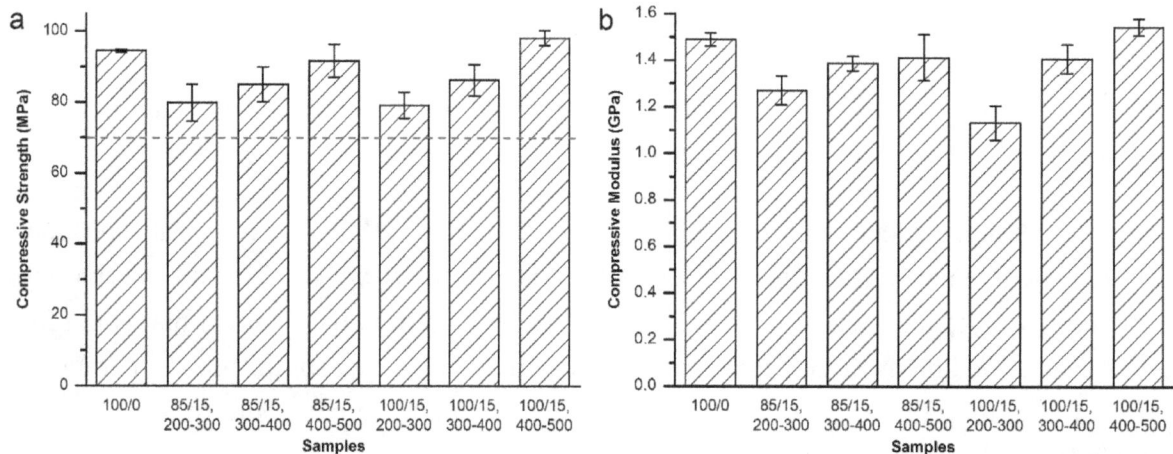

Figure 5. (a) Compressive strength and (b) compressive modulus of the MC modified Mendec® Spine bone cement.

As shown in Figure 5a, the compressive strength of the MC modified Mendec® Spine bone cement met the requirement of ISO 5833-2002 (red dash line in Figure 5a). Wherein, the direct addition of 200–300 μm MC particles achieved the best effect that the compressive modulus decreased by 24.0% (Figure 5b), and was statistically different from each of the other groups. Although the equivalent replacement using the same particle size range also obtained obvious down-regulatory effect, the direct addition would be more convenient.

Figure 6 shows the bending strength and modulus of the Mendec® Spine bone cement modified by 200–300 μm MC particles. Both specimens that modified by equivalent replacement and direct addition methods using were tested. The results show the bending strength and modulus slightly decreased by MC addition but met the requirement of ISO 5833-2002 (red dash lines in Figure 6).

As a result, direct addition of 15 wt% MC particles with 200–300 μm particle size was considered to be the best resolution for the modification of the Mendec® Spine bone cement.

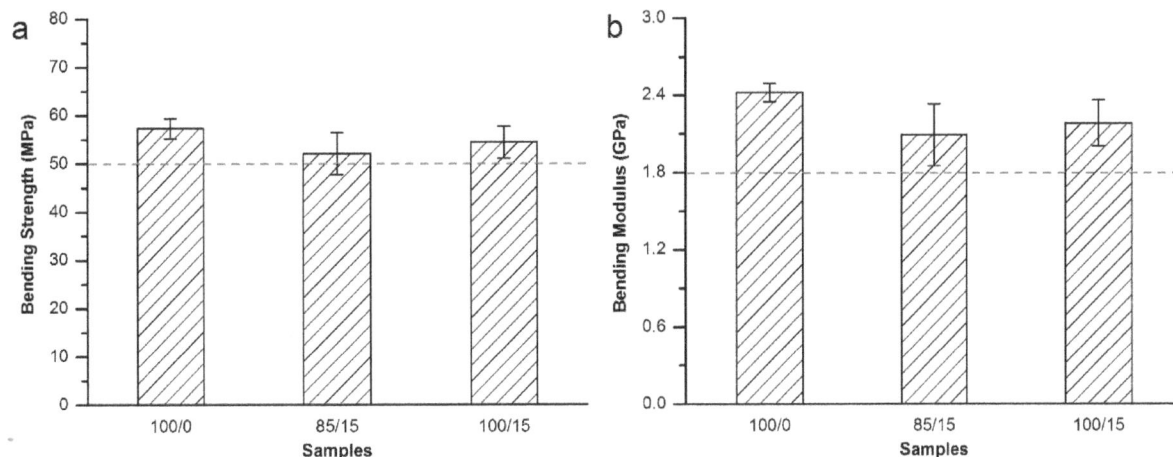

Figure 6. (**a**) Bending strength and (**b**) bending modulus of the MC modified Mendec® Spine bone cement.

3.3.3. Mechanical Properties of the Modified Spineplex™ Bone Cement

Similar to the study process of the Mendec® Spine bone cement, six experimental groups including three particle size ranges and two addition methods were investigated. The compressive strength and modulus are shown in Figure 7.

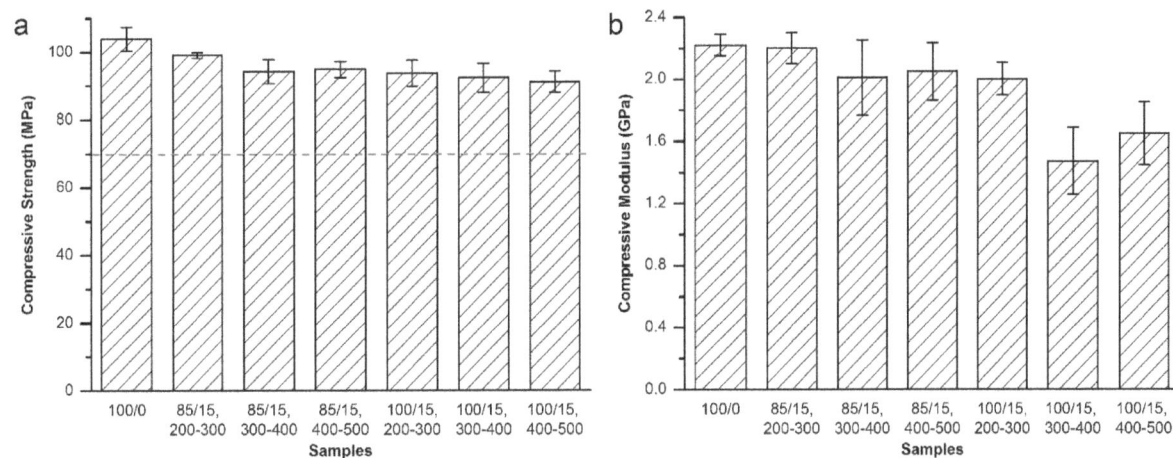

Figure 7. (**a**) Compressive strength and (**b**) compressive modulus of the MC modified Spineplex™ bone cement.

As shown in Figure 7a, the compressive strength of the Spineplex™ bone cement modified by MC particles slightly decreased, and met the requirement of ISO 5833-2002 (red dash line in Figure 7a). Figure 7b shows direct addition of MC with 300–400 μm particle size obtained the best modification effect that the compressive modulus was down-regulated by 33.8%, and was statistically different from each of the other groups, except the (100/15, 400–500) group, which also obtained obvious down-regulation effect in comparison with the control group.

Figure 8 shows the bending strength and modulus of the Spineplex™ bone cement modified by direct addition of the MC particles. Specimens modified by 300–400 μm and 400–500 μm MC particles were tested. The results show the bending strength and modulus decreased a little after the modification but met the requirement of ISO 5833-2002 (red dash line in Figure 8).

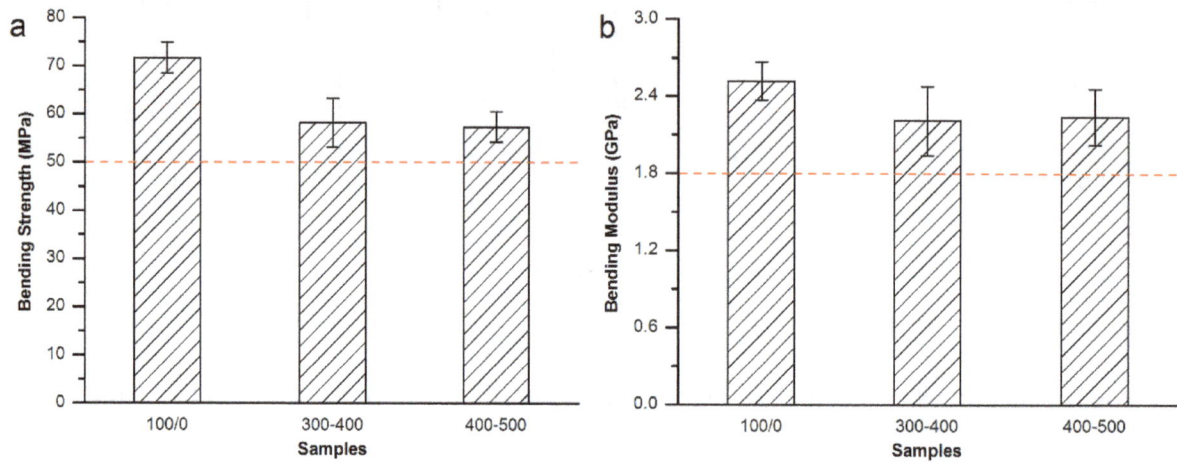

Figure 8. (a) Bending strength and **(b)** bending modulus of the MC modified Spineplex™ bone cement.

As a result, direct addition of 15 wt% MC particles with 300–400 μm particle size was considered to be the best resolution for the modification of the Spineplex™ bone cement.

3.4. Maximum Temperature and Setting Time

The maximum temperature comparisons between the unmodified bone cements and their perspective optimal modification group are shown in Figure 9.

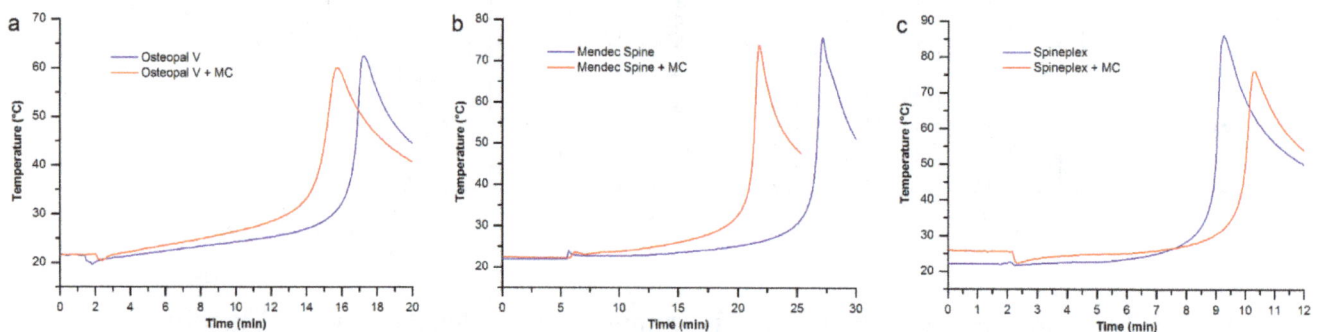

Figure 9. The maximum temperature comparisons between the unmodified and modified bone cements: **(a)** Osteopal® V bone cement; **(b)** Mendec® Spine bone cement; and **(c)** Spineplex™ bone cement.

For each bone cement product investigated in this study, the maximum temperature of the modified bone cement decreased compared to its original product. Because the added MC particles absorbed a portion of heat generated by the polymerization of the PMMA bone cements. The lower maximum temperature is beneficial for clinical applications, since such low temperature could reduce damage on tissues near the bone cement caused by the heat of polymerization.

Table 4 lists the setting time of the unmodified and modified bone cements. The addition of MC particles took effects on these bone cements. Wherein, the setting time shortened for Osteopal® V and

Mendec® Spine bone cements after the modification, while the setting time of Spineplex™ became longer. In regard to specific setting time for each bone cement, Osteopal® V and Mendec® Spine bone cements had overlong setting time, while Spineplex™ bone cement set too fast. A change in setting time for all bone cements, by some modification, would make it more convenient for clinical use by a surgeon.

Table 4. Setting time of the original bone cement products and modified bone cements.

Bone cements	Osteopal® V	Mendec® Spine	Spineplex™
Original product	16'44"	26'36"	9'02"
Modified by MC particles	14'51"	21'18"	10'01"

3.5. Processing Times for the Modified Bone Cements

Processing times for each bone cement, before and after the modification, are shown in Figure 10. The processing times for each bone cement varied a little after the modification by MC particles, and the variation was 0.5–1 min for each phase. Such small variation in the processing times makes no changes to the operating habits of surgeons.

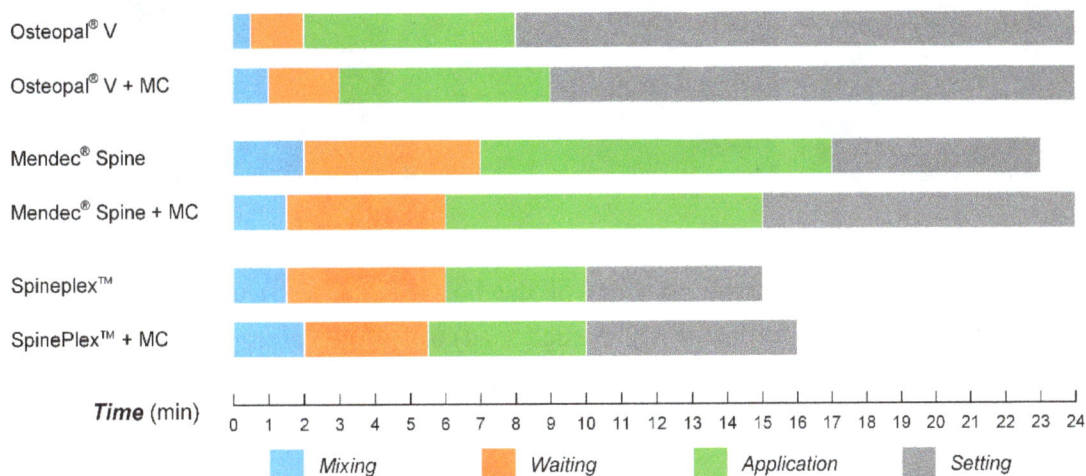

Figure 10. Processing times for the bone cements before and after the modification by MC.

Figure 11 takes Osteopal® V bone cement as an example to show operation process and processing times of the MC modified PMMA bone cement.

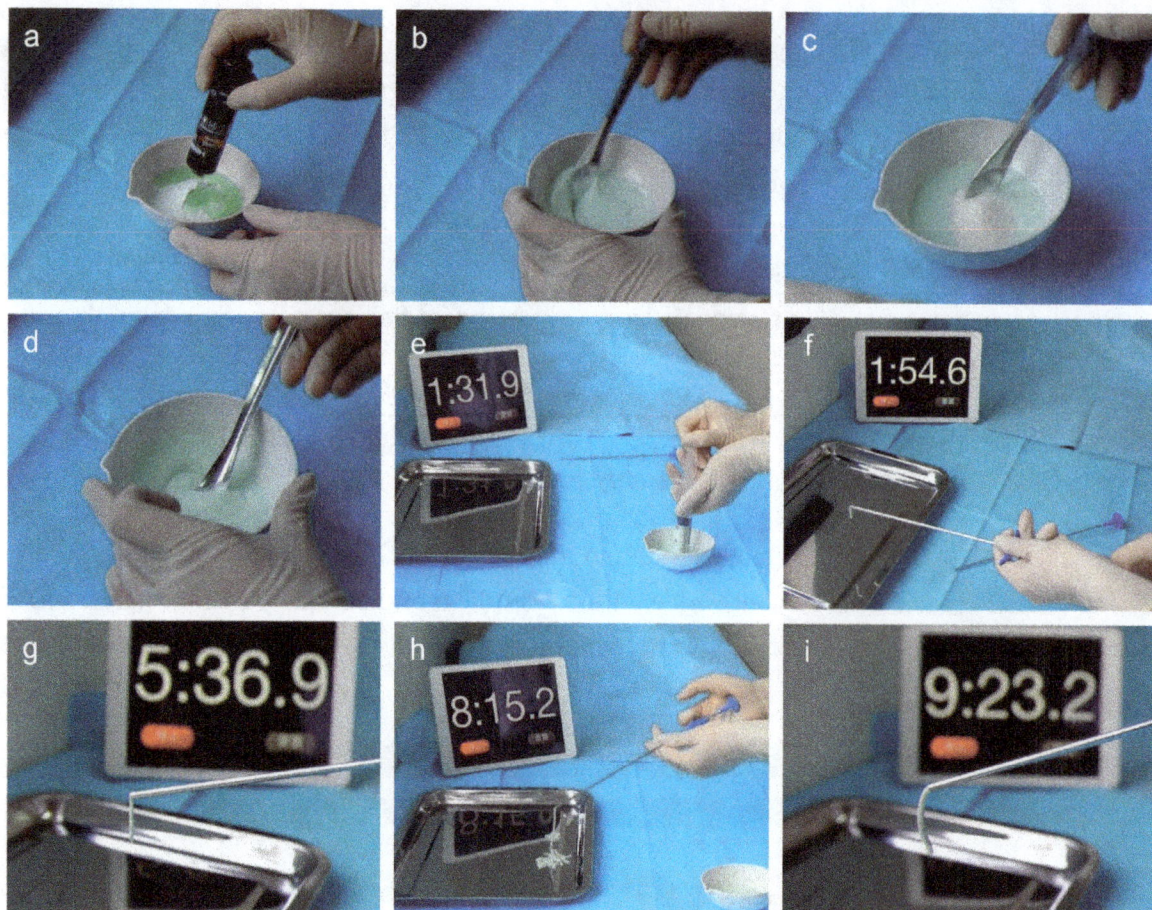

Figure 11. Operation process and processing times of the MC modified Osteopal® V bone cement: (**a**) mixing powder and liquid parts of the bone cement; (**b**) uniformly mixing powder and liquid parts; (**c**) addition of MC particles; (**d**) uniformly mixing all components; (**e**) extracting flowing bone cement by a syringe; (**f**) injection of the bone cement into a bone filler device; (**g**) earlier stage of the bone cement; (**h**) middle stage of the bone cement; and (**i**) later stage of the bone cement.

3.6. Cytocompatibility Improvement of the Modified Bone Cements

Cytocompatibility improvement of the MC modified bone cements were evaluated by proliferation quantification and attachment observation of MC3T3-E1 cells on the unmodified and MC modified bone cements. The proliferation of the cells on the bone cements are shown in Figure 12.

For both Osteopal® V and Mendec® Spine bone cements, cells proliferated well on each bone cement. Cell count on the MC modified bone cement was significantly higher than that on the unmodified original bone cement, with regard to either Osteopal® V or Mendec® Spine bone cement. At day 5 and 7, there were statistically significant differences between the MC incorporated group and the modified group, as well as between the MC incorporated group and the blank control. Cell counts for the unmodified group and the blank control group were closed without statistical differences at day 5 and 7, for both PMMA bone cement products, since pure PMMA bone cements and well-plates were all bioinert materials that had no effect on cell proliferation. The result indicated that the modification by using MC largely improved cytocompatibility of the PMMA bone cements, and the contrast agent, ZrO_2 or $BaSO_4$, did not affect such improvement effects.

Figure 13 Shows cell morphology on the Osteopal® V bone cement before and after the modification. Figure 13b and 13d are the amplification of the center areas (noted by dash boxes) of Figure 13a and 13c, respectively.

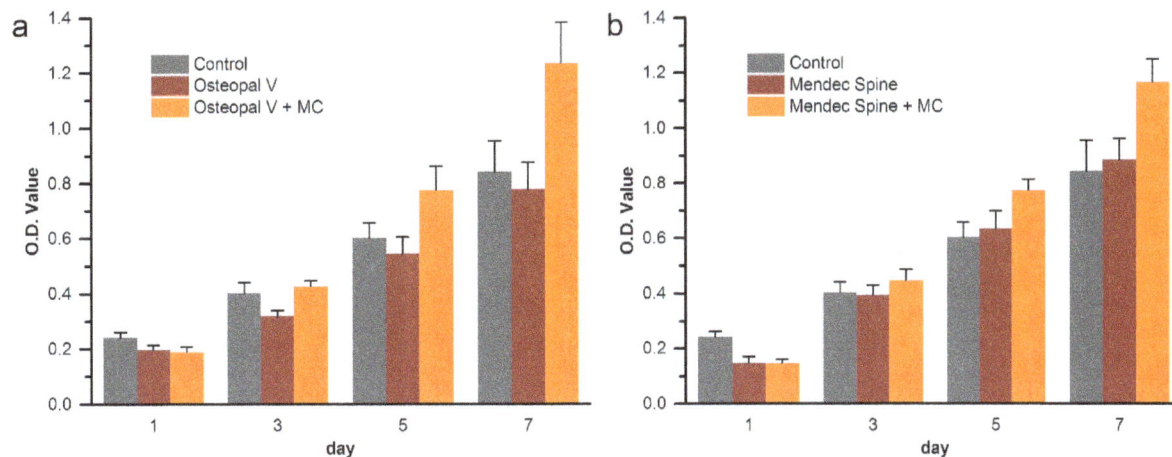

Figure 12. Cell proliferation on (**a**) Osteopal® V and (**b**) Mendec® Spine bone cements.

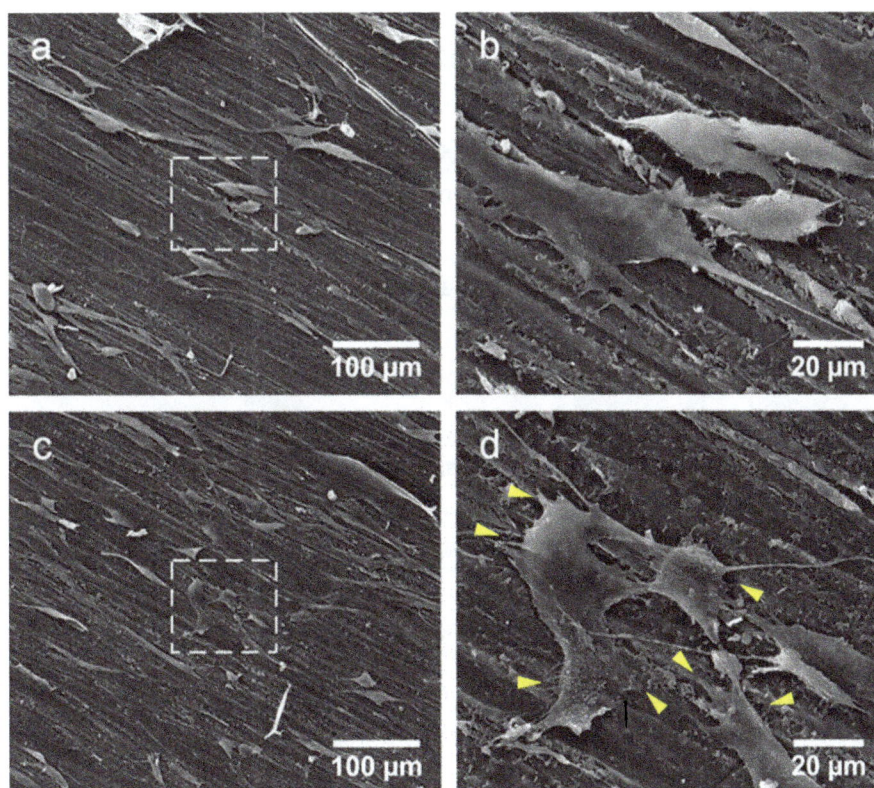

Figure 13. Cell observations on the bone cements before and after the MC modification by SEM: (**a**) cells on unmodified bone cement; (**b**) amplification of the center of Figure 13a; (**c**) cells on MC modified bone cement: and (**d**) amplification of the center of Figure 13c.

As shown in Figure 13, cells grew well on the bone cements and filopodia stretched out to anchor the cells on the bone cements. By comparing cells on the unmodified and modified bone cements, there were no differences on cell counts, which were in conformity with histograms shown in Figure 12. From the detail of the cell morphology shown in Figure 13b and 13d, it can be seen that a large number of filopodia stretched out from the cells on the MC modified bone cement (noted by yellow arrows), while the cells on the unmodified bone cement had less filopodia. The cytocompatibility of the MC modified PMMA bone cement was better than that of the unmodified original bone cement, and cell adhesion would be improved by such modification. The results indicates that the modification of PMMA bone cements by addition of MC could improve its cytocompatibility, which is beneficial for the formation of good osteointegration between the bone cement and the host bone in clinical applications.

4. Discussion

The spine is the load bearing structure in the human skeleton, and vertebral bodies are the basic structural units. For upright walking human beings, the major direction of the loading on the vertebral body is compressive force in the vertical direction, including compressive force from above lower endplate and support force from bottom upper endplate. Therefore, the compressive strength and modulus are key mechanical factors for those bone cements used for PVP and PKP. Overhigh compressive modulus of the bone cement produces overhigh stiffness of the PVP or PKP treated segment, which resulting in stress concentration at the segment, and would easily cause secondary fracture on adjacent vertebral bodies and endplate near the surgical segment [9,10].

Bioinert is another disadvantage of the PMMA bone cement, since osteocytes cannot grow into the bioinert PMMA, it is unable to form stable osteointegration between the bone cement and the host bone at the implant site [13–15]. As described by the analysis in the introduction section, aseptic loosening or even dislodgement of the bone cement are very dangerous for patients, as a free hard block may press on the spinal nerve to produce hazardous results [22].

Many efforts were made to improve the mechanical properties and biocompatibility of the PMMA bone cements for PVP and PKP. In light of above-mentioned disadvantages, these studies were focused on down-regulation of the compressive modulus, as well as improvement of the biocompatibility of the PMMA bone cement.

As the main inorganic component of natural bone tissue, HA was popular in the modification studies on PMMA bone cements. Many studies used HA and element-doped HA, such as strontium-doped HA to modify the PMMA bone cement. However, in some studies, the addition of HA largely decreased compressive strength of the bone cement that cannot meet the requirement of ISO 5833-2002 [23]; in some other studies, the compressive modulus even largely increased after the addition of HA [24]. Moreover, the addition of HA into PMMA bone cement did not exhibit improved biocompatibility [25].

Introduction of a biodegradable component was another modification idea. For example, chitosan and sodium hyaluronate were studied to form porous structure by degradation [26,27]. However, the compressive strength of the bone cement also decreased with the degradation of the biodegradable component, and became much lower than the lower limit specified by ISO 5833-2002 [26,27].

Modification of MMA monomer was tried by some researchers to down-regulate the compressive modulus of the PMMA bone cements. For example, N-methyl-pyrrolidone monomer and linoleic acid were, respectively, used to partially replace the MMA monomer in the polymerization of the PMMA bone cement. However, with the down-regulation of the compressive modulus, the compressive strength also decreased to be much lower than the requirement of ISO 5833-2002 [28,29].

In summary, previous studies on the modification of PMMA bone cement did not obtain a perfect solution that both down-regulated compressive modulus without affecting the compressive strength, and improved biocompatibility of the PMMA bone cement. In this study, a biomimetic material MC with good biocompatibility and osteogenic activity was used for the modification of the PMMA bone cement. MC was compatible with the PMMA and could be homogeneously dispersed within the PMMA bone cement. The dispersed MC particles broke integrality of the polymerized bone cement, and were able to regulate mechanical properties by verifying addition amounts, particle size range, and addition method of the MC particles. Through a series of experiments, both of mechanical properties and cytocompatibilities of three commonly used PMMA bone cements for PVP and PKP were successfully improved by addition of different MC particles with different addition methods. However, related mechanical properties regulation mechanisms need further investigations, and the clinical outcomes of the modification need long-term clinical observations.

5. Conclusions

Biomimetic MC with good osteogenic activity was added to commercially available PMMA bone cement products to improve both the mechanical properties and the cytocompatibility in this study. As the compressive strength of the modified bone cements remained well, the compressive elastic modulus were significantly down-regulated by the MC. Meanwhile, the adhesion and proliferation of pre-osteoblasts on the modified bone cements were improved compared with cells on those unmodified. The results are beneficial for both reducing the pressure on the adjacent vertebral bodies, and the osteointegration formation between the bone cement and the host bone tissue in clinical applications. Moreover, the modification of the PMMA bone cements by adding MC did not much influence the injectability and processing times of the cement. As a result, improvement of PMMA bone cements by incorporating MC particles is an effective and easy-to-operate clinical approach for improving the quality of the surgery and reducing complications after PVP and PKP.

Acknowledgments

This work was in part supported by the National Basic Research Program of China funded by the Ministry of Science and Technology of China (2011CB606205), and the National Natural Science Fund funded by the National Natural Science Foundation of China (21371106, 51402167).

Author Contributions

Hong-Jiang Jiang conceived the study and drafted the manuscript. Jin Xu and Yun Cui performed the experiments and analyzed the data. Zhi-Ye Qiu performed the experiments and drafted the manuscript. Xin-Long Ma participated in the experiment design and analyzed the data. Zi-Qiang Zhang analyzed the data. Xun-Xiang Tan participated in the experiment design. Fu-Zhai Cui conceived the study and participated in the experiment design. All the authors read and approved the final manuscript.

Conflicts of Interest

The authors declare no conflict of interest.

References

1. Old, J.L.; Calvert, M. Vertebral compression fractures in the elderly. *Am. Fam. Physician* **2004**, *69*, 111–116.

2. Furstenberg, C.H.; Grieser, T.; Wiedenhofer, B.; Gerner, H.J.; Putz, C.M. The role of kyphoplasty in the management of osteogenesis imperfecta: Risk or benefit? *Eur. Spine J.* **2010**, *19* (Suppl. 2), S144–S148.

3. Gu, Y.F.; Li, Y.D.; Wu, C.G.; Sun, Z.K.; He, C.J. Safety and efficacy of percutaneous vertebroplasty and interventional tumor removal for metastatic spinal tumors and malignant vertebral compression fractures. *AJR Am. J. Roentgenol.* **2014**, *202*, W298–W305.

4. Banse, X.; Sims, T.J.; Bailey, A.J. Mechanical properties of adult vertebral cancellous bone: Correlation with collagen intermolecular cross-links. *J. Bone Miner. Res.* **2002**, *17*, 1621–1628.

5. Hou, F.J.; Lang, S.M.; Hoshaw, S.J.; Reimann, D.A.; Fyhrie, D.P. Human vertebral body apparent and hard tissue stiffness. *J. Biomech.* **1998**, *31*, 1009–1015.

6. Morgan, E.F.; Bayraktar, H.H.; Keaveny, T.M. Trabecular bone modulus-density relationships depend on anatomic site. *J. Biomech.* **2003**, *36*, 897–904.

7. Kurtz, S.M.; Villarraga, M.L.; Zhao, K.; Edidin, A.A. Static and fatigue mechanical behavior of bone cement with elevated barium sulfate content for treatment of vertebral compression fractures. *Biomaterials* **2005**, *26*, 3699–3712.

8. Jasper, L.E.; Deramond, H.; Mathis, J.M.; Belkoff, S.M. Material properties of various cements for use with vertebroplasty. *J. Mater. Sci. Mater. Med.* **2002**, *13*, 1–5.

9. Grados, F.; Depriester, C.; Cayrolle, G.; Hardy, N.; Deramond, H.; Fardellone, P. Long-term observations of vertebral osteoporotic fractures treated by percutaneous vertebroplasty. *Rheumatology (Oxford)* **2000**, *39*, 1410–1414.

10. Trout, A.T.; Kallmes, D.F.; Layton, K.F.; Thielen, K.R.; Hentz, J.G. Vertebral endplate fractures: An indicator of the abnormal forces generated in the spine after vertebroplasty. *J. Bone Miner. Res.* **2006**, *21*, 1797–1802.

11. Burton, A.W.; Mendoza, T.; Gebhardt, R.; Hamid, B.; Nouri, K.; Perez-Toro, M.; Ting, J.; Koyyalagunta, D. Vertebral compression fracture treatment with vertebroplasty and kyphoplasty: Experience in 407 patients with 1,156 fractures in a tertiary cancer center. *Pain Med.* **2011**, *12*, 1750–1757.

12. Trout, A.T.; Kallmes, D.F.; Kaufmann, T.J. New fractures after vertebroplasty: Adjacent fractures occur significantly sooner. *AJNR Am. J. Neuroradiol.* **2006**, *27*, 217–223.

13. Sugino, A.; Miyazaki, T.; Kawachi, G.; Kikuta, K.; Ohtsuki, C. Relationship between apatite-forming ability and mechanical properties of bioactive PMMA-based bone cement modified with calcium salts and alkoxysilane. *J. Mater. Sci. Mater. Med.* **2008**, *19*, 1399–1405.

14. Portigliatti-Barbos, M.; Rossi, P.; Salvadori, L.; Carando, S.; Gallinaro, M. Bone-cement interface: A histological study of aseptic loosening in twelve prosthetic implants. *Ital. J. Orthop. Traumatol.* **1986**, *12*, 499–505.

15. Mann, K.A.; Miller, M.A.; Cleary, R.J.; Janssen, D.; Verdonschot, N. Experimental micromechanics of the cement-bone interface. *J. Orthop. Res.* **2008**, *26*, 872–879.

16. Cui, F.Z.; Li, Y.; Ge, J. Self-assembly of mineralized collagen composites. *Mater. Sci. Eng.: R: Rep.* **2007**, *57*, 1–27.

17. Zhang, W.; Liao, S.S.; Cui, F.Z. Hierarchical Self-Assembly of Nano-Fibrils in Mineralized Collagen. *Chem. Mater.* **2003**, *15*, 3221–3226.

18. Liao, S.S.; Cui, F.Z. *In vitro* and *in vivo* degradation of mineralized collagen-based composite scaffold: Nanohydroxyapatite/collagen/poly(*L*-lactide). *Tissue Eng.* **2004**, *10*, 73–80.

19. Liao, S.S.; Guan, K.; Cui, F.Z.; Shi, S.S.; Sun, T.S. Lumbar spinal fusion with a mineralized collagen matrix and rhBMP-2 in a rabbit model. *Spine (Phila Pa 1976)* **2003**, *28*, 1954–1960.

20. Liao, S.S.; Cui, F.Z.; Zhang, W.; Feng, Q.L. Hierarchically biomimetic bone scaffold materials: Nano-HA/collagen/PLA composite. *J. Biomed. Mater. Res. B Appl. Biomater.* **2004**, *69*, 158–165.

21. International Organization for Standardization. *Implants for Surgery—Acrylic Resin Cements*; ISO 5833:2002(E); International Organization for Standardization: Geneva, Switzerland, 2002.

22. Tsai, T.T.; Chen, W.J.; Lai, P.L.; Chen, L.H.; Niu, C.C.; Fu, T.S.; Wong, C.B. Polymethylmethacrylate cement dislodgment following percutaneous vertebroplasty: A case report. *Spine (Phila Pa 1976)* **2003**, *28*, E457–E460.

23. Lam, W.; Pan, H.B.; Fong, M.K.; Cheung, W.S.; Wong, K.L.; Li, Z.Y.; Luk, K.D.; Chan, W.K.; Wong, C.T.; Yang, C.; Lu, W.W. *In vitro* characterization of low modulus linoleic acid coated strontium-substituted hydroxyapatite containing PMMA bone cement. *J. Biomed. Mater. Res. B Appl. Biomater.* **2011**, *96*, 76–83.

24. Hernandez, L.; Gurruchaga, M.; Goni, I. Injectable acrylic bone cements for vertebroplasty based on a radiopaque hydroxyapatite. Formulation and rheological behaviour. *J. Mater. Sci. Mater. Med.* **2009**, *20*, 89–97.

25. Hernandez, L.; Parra, J.; Vazquez, B.; Bravo, A.L.; Collia, F.; Goni, I.; Gurruchaga, M.; San Roman, J. Injectable acrylic bone cements for vertebroplasty based on a radiopaque hydroxyapatite. Bioactivity and biocompatibility. *J. Biomed. Mater. Res. B Appl. Biomater.* **2009**, *88*, 103–114.

26. Kim, S.B.; Kim, Y.J.; Yoon, T.L.; Park, S.A.; Cho, I.H.; Kim, E.J.; Kim, I.A.; Shin, J.W. The characteristics of a hydroxyapatite-chitosan-PMMA bone cement. *Biomaterials* **2004**, *25*, 5715–5723.

27. Boger, A.; Bohner, M.; Heini, P.; Verrier, S.; Schneider, E. Properties of an injectable low modulus PMMA bone cement for osteoporotic bone. *J. Biomed. Mater. Res. B Appl. Biomater.* **2008**, *86*, 474–482.

28. Boger, A.; Wheeler, K.; Montali, A.; Gruskin, E. NMP-modified PMMA bone cement with adapted mechanical and hardening properties for the use in cancellous bone augmentation. *J. Biomed. Mater. Res. B Appl. Biomater.* **2009**, *90*, 760–766.

29. Lopez, A.; Mestres, G.; Karlsson Ott, M.; Engqvist, H.; Ferguson, S.J.; Persson, C.; Helgason, B. Compressive mechanical properties and cytocompatibility of bone-compliant, linoleic acid-modified bone cement in a bovine model. *J .Mech. Behav. Biomed. Mater.* **2014**, *32*, 245–256.

Bioactive Wollastonite-Diopside Foams from Preceramic Polymers and Reactive Oxide Fillers

Laura Fiocco [1], **Hamada Elsayed** [1], **Letizia Ferroni** [2], **Chiara Gardin** [2], **Barbara Zavan** [2] **and Enrico Bernardo** [1,*]

[1] Department of Industrial Engineering, University of Padova, Via Marzolo 9, Padova 35131, Italy; E-Mails: laurafiocco@hotmail.com (L.F.); elsisy_chem@yahoo.com (H.E.)

[2] Department of Biomedical Sciences, University of Padova, Via Ugo Bassi 58/B, Padova 35131, Italy; E-Mails: letizia.ferroni@unipd.it (L.F.); chiara.gardin@unipd.it (C.G.); barbara.zavan@unipd.it (B.Z.)

* Author to whom correspondence should be addressed; E-Mail: enrico.bernardo@unipd.it;

Academic Editor: Andrew J. Ruys

Abstract: Wollastonite ($CaSiO_3$) and diopside ($CaMgSi_2O_6$) silicate ceramics have been widely investigated as highly bioactive materials, suitable for bone tissue engineering applications. In the present paper, highly porous glass-ceramic foams, with both wollastonite and diopside as crystal phases, were developed from the thermal treatment of silicone polymers filled with CaO and MgO precursors, in the form of micro-sized particles. The foaming was due to water release, at low temperature, in the polymeric matrix before ceramic conversion, mainly operated by hydrated sodium phosphate, used as a secondary filler. This additive proved to be "multifunctional", since it additionally favored the phase development, by the formation of a liquid phase upon firing, in turn promoting the ionic interdiffusion. The liquid phase was promoted also by the incorporation of powders of a glass crystallizing itself in wollastonite and diopside, with significant improvements in both structural integrity and crushing strength. The biological characterization of polymer-derived wollastonite-diopside foams, to assess the bioactivity of the samples, was performed by means of a cell culture test. The MTT assay and LDH activity tests gave positive results in terms of cell viability.

Keywords: polymer-derived ceramics; bioactivity; wollastonite; diopside; glass-ceramic

1. Introduction

The technology of polymer-derived ceramics (PDCs) is among the most novel approaches for the synthesis and shaping of advanced ceramics. In the vast range of polymeric precursors, silicone resins are undoubtedly widely explored and exploited thanks to their low cost, large availability and easy handling [1]. The synthesis of many types of silicate ceramics can be easily achieved by the addition of metal oxide precursors, in the form of micro- or nano-sized particles. Highly phase pure ceramics can be obtained at relatively low temperatures, due to the high reactivity of the metal oxide precursors with the particularly defective network of the amorphous silica, left as a ceramic residue of oxidative decomposition of silicones [2,3].

In the field of bioceramics, Ca-silicates and Ca-Mg silicates have recently received a growing interest for their bioactivity properties, according to their ability to stimulate body tissues to repair themselves, in particular for bone ingrowth [4–9]. Silicone/fillers mixtures do not only allow one to get these peculiar bioactive formulations, but also facilitate the shaping of the ceramic components in the form of highly porous bodies, which are extremely useful, especially in the field of scaffolds for bone regeneration [10,11]. As an example, porous akermanite ($Ca_2MgSi_2O_7$) was successfully fabricated from preceramic polymers [12], as well as porous wollastonite ($CaSiO_3$) [13,14] and foamed wollastonite-diopside glass ceramic ($CaSiO_3$-$CaMgSi_2O_6$) [15]. Concerning the shaping techniques, different methods can be applied, such as warm-pressing of composite powders mixed with sacrificial PMMA microbeads, evolution of CO_2 previously entrapped in the polymer matrix by supercritical CO_2-assisted extrusion, 3D printing of porous scaffolds from direct extrusion of preceramic pastes and foaming by water release from specific hydrated fillers [12–15].

While a high phase purity is usually achievable in binary systems derived from preceramic polymers, such as Ca-silicates, ternary systems generally imply some difficulties, due to the potential formation of undesired binary compounds instead of the expected ternary compounds. As described in a couple of previous papers, the problem may be solved by providing a liquid phase upon firing, which could promote the ionic interdiffusion, operating with specific fillers [16]. A fundamental example is that of hydrated sodium borate, also known as borax ($Na_2B_4O_7 \cdot 10H_2O$) included in the formulations for akermanite ($Ca_2MgSi_2O_7$) [12] and wollastonite-diopside ceramics [15]. The additive formed a borate liquid phase upon firing and helped the crystallization of the desired phases. The borate liquid phase, after cooling at room temperature, remained as a glass phase, so that the resulting product could be seen as a sort of "polymer-derived glass-ceramic". Borax could be seen actually as a multifunctional filler, since its use in a liquid silicone could be exploited also for an abundant and uniform foaming, due to the water release associated with the dehydration reaction, occurring at only 350 °C. The cross-linking of the polymer stabilized the porosity, maintained also after the conversion of the polymer into amorphous silica and the formation of silicates [12–15]. It must be noted that $Mg(OH)_2$, used as the MgO precursor for Ca-Mg silicates, may contribute to the foaming, but its impact is much lower than that of borax [12].

Although the addition of borax is undeniably significant for the obtainment of glass-ceramic samples with a specific phase assemblage and with a homogeneous cellular structure, the effect on the biocompatibility of the same samples is still controversial. Several studies highlighted a concern associated with borate bioactive glasses, due to the potential toxicity of boron released in the solution as borate ions $(BO_3)^{3-}$ [17,18]. As an example, the well-known borate bioglass 13-93B3 was found to be toxic to murine MLO-A5 osteogenic cells *in vitro*, above a boron threshold concentration of 0.65 mmol in the cell culture medium, while it supported the proliferation and growth of the cells below that concentration [19]. However, the same scaffolds did not show toxicity to cells *in vivo* and supported new tissue infiltration when implanted in rats [20–23]. Other boron-containing glasses are reported to be biocompatible and bioactive [24,25].

The materials described in previous papers [12,15] have a low amount of boron, but it should be remarked that boron was reasonably concentrated in the glass phase between silicate crystals. At present, the biological characterization of wollastonite-diopside porous glass-ceramics, obtained by borax addition in silicone-based mixtures, is still in progress, but it confirms the controversial impact of the specific element. In fact, dissolution studies in simulated body fluid (SBF) proved the positive behavior of the material in terms of bioactivity and ion release, while a 24 h *in vitro* cell culture test showed that the material was not suitable for cell living and proliferation.

In the present paper, we discuss a further development concerning highly porous wollastonite-diopside "polymer-derived glass-ceramics", based on the replacement of borax with sodium phosphate dibasic heptahydrate ($Na_2HPO_4 \cdot 7H_2O$), aimed at overcoming the above-described difficulties arising from the presence of boron. The selected filler, like borax, is multifunctional, *i.e.*, it contributes to both foaming and forming a liquid phase upon firing, as illustrated by Figure 1.

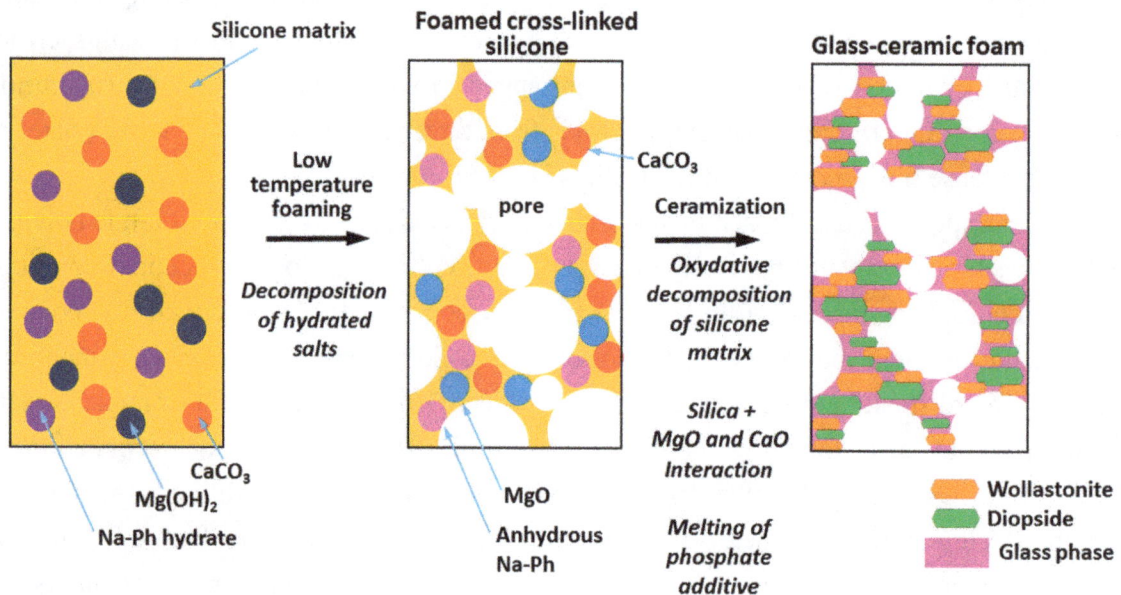

Figure 1. Scheme for the obtainment of wollastonite-diopside "polymer-derived glass-ceramic" foams, according to the dual role of hydrated sodium phosphate filler (Na-Ph hydrate).

Like in the previously developed wollastonite-diopside ceramics [15], the addition of a further filler, in the form of powders of a glass crystallizing into wollastonite and diopside, will be discussed in order to optimize the integrity of samples. In fact, the ceramization step does not modify the macro-porosity formed in the low-temperature foaming step, but it implies the formation of micro-cracks, caused by internal stresses. The glass addition is essentially conceived to reduce the cracks, enhancing the stress relaxation operated by the liquid phase, upon firing, with no impact on foaming and phase development.

Although preliminary, the results of a five-day cell culture test, on phosphate-modified wollastonite-diopside ceramics, indicate a good biocompatibility, independent of the glass addition.

2. Experimental Procedure

2.1. Starting Materials

Two commercially available silicones, H62C and MK (Wacker-Chemie GmbH, Munich, Germany), were considered as silica precursors, with a yield of 58 wt% and 84 wt%, respectively [2]. CaO and MgO precursors consisted of $CaCO_3$ (Sigma Aldrich, Gillingham, UK) and $Mg(OH)_2$ (Industrie Bitossi, Vinci, Italy), respectively, in the form of powders with a diameter below 10 µm. The amounts of silicones and precursors for CaO and MgO were calibrated in order to match the $CaO-MgO-SiO_2$ molar proportion of 2-1-3, corresponding to an equimolar mixture of wollastonite ($CaSiO_3$ or $CaO \cdot SiO_2$, $CaO-MgO-SiO_2$ molar proportion of 1-0-1) and diopside ($CaMgSi_2O_6$ or $CaO \cdot MgO \cdot 2SiO_2$, $CaO-MgO-SiO_2$ molar proportion of 1-1-2).

Sodium phosphate dibasic heptahydrate ($Na_2HPO_4 \cdot 7H_2O$, Sigma Aldrich, Gillingham, UK) was used as additional filler. Finally, a powdered Ca/Mg-rich silicate glass with a particle size <60 µm (mean diameter ~5 µm—known as G20CaII glass [15]), was added. The chemical composition of the glass additive is reported in Table 1. The molar proportions between CaO, MgO and SiO_2 roughly correspond to those of the desired mixture of wollastonite and diopside, with alkali oxides used as fluxes. The use of Li_2O, in addition to Na_2O, is in agreement with recent findings concerning the positive effect of this oxide added in formulations of bioglasses, previously involving only sodium oxide [26,27].

Table 1. Chemical composition of the glass additive used in silicone-based mixtures.

Composition (% mol)				
SiO_2	CaO	MgO	Na_2O	Li_2O
55.3	22.0	12.0	9.0	1.7

2.2. Preparation of Foams

H62C was first dissolved in isopropanol (10 mL for 10 g of final ceramic) and then mixed with micro-sized fillers, including sodium phosphate, in the as-received, hydrated form (the quantity of salt was 10 wt% of the theoretical ceramic yield of the other components, corresponding to 5 wt% of anhydrous salt). Selected samples included also glass powders (10 wt% of the theoretical ceramic yield of the other components). The mixing was performed under magnetic stirring, followed by sonication for 10 min, which allowed obtaining stable and homogeneous dispersions. The mixtures were poured into large glass containers and dried at 60 °C overnight.

After first drying, the mixtures were in the form of thick pastes, later manually transferred into cylindrical Al molds and then subjected to a foaming treatment at 350 °C in air for 30 min. Cylindrical samples, 10 mm in diameter and 7–8 mm in height, were obtained from the foams. The top surfaces were polished with abrasive paper. The samples (after removal from Al molds) were fired at 1100 °C for 1 h, using a heating rate of 2 °C/min.

2.3. Preparation of Pellets

Monolithic pellets were prepared using the MK mixed with $Mg(OH)_2$ and $CaCO_3$ micro-particles, anhydrous sodium phosphate (the same salt cited above, after preliminary dehydration at 450 °C, with a heating rate of 5 °C/min, for 1 h) and glass additive. MK was dissolved in isopropanol (10 mL for 10 g of final ceramic) and then mixed with the fillers. Stable and homogeneous dispersions in isopropanol were obtained using the same conditions applied for the H62C-based mixtures and left to dry overnight at 60 °C.

After drying, the silicone-based mixtures were in the form of solid fragments, later converted into fine powders by ball milling at 350 rpm for 30 min. The powders were cold-pressed in a cylindrical steel die applying a pressure of 20 MPa for 1 min, without using any additive. Specimens of 0.5 g, 16.6 mm in diameter and approximately 1.7 mm in thickness, were obtained. For comparison purposes, pellets of glass-free formulation were also prepared. The cold-pressed samples were fired at 1100 °C for 1 h, using a heating rate of 2 °C/min.

2.4. Cell Culture and Seeding

For cell culture studies, samples were cut to 10 mm × 10 mm × 5 mm and fixed to 48-well plates. The entire well plates where then sterilized. Human fibroblasts were seeded at a density of 4×10^5 cells/piece in cDMEM, which consisted of Dulbecco's Modified Eagle Medium (DMEM) (Lonza S.r.l., Milano, Italy), supplemented with 10 vol% fetal bovine serum (FBS) (Bidachem S.p.A., Milano, Italy) and 1 vol% penicillin/streptomycin (P/S) (EuroClone, Milano, Italy). The 3D cultures were incubated at 37 °C and 5% CO_2 for 7 days, with media changes every 2 days.

2.5. Analysis of Cell Viability

The cell proliferation rate was evaluated after 3 and 7 days from seeding with the MTT (methylthiazolyl-tetrazolium)-based proliferation assay, performed according to the method of Denizot and Lang with minor modifications [28]. Briefly, samples were incubated for 3 h at 37 °C in 1 mL of 0.5 mg/mL MTT solution prepared in phosphate buffered saline (PBS) (Euroclone). After removal of the MTT solution by pipette, 0.5 mL of 10% DMSO in isopropanol was added to extract the formazan in the samples for 30 min at 37 °C. For each sample, absorbance values at 570 nm were recorded in duplicate on 200 µL aliquots deposited in microwell plates using a multi-label plate reader (Victor 3, Perkin Elmer, Milano, Italy).

Lactate Dehydrogenase Activity (LDH activity) was measured using a specific LDH Assay Kit (SigmaAldrich, St. Louis, MO, USA) according to the manufacturer's instructions. All conditions were tested in duplicate. The culture medium was reserved to determine extracellular LDH. The intracellular LDH was estimated after cells lysis with the assay buffer contained in the kit. All samples were incubated with a supplied reaction mixture, resulting in a product whose absorbance was measured at 450 nm using a Victor 3 multi-label plate reader.

For SEM imaging, fibroblasts grown on samples for 3 and 7 days were fixed in 2.5% glutaraldehyde in 0.1 M cacodylate buffer for 1 h, then progressively dehydrated in ethanol. Control and treated Ti discs without cells were also examined.

2.6. Statistical Analysis

t-tests were used to determine significant differences ($p < 0.05$). Repeatability was calculated as the standard deviation of the difference between measurements. All testing was performed in SPSS 16.0 software (SPSS Inc., Chicago, IL, USA) (license of the University of Padua, Padua, Italy).

2.7. Characterization

Microstructural characterizations were performed by optical stereomicroscopy (AxioCam ERc 5 s Microscope Camera, Carl Zeiss Microscopy, Thornwood, NY, USA) and scanning electron microscopy (FEI Quanta 200 ESEM, Eindhoven, The Netherlands) equipped with energy dispersive spectroscopy (EDS).

The crystalline phases were identified by means of X-ray diffraction on powdered samples (XRD; Bruker AXS D8 Advance, Bruker, Germany—CuKα radiation, 0.15418 nm, 40 kV–40 mA, 2θ = 15°–70°, step size = 0.05°, 2 s counting time), supported by data from the PDF-2 database (Powder Diffraction File, ICDD-International Center for Diffraction Data, Newtown Square, PA, USA) and the Match! program package (Crystal Impact GbR, Bonn, Germany).

The bulk density of the foams was determined from the weight-to-volume ratio, using a caliper and a digital balance. The true density of the samples was measured by means of a gas pycnometer (Micromeritics AccuPyc 1330, Norcross, GA, USA), operating with He gas on finely-milled samples.

The crushing strength of foams was measured at room temperature, by means of an Instron 1121 UTM (Instron Danvers, MA, USA) operating with a cross-head speed of 1 mm/min. Each data point represents the average value of 5–10 individual tests.

3. Results and Discussion

3.1. Foaming and Phase Development

Figure 2a testifies to the very homogeneous foaming achieved according to the approach described in Figure 1. Many interconnections between adjacent pores were visible from both top and side views, as proof of the open porosity. The morphology of the newly obtained foams is comparable to that of previous wollastonite-diopside polymer-derived ceramics foamed by decomposition of borax, although the amount of foaming additive had to be drastically revised. The effect of 10 wt% hydrated Na-phosphate, in other words, roughly corresponded to that 3 wt% borax (samples with a lower content of phosphate salt, exhibiting a much less abundant and uniform foaming, are not discussed here for the sake of brevity) in previous experiments [15].

Like borax, the phosphate salt did not contribute to the formation of any crystal phase. In particular, Figure 3a (upper pattern) shows that the expected silicate phases, *i.e.*, wollastonite (PDF#42-0547) and diopside (PDF#86-0932), effectively formed at 1100 °C from H62C silicone and oxide precursors, with only minor traces of akermanite (PDF#83-1815) and merwinite ($Ca_3MgSi_2O_8$; PDF#74-0382).

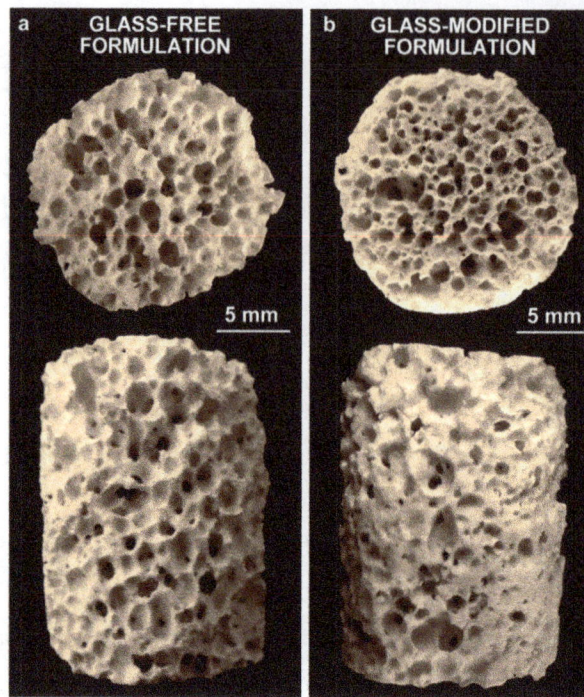

Figure 2. Morphology of the foams (top and side views): (**a**) glass-free formulation; (**b**) glass-modified formulation (10 wt% glass).

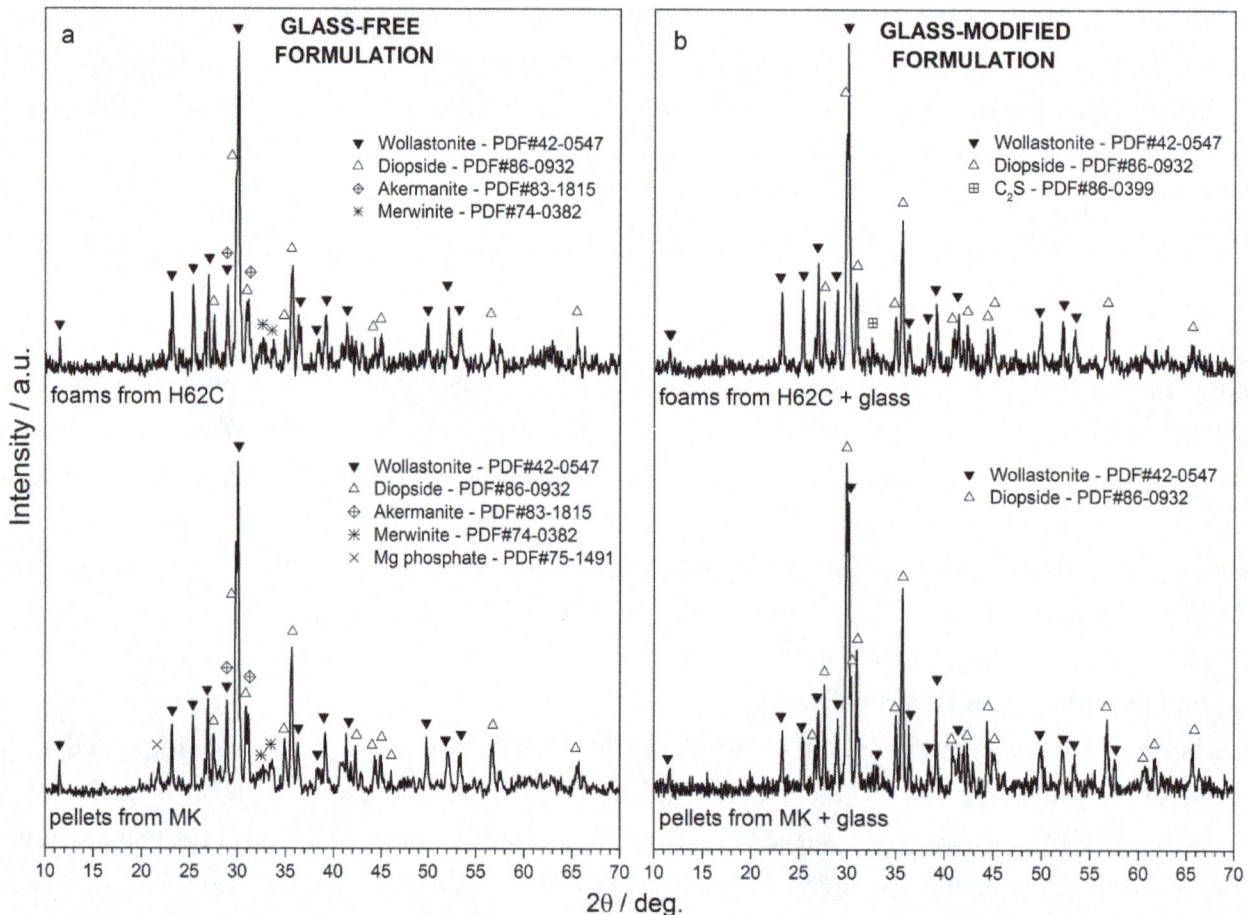

Figure 3. X-ray diffraction pattern of polymer-derived glass-ceramic samples (foams from H62C, pellets from MK): (**a**) glass-free formulations; (**b**) glass-modified formulations.

The similarity with the previous wollastonite-diopside foams, developed with borax, was further confirmed by the physical and mechanical data reported in Table 2. Bulk density, the amount of open porosity and crushing strength were practically identical. The crushing strength (approximately 1.5 MPa), in particular, was quite low, considering the high crystallinity inferable from the diffraction pattern (the absence of an "amorphous halo" suggested a limited amount of glass phase, mostly attributable to sodium phosphate).

Table 2. Physical and mechanical properties of polymer-derived wollastonite-diopside foams.

Foam Formulation	Bulk Density (g/cm^3)	Open Porosity (%)	Crushing Strength (MPa)
H62C + fillers (borax) *	0.73 ± 0.02	77.0	1.8 ± 0.3
H62C + fillers (Na-phosphate)	0.70 ± 0.02	76.5	1.4 ± 0.1
H62C + fillers + 10 wt% glass (Na-phosphate)	0.63 ± 0.10	79.4	3.1 ± 0.7

* Data from Fiocco *et al.* [15].

As illustrated by Figure 4a–c, the foamed samples from glass-free formulation exhibited a large number of microcracks, which could be due to the development of internal stresses upon ceramization. These stresses could be attributed to multiple factors, such as gas release from the polymer-to-ceramic conversion of silicones, decomposition of calcium carbonate (used as CaO precursor) and volume changes associated with the crystallization of silicates, visible as small granules in Figure 4c.

Despite a slightly less homogeneously distributed macro-porosity and mean diameter (Figure 2b), with respect to the samples from the glass-free formulation (Figure 2a), foams developed with glass powders as additional fillers exhibited an improvement in the structural integrity (Figure 4d–f). The viscous flow, due to the softening of glass particles, likely overlapped with that of the liquid phase offered by sodium phosphate and caused some stress relaxation. The formation of elongated crystals, shown in Figure 4f, could be seen as proof of enhanced flow. The crystals can be practically attributed only to wollastonite and diopside, considering the upper pattern of Figure 1b, showing only very small traces of dicalcium silicate (C$_2$S, Ca$_2$SiO$_4$ or 2CaO·SiO$_2$; PDF#86-0399) in addition to the well-defined peaks of the desired phases.

As reported in Table 2, both bulk density and the amount of open porosity were not affected by the glass addition. However, the glass addition was more effective, owing to the reduction of cracks, in the improvement of the mechanical strength, which increased from 1.4 ± 0.1 (for foams without glass) up to 3.1 ± 0.7 (for foams added with the 10 wt% of glass).

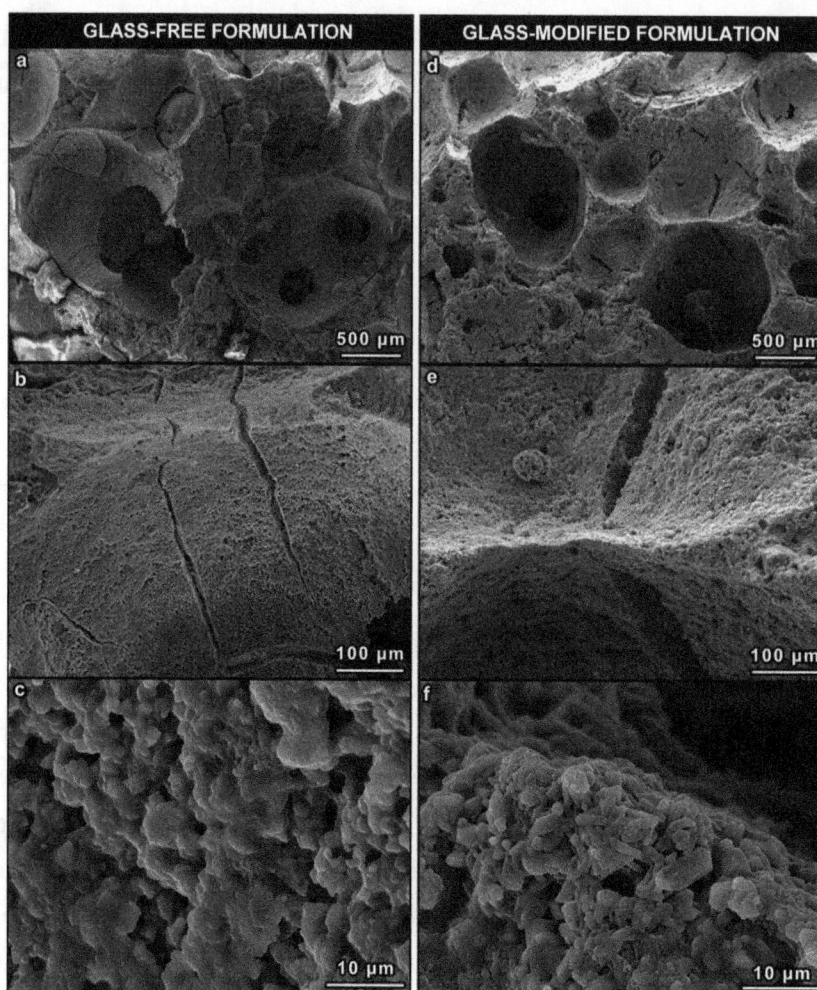

Figure 4. Higher magnification details of the foams: (**a–c**) glass-free formulation; (**d–f**) glass-modified formulation (10 wt% glass).

3.2. Impacts of Preceramic Polymer and Glass on Phase Development

Cell culture tests are generally easier to perform with flat samples, instead of foamed samples. For the specific purpose of preparing disc samples, H62C was replaced by MK. The solid silicone allowed an easy shaping of pellets by cold pressing of powdered silicone-fillers mixtures. The amount of MK was obviously calibrated, keeping the reference CaO-MgO-SiO$_2$ molar proportion, considering the different yield of silica, compared to H62C; since no foaming was expected, sodium phosphate was used in anhydrous form.

The lower pattern of Figure 3a clearly shows that the change in the preceramic polymer had no practical impact on the phase development, except for the formation of traces of magnesium phosphate (Mg$_3$P$_2$O$_8$; PDF#75-1491). This phosphate phase, together with akermanite and merwinite, completely disappeared in an MK-based formulation comprising glass particles, as shown in the lower pattern of Figure 3b. The "purifying" effect of the glass additive (an enhanced content of liquid phase promotes the interdiffusion), found for H62C, was confirmed in the system based on MK.

An additional discussion, concerning the phase development, can be done on the basis of semi-quantitative analysis provided by the Match! (Crystal Impact GbR, Bonn, Germany) program package, already employed for phase identification. Considering wollastonite and diopside, as a first approximation, as the only crystal phases, the program package could predict several weight ratios, reported in Table 3, corresponding to the best matching between experimental and theoretical diffraction patterns, depending on the formulation. In an ideal ceramic with wollastonite and diopside in equivalent molar amounts (molar ratio equal to one), the theoretical wollastonite/diopside weight balance would be equal to 35/65; from Table 3, we can easily note that the best agreement with the theoretical weight balance was provided by glass-modified formulation, based on both H62C and MK polymers.

As previously mentioned, the glass additive was proven to crystallize, alone, in wollastonite and diopside [15]. Considering the chemical composition (Table 1), we estimated a certain weight balance between the crystalline and amorphous phase, in the hypothesis of CaO included only in wollastonite and diopside, in equivalent molar content, as reported in Table 4. Repeating the same calculation, on the basis of the weight balances reported in Table 3, for polymer-based mixtures (Table 4, again) we can note that: (i) the amount of glass phase, in the glass-free formulation, is only slightly above that expected from the sodium phosphate additive (5 wt%); and (ii) the addition of glass did not "dilute" the crystallization, wollastonite and diopside being formed not only by polymer-filler reactions, but also by glass devitrification.

Table 3. Wollastonite-diopside weight balances according to the semi-quantitative X-ray diffraction analysis provided by the Match! program package.

	Formulations	Wollastonite (wt%)	Diopside (wt%)
Theoretical	$CaO \cdot SiO_2 + CaO \cdot MgO \cdot 2SiO_2$	35	65
1	H62C + fillers	56	44
2	H62C + fillers + 10 wt% glass	40	60
3	MK + fillers	49	51
4	MK + fillers + 10 wt% glass	42	58

Table 4. Semi-quantitative analysis of the weight balance between crystalline and amorphous phases.

	Formulations	Crystalline Phase (wt%)	Amorphous Phase (wt%)
	Pure Ca/Mg-rich glass	66	34
1	H62C + fillers	88	12
2	H62C + fillers + 10 wt% glass	98	2
3	MK + fillers	92	8
4	MK + fillers + 10 wt% glass	96	4

The calculations in Table 4 are only indicative (a more precise phase quantification, based on specific software packages, is in progress), but we can certainly say that silicone/fillers mixtures and the adopted Ca/Mg-rich glass have an intrinsic, very significant "compatibility"; one system had a great potential in supporting the other. Going back to foams from H62C, the increase of the liquid phase formed upon firing could be achieved by a simple increase of the amount of sodium phosphate additive, but with the risks of coarsening and/or viscous collapse of the cellular structure, upon firing, due to the dilution of the fraction leading to wollastonite and diopside. The glass additive represented a valid alternative, offering a "transient liquid phase", mostly transformed in the desired crystal phases. The tests with MK, despite providing pellets for cell tests, are promising for the application of shaping techniques based on this specific polymer (foaming by release of CO_2, embedded upon supercritical CO_2-assisted extrusion [14]) or on MK/H62C mixtures (scaffolds from fused deposition of silicone-based pastes [13]).

3.3. In Vitro Biological Characterization

As previously stated, a preliminary biological study, *i.e.*, the MTT assay, was performed on MK-derived pellets. The graph in Figure 5a shows that an increase in cell viability was observed passing from Day 3–7 for both the formulations (*i.e.*, glass-free and glass-modified), implying that the fibroblast surviving at Day 3 might have duplicated and proliferated up to Day 7. Interestingly, the incorporation of glass seemed to make the pellets generally even more biocompatible.

The successful tests on pellets stimulated the application of the MTT assay on H62C-derived foams, having a morphological organization closer to that of natural bones. As summarized in Figure 5b, at Day 3, cell viability looked higher in the glass-modified foams, as already seen in Figure 5a, while at Day 7, cells on the glass-free foams were more proliferated. From this observation, the addition of glass in the formulation of the foams did not lead to a clear improvement in cell viability at Day 7, but only contributed to increasing the biocompatibility at Day 3.

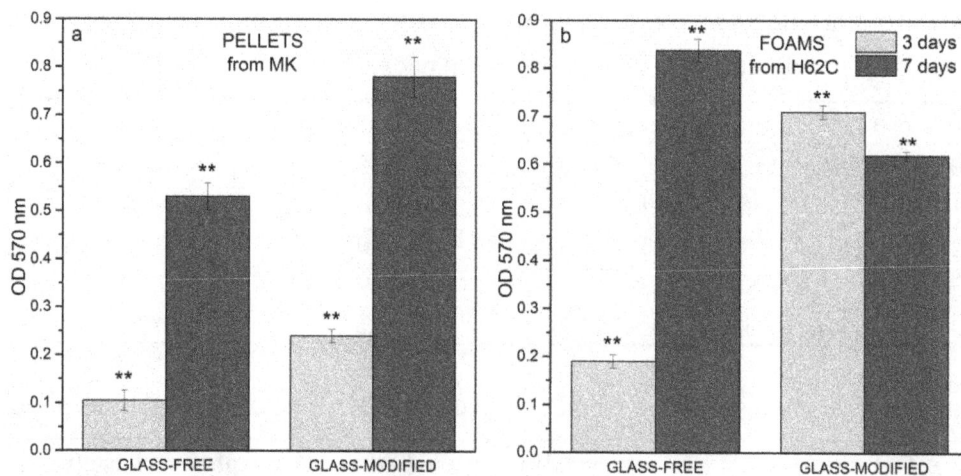

Figure 5. MTT assay: (**a**) pellets, 3–7 days; (**b**) foams, 3–7 days. Significant difference * ($p < 0.05$); ** ($p < 0.01$); *** ($p < 0.001$).

Comparing the behavior of cells seeded on pellets and on foams, with regards to glass-free formulation, the foams allowed a more extensive cell viability; concerning the glass-modified formulation, the foams showed an improvement in viability only at Day 3.

In order to overcome the controversial results of the MTT assay obtained for pellets and foams, the LDH activity assay was also performed on the cells. Figure 6a shows the intracellular LDH activity of the cells seeded on pellets: the graph proves that cells were able to produce metabolites, with improved results after seven days from seeding. As reported in Figure 6b, extracellular LDH activity was also measured on the culture medium: the graph confirms that metabolites were secreted by the same cells.

Even if the results of intracellular and extracellular LDH activity assays were not perfectly in agreement with each other, it can be observed that the incorporation of glass, which was effective in improving the mechanical behavior of the foams and the phase assemblage, was not detrimental to cell survival and proliferation.

SEM images of the foams, shown in Figure 7, were taken after three and seven days from fibroblast seeding. After three days (Figure 7a,b), fibroblasts were found to be alive and spread on the surface of the samples, of both glass-free and glass-modified formulations; in particular, they had a more elongated profile when seeding on glass-modified foams (Figure 7b). After seven days, cells had colonized the surface of the foams, still demonstrating elongated profiles, as shown in Figure 7c,d for glass-modified samples. Moreover, the formation of hydroxyapatite precipitates (nodules in Figure 7c,d) was observed, giving further evidence of the biocompatibility of the material.

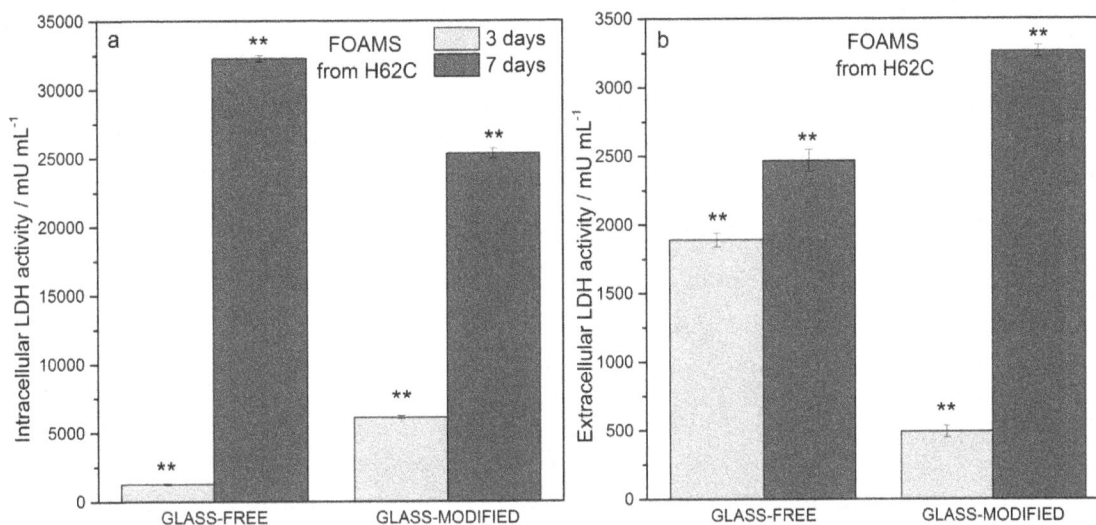

Figure 6. LDH activity assay. (**a**) Intracellular LDH activity, foams, 3–7 days; (**b**) Extracellular LDH activity, foams, 3–7 days. Significant difference * ($p < 0.05$); ** ($p < 0.01$); *** ($p < 0.001$).

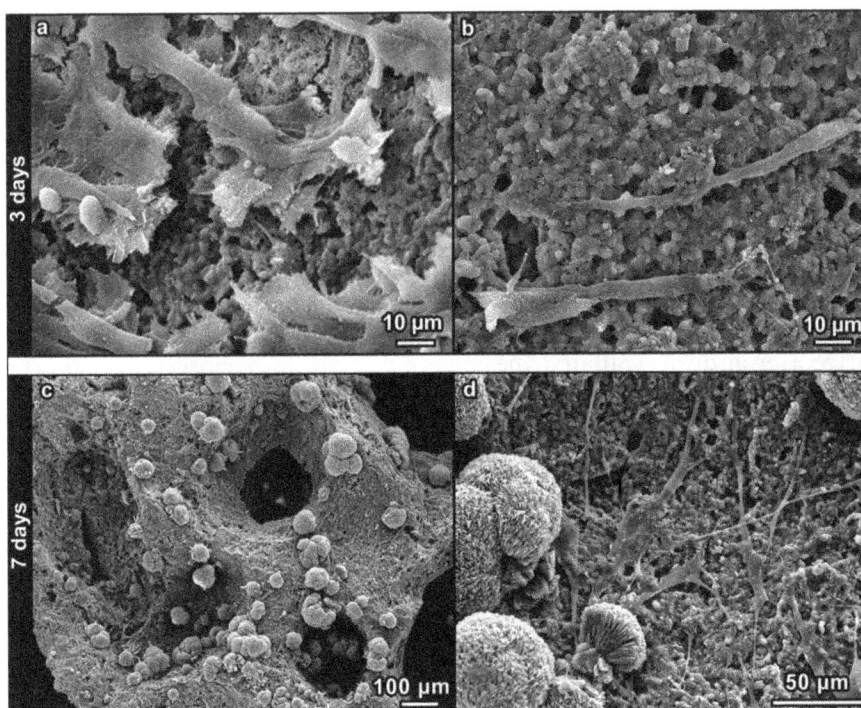

Figure 7. SEM images after cell culture on foams: (**a**) glass-free formulation, three days; (**b**) glass-modified formulation, three days; (**c,d**) glass-modified formulation, seven days.

4. Conclusions

We may conclude that:

- Wollastonite-diopside ceramics can be fabricated by firing mixtures based on preceramic polymers, in the form of silicone resins (acting as silica sources), mixed with powdered metal oxide precursors;
- The choice of silicone polymers with different natures and chemistry (liquid H62C, solid MK) does not affect the ceramic product in terms of main phase assemblage;
- A liquid silicone can be easily foamed by water release, in turn due to the decomposition of hydrated sodium phosphate; the ceramic conversion implies the transformation of the silicone foam into a glass-ceramic foam, incorporating silicate crystals embedded in the glass phase provided by the same phosphate additive;
- The liquid phase developed upon firing can be increased by the introduction of a glass filler; the positive impact on the structural integrity of samples is not accompanied by any change in the phase assemblage, operating with a glass crystallizing itself in wollastonite and diopside;
- Both dense and foamed wollastonite-diopside ceramic samples showed positive results in terms of cell viability, according to the MTT assay and LDH activity tests; the incorporation of glass in the formulations proved not to be detrimental to cell survival and proliferation;
- While the incorporation of glass in the formulation was not crucial for viability at Day 7, it was definitively effective at improving the biocompatibility of the samples throughout the cell culture period up to Day 3.

Acknowledgments

The authors acknowledge the University of Padova for funding in the framework of the project "BIOBONE: Design, prototyping and validation of advanced BIOceramics for BONE tissue engineering". The authors also acknowledge Juliana Kelmy Macàrio de Faria Daguano Daguano (Centro de Engenharia, Modelagem e Ciências Sociais Aplicadas, Federal University of ABC, Brazil) and Viviane Oilveira Soares (Departamento de Ciências, State University of Maringá, Brazil) for supplying the Ca/Mg-rich glass.

Author Contributions

For this paper, Enrico Bernardo formulated the research ideas, supervised the experiment on dense and porous glass-ceramic materials and planned the structure of the article. Laura Fiocco performed the general experimentation, except the preparation and X-ray diffraction characterization of pellets, which were done by Hamada Elsayed, and preliminary cell tests (including statistical analysis), which were done by Letizia Ferroni and Chiara Gardin, under the supervision of Barbara Zavan. The paper was written and edited by Laura Fiocco, Enrico Bernardo and Barbara Zavan.

Conflicts of Interest

The authors declare no conflict of interest.

References

1. Colombo, P.; Mera, G.; Riedel, R; Sorarù, G.D. Polymer-derived ceramics: 40 years of research and innovation in advanced ceramics. *J. Am. Ceram. Soc.* **2010**, *93*, 1805–1837.

2. Colombo, P.; Bernardo, E.; Parcianello, G. Multifunctional advanced ceramics from preceramic polymers and nano-sized active fillers. *J. Eur. Ceram. Soc.* **2013**, *33*, 453–469.

3. Bernardo, E.; Fiocco, L.; Parcianello, G.; Storti, E.; Colombo, P. Advanced ceramics from preceramic polymers modified at the nano-scale: A review. *Materials* **2014**, *7*, 1927–1956.

4. De Aza, P.N.; Guitian, F.; de Aza. Bioactivity of wollastonite ceramics: *In vitro* evaluation. *Scr. Metall. Mater.* **1994**, *31*, 1001–1005.

5. Lin, K.; Zhai, W.; Ni S.; Chang, J.; Zeng, Y.; Qian, W. Study of mechanical property and *in vitro* biocompatibility of $CaSiO_3$ ceramics. *Ceram. Inter.* **2005**, *31*, 323–326.

6. Wu, C.; Chang, J. Degradation, bioactivity and cytocompatibility of diopside, akermanite and bredigite ceramics. *J. Biomed. Mater. Res.-B Appl. Biomater.* **2007**, *83*, 153–160.

7. Ventura, J.M.G.; Tulyaganov, D.U.; Agathopoulos, S.; Ferreira, J.M.F. Sintering and crystallization of akermanite-based glass–ceramics. *Mater. Lett.* **2006**, *60*, 1488–1491.

8. Nonami, T.; Tsutsumi, S. Study of diopside ceramics for biomaterials. *J. Mater. Sci. Mater. Med.* **1999**, *10*, 475–479.

9. Wu, C.; Ramaswamy, Y.; Zreiqat, H. Porous diopside ($CaMgSi_2O_6$) scaffold: A promising bioactive material for bone tissue engineering. *Acta Biomater.* **2010**, *6*, 2237–2245.

10. Jones, J.R.; Hench, L.L. Regeneration of trabecular bone using porous ceramics. *Curr. Opin. Solid State Mater. Sci.* **2003**, *7*, 301–307.

11. Jones, J.R.; Lee, P.D.; Hench, L.L. Hierarchical porous materials for tissue engineering. *Philos. Trans. R. Soc. A* **2006**, *364*, 263–281.

12. Bernardo, E.; Carlotti, J.-F.; Dias, P.M.; Fiocco, L.; Colombo, P.; Treccani, L.; Hess, U.; Rezwan, K. Novel akermanite-based bioceramics from preceramic polymers and oxide fillers. *Ceram. Int.* **2014**, *40*, 1029–1035.

13. Bernardo, E.; Colombo, P.; Dainese, E.; Lucchetta, G.; Bariani, P.F. Novel 3D wollastonite-based scaffolds from preceramic polymers containing micro- and nano-sized reactive particles. *Adv. Eng. Mater.* **2012**, *14*, 269–274.

14. Bernardo, E.; Parcianello, G.; Colombo, P.; Matthews, S. Wollastonite foams from an extruded preceramic polymer mixed with $CaCO_3$ microparticles assisted by supercritical carbon dioxide. *Adv. Eng. Mater.* **2013**, *5*, 60–65.

15. Fiocco, L.; Elsayed, H.; Bernardo, E.; Daguano, J.K.M.F.; Soares, V.O. Silicone resins mixed with active oxide fillers and Ca-Mg Silicate glass as alternative/integrative precursors for wollastonite-diopside glass-ceramic foams. *J. Non-Cryst. Sol.* **2015**, *416*, 44–49.

16. Bhattacharya, G.; Zhang, S.; Jayaseelan, D.D.; Lee, W.E. Mineralizing effect of $Li_2B_4O_7$ and $Na_2B_4O_7$ on magnesium aluminate spinel formation. *J. Am. Ceram. Soc.* **2007**, *90*, 97–106.

17. Rahaman, M.N.; Day, D.E.; Bal, B.S.; Fu, Q.; Jung, S.B.; Bonewald, L.F.; Tomsia, A.P. Bioactive glass in tissue engineering. *Acta Biomater.* **2001**, *7*, 2355–2373.

18. Hoppe, A.; Güldal, N.S.; Boccaccini, A.R. A review of the biological response to ionic dissolution products from bioactive glasses and glass-ceramics. *Biomaterials* **2011**, *32*, 2757–2774.

19. Fu, Q.; Rahaman, M.N.; Bal, B.S.; Bonewald, L.F.; Kuroki, K.; Brown, R.F. Silicate, borosilicate and borate bioactive glass scaffolds with controllable degradation rates for bone tissue engineering applications. II. *In vitro* and *in vivo* biological evaluation. *J. Biomed. Mater. Res. A* **2010**, *95*, 172–179.

20. Gorustovich, A.A.; Lopez, J.M.P.; Guglielmotti, M.B.; Cabrini, R.L. Biological performance of boron-modified bioactive glass particles implanted in rat tibia bone marrow. *Biomed. Mater.* **2006**, *1*, 100–105.

21. Jung, S.B.; Day, D.E.; Brown, R F.; Bonewald, L.F. Potential toxicity of bioactive borate glasses *in-vitro* and *in-vivo*. In *Advances in Bioceramics and Porous Ceramics V*; Narayan, R., Colombo, R.P., Halbig, M., Mathur, S., Eds.; John Wiley & Sons, Inc.: Hoboken, NJ, USA, 2012.

22. Baino, F.; Brovarone, C.V. Three-dimensional glass-derived scaffolds for bone tissue engineering: Current trends and forecasts for the future. *J. Biomed. Mater. Res. A* **2010**, *97A*, 514–535.

23. Kaur, G.; Pandey, O.P.; Singh, K.; Homa, D.; Scott, B.; Pickrell, G. A review of bioactive glasses: Their structure, properties, fabrication, and apatite formation. *J. Biomed. Mater. Res. A* **2013**, *102*, 254–274.

24. Rahaman, M.N.; Liang, W.; Day, D.E.; Marion, N.W.; Reilly, G.C.; Mao, J.J. Preparation and bioactive characteristics of porous borate glass substrates. *Ceram. Eng. Sci. Proc.* **2005**, *26*, 3–10.

25. Kaur, G.; Pickrell, G.; Kimsawatde, G.; Homa, D.; Allbee, H.A.; Sriranganathan, N. Synthesis, cytotoxicity, and hydroxypatite formation in 27-Tris-SBF for sol-gel based $CaO-P_2O_5-SiO_2-B_2O_3-ZnO$ bioactive glasses. *Sci. Rep.* **2014**, *4*, 1–14.

26. Khorami, M.; Hesaraki, S.; Behnamghader, A.; Nazarian, H.; Shahrabi, S. *In vitro* bioactivity and biocompatibility of lithium substituted 45S5 bioglass. *Mat. Sci. Eng. C* **2011**, *31*, 1584–1592.

27. Miguez-Pacheco, V.; Büttner, T.; Maçon, A.L.B.; Jones, J.R.; Fey, T.; de Ligny, D.; Greil, P.; Chevalier, J.; Malchere, A.; Boccaccini, A.R. Development and characterization of lithium-releasing silicate bioactive glasses and their scaffolds for bone repair. *J. Non-Cryst. Sol.* **2015**, in press, doi:10.1016/j.jnoncrysol.2015.03.027.

28. Denizot, F.; Lang, R. Rapid colorimetric assay for cell growth and survival. Modifications to the tetrazolium dye procedure giving improved sensitivity and reliability. *J. Immunol. Methods* **1986**, *89*, 271–277.

8

Disassembly Properties of Cementitious Finish Joints Using an Induction Heating Method

Jaecheol Ahn [1,*], Takafumi Noguchi [2] and Ryoma Kitagaki [2]

[1] Department of Architecture, Dong-A University, 550 Beon-gil Saha-gu, Busan 604-714, Korea

[2] Department of Architecture, the University of Tokyo, Hongo 7-3-1, Tokyo 113-8654, Japan;
    E-Mails: noguchi@bme.arch.t.u-tokyo.ac.jp (T.N.); ryoma@bme.arch.t.u-tokyo.ac.jp (R.K.)

* Author to whom correspondence should be addressed; E-Mail: jcan222@donga.ac.kr;

Academic Editor: Klara Hernadi

Abstract: Efficient maintenance and upgrading of a building during its lifecycle are difficult because a cementitious finish uses materials and parts with low disassembly properties. Additionally, the reuse and recycling processes during building demolition also present numerous problems from the perspective of environmental technology. In this study, an induction heating (IH) method was used to disassemble cementitious finish joints, which are widely used to join building members and materials. The IH rapidly and selectively heated and weakened these joints. The temperature elevation characteristics of the cementitious joint materials were measured as a function of several resistor types, including wire meshes and punching metals, which are usually used for cementitious finishing. The disassembly properties were evaluated through various tests using conductive resistors in cementitious joints such as mortar. When steel fiber, punching metal, and wire mesh were used as conductive resistors, the cementitious modifiers could be weakened within 30 s. Cementitious joints with conductive resistors also showed complete disassembly with little residual bond strength.

Keywords: induction heating; cementitious joint; mortar; steel fiber; conductive resistor; disassembly property

1. Introduction

The building production technologies that have been realized to date include those for the composition of building materials and structural members, as well as the simplification of assembly processes. However, because of the low disassembly properties of individual materials and parts, efficient maintenance and upgrading of buildings during the course of their lifecycles have presented challenges. Furthermore, the reuse and recycling processes during the demolition of buildings also involve numerous problems from the perspective of environmental technology [1,2]. In Japan, the impurity contents of finishing materials have been greater than 10% of total concrete waste, even if the preceding collection of demolished finishing materials was conducted at demolition site of reinforced concrete structure buildings. This affects the quality of building material recycling as a result [3].

In this study, induction heating was used to determine the disassembly properties of a cementitious joint widely used to join building members and materials. The induction heating technique has high energy efficiency [4–8]. Hence, it is expected to provide local heating only for building members and materials that are in contact. Therefore, this technique is highly regarded as an environment-friendly technique for a forward-backward process. In particular, because wire mesh and rebar are widely used as the conductive materials of the heating source for induction heating in existing buildings, they are judged to be capable of facilitating the easy specification of disassembly properties without the deterioration of the assembly properties.

Therefore, the purpose of this study was to specify disassembly properties that enable field disassembly while maintaining assembly properties such as workability and bond strength, using a conductive resistor at a cementitious joint. In addition, basic data for the development of an optimal forward-backward process production system were proposed by selecting a conductive resistor with outstanding temperature elevation characteristics in response to induction heating.

2. Separation Model for Cementitious Joints during Induction Heating

Cementitious materials are likely to decrease in strength because of the dehydration and metamorphosis of the inner hydrate and the accompanying increase in the pore volume when subjected to heating at 300 °C [9,10]. If the cement paste, the main component of mortar, is dehydrated and weakened by heating the adhered mortar, the attached material and adhesive mortar can be efficiently separated. In particular, the induction heating method [11] can be used to selectively heat only the desired part with outstanding heat efficiency. Hence, it is deemed capable of improving the cementitious material's temperature elevation characteristics in the most economical way.

Therefore, if a conductive resistor such as steel fiber or a wire fabric is mixed into the adhered mortar, the disassembly property can be specified, while maintaining the assembly properties of the mortar-like bond strength and workability. Figure 1 shows the disassembly mechanism of the induction heating method, and Figure 2 shows a separation model of the finishing items attached as cementitious material.

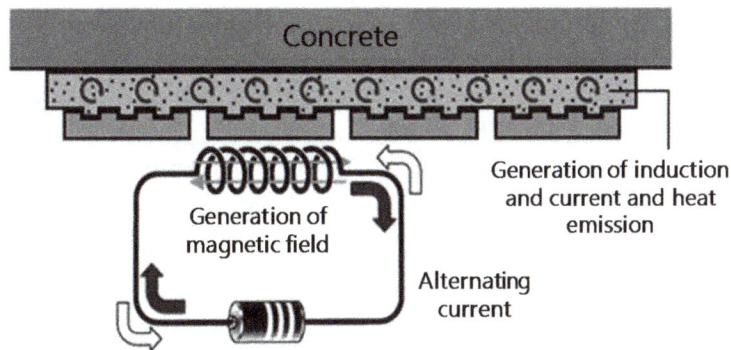

Figure 1. Separation mechanism of a cementitious joint using induction heating method.

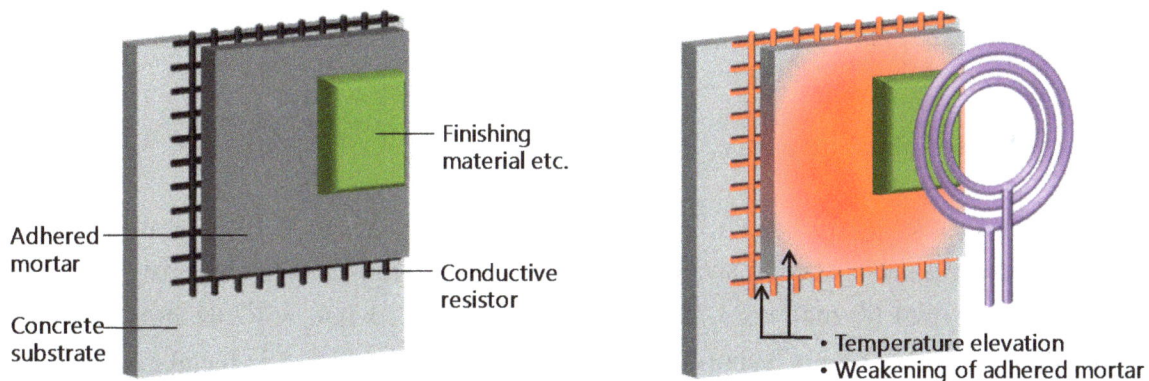

Figure 2. Separation model of finishing items.

3. Temperature Elevation Characteristics of Conductive Resistors under Induction Heating

3.1. Overview of Experiment

In this study, a conductive resistor with high energy efficiency was selected by analyzing the temperature elevation characteristics of induction heating according to the type of conductive resistor. The efficiency of the induction heating was anticipated to be significantly affected by the cross-sectional area of the conductive resistor, and the wire fabric was expected to vary according to the adhered shape of the contact points of the horizontal and vertical wires [12]. The temperature elevation characteristics as a function of the radio frequency and heating distance of the conductive resistor were also analyzed.

3.2. Experiment Factors

Wire, stainless metal mesh, and punching metal were used as conductive resistors. After adjusting the radio frequency power to 1, 2.25, and 4 kW, variations in the temperature elevation characteristics according to the type of conductive resistor were analyzed, and the results are listed in Table 1.

Table 1. Experimental factors for temperature elevation.

Specimen Symbol	Conductive Resistor	
	Metal Mesh and Punching Metal	**Size**
F4.0	Wire fabric (zinc plate)	ø1.60 mm × P4.75 mm (4 Mesh)
F10		ø0.55 mm × P1.99 mm (10 Mesh)
S4.0	Stainless fabric	ø1.60 mm × P4.75 mm (4 Mesh)
S10		ø0.53 mm × P2.01 mm (10 Mesh)
Fp0.5–ø3	Steel punching metal	T0.5 mm–ø3 mm × P5 mm (Opening rate 32.4%)
Fp1.0–ø3		T1.0 mm–ø3 mm × P5 mm (Opening rate 32.4%)
Fp1.0–ø5		T1.0 mm–ø5 mm × P6 mm (Opening rate 62.5%)
Sp0.5–ø3	Stainless punching metal	T0.5 mm–ø3 mm × P5 mm
Sp1.0–ø3		T1.0 mm–ø3 mm × P5 mm

Radio frequency power: 1 kW, 2.25 kW, 4.0 kW.

3.3. Production of Materials and Specimen

For the metal mesh, a ø1.60 mm × P4.75 mm 4.0 mesh or ø0.55 mm × P1.99 mm 10 mesh of zinc-plated wire fabric was used, where ø and P denote the diameter (mm) and pitch (mm), respectively. For the stainless mesh, a ø1.60 mm × P4.75 mm 4.0 mesh or ø0.53 mm × P2.01 mm 10 mesh of SUS304 (18Cr-8Ni) was used. Figure 3 shows the wire fabric and stainless fabric metal meshes.

The punching metal material was the same as that of the metal mesh. For the punching metal, either a steel with a 32.4% (ø3 mm × P5 mm) opening rate and a thickness (T) of T0.5 or T1.0 mm or one with a 62.5% (ø5 mm × P6 mm) opening rate and a thickness of T1.0 mm was used. For the stainless punching metal, a plate with a 32.4% (ø3 mm × P5 mm) opening rate and a thickness of T1.0 mm was used. The hole was punched at the top of a regular triangle. Figure 4 shows the punching metals used in this study.

ø1.60 mm × P4.75 mm   ø0.55 mm × P1.99 mm   ø1.60 mm × P4.75 mm   ø0.53 mm × P2.01 mm
(Wire Fabric, 4.0 mesh)   (Wire Fabric, 10 mesh)   (SUS304, 4.0 mesh)   (SUS304, 10 mesh)

Figure 3. Wire fabric and stainless fabric.

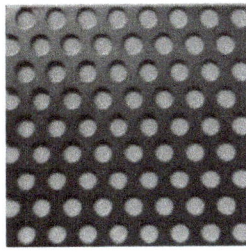  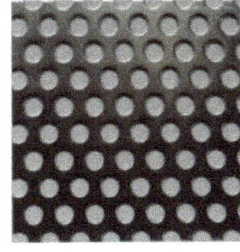

T1.0mm–ø3 mm × P5 mm   T1.0mm–ø5 mm × P6 mm   T1.0 mm–ø3 mm × P5 mm
(Steel, opening rate 32.4%)   (Steel, opening rate 62.5%)   (SUS304, opening rate 32.4%)

Figure 4. Punching metal.

3.4. Induction Heating

An experimental device with a 100 kHz (60 to 120 kHz operating frequency) and 5 kW maximum radio frequency was used for the induction heating. The power was stable within a range of ±2%, as shown in Figure 5. The power was adjusted using the direct current (DC) to DC voltage relationship. For the radio frequency power, an auto-track method was adopted, with a resonance frequency determined by the resonance condenser inside the adapter, as well as the inductances of the power lead and heating induction coil. The heating induction coil had a ø10 mm with an external diameter of ø120 mm. Each temperature measurement was conducted using an infrared thermometer (thermographic camera) in the temperature range of 0 °C–800 °C.

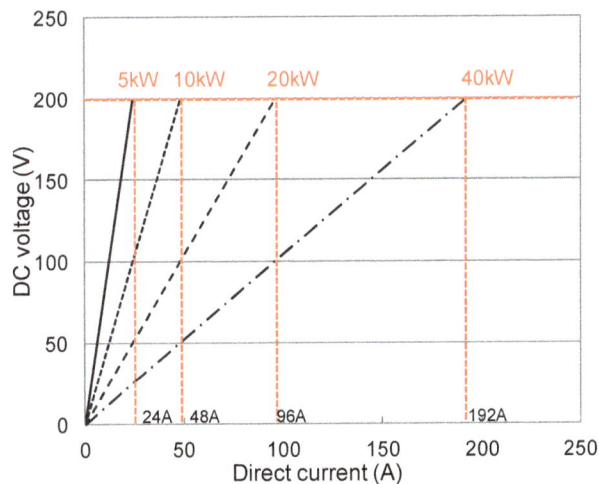

Figure 5. Induction heating power.

3.5. Temperature Elevation Characteristics of Conductive Resistor by Induction

The temperature elevation characteristics of the conductive resistors during induction heating are shown in Figures 6 and 7 as a function of the type of resistor (metal mesh and punching metal), power consumption, and heating distance. The experiments were performed with a basic frequency of 100 kHz, and three different power levels were used: 1 kW (DC voltage 100 V × DC current 10 A), 2.25 kW (DC voltage 150 V × DC current 15 A), 4 kW (DC voltage 200 V × DC current 20 A). For temperatures above 800 °C, which could not be measured, the value was set to 800 °C. To measure the overall temperature elevation characteristics, heating gradually progressed toward the inside as a function of time to make heating the outside easier than heating the center. In addition, the temperature elevation characteristics varied depending on the type and cross-sectional area of the conductive resistor.

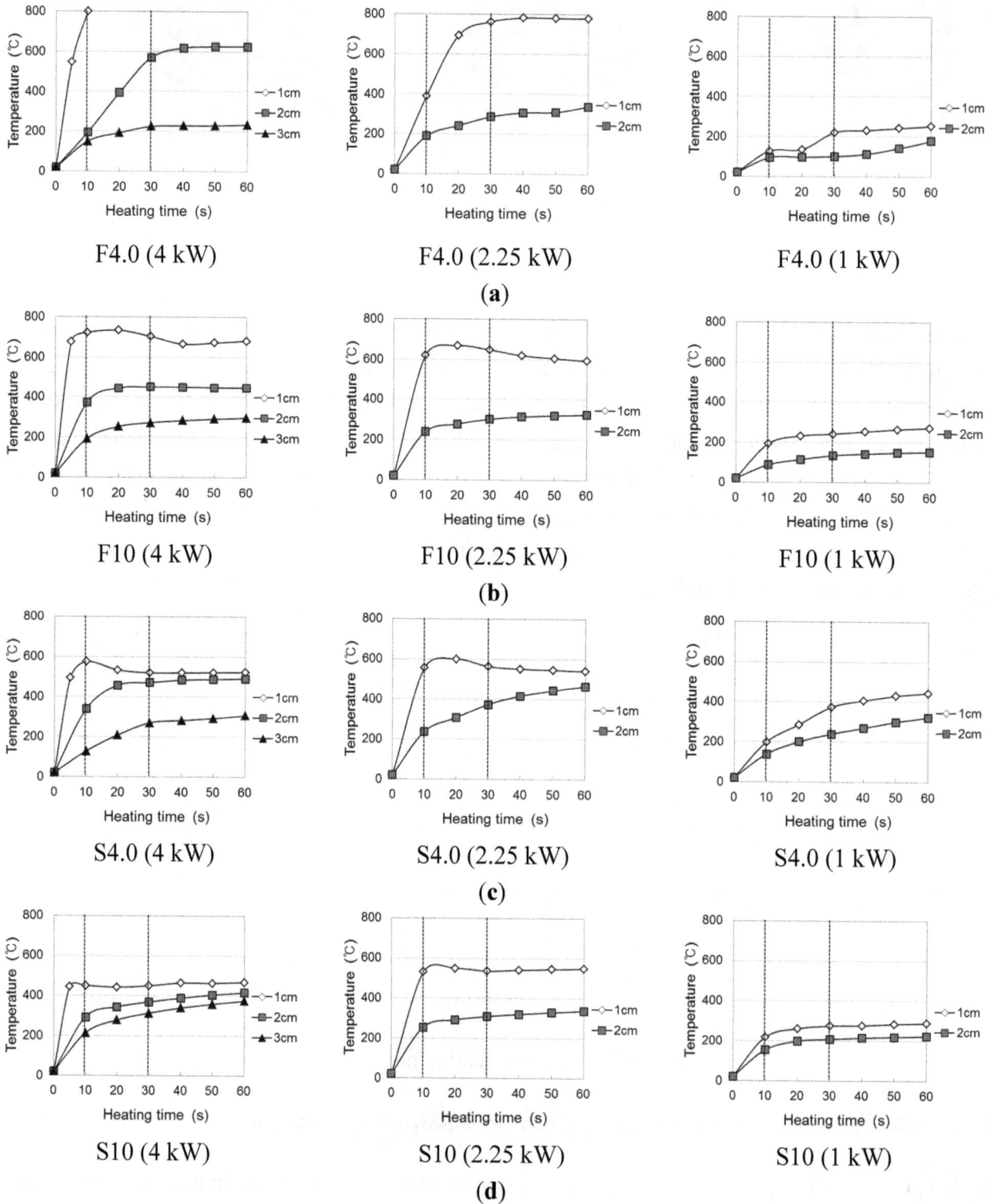

Figure 6. Temperature elevation of induction heating of conductive resistor (Metal mesh). (**a**) Wire fabric of 4 mesh; (**b**) wire fabric of 10 mesh; (**c**) stainless fabric of 4 mesh; (**d**) stainless fabric of 10 mesh.

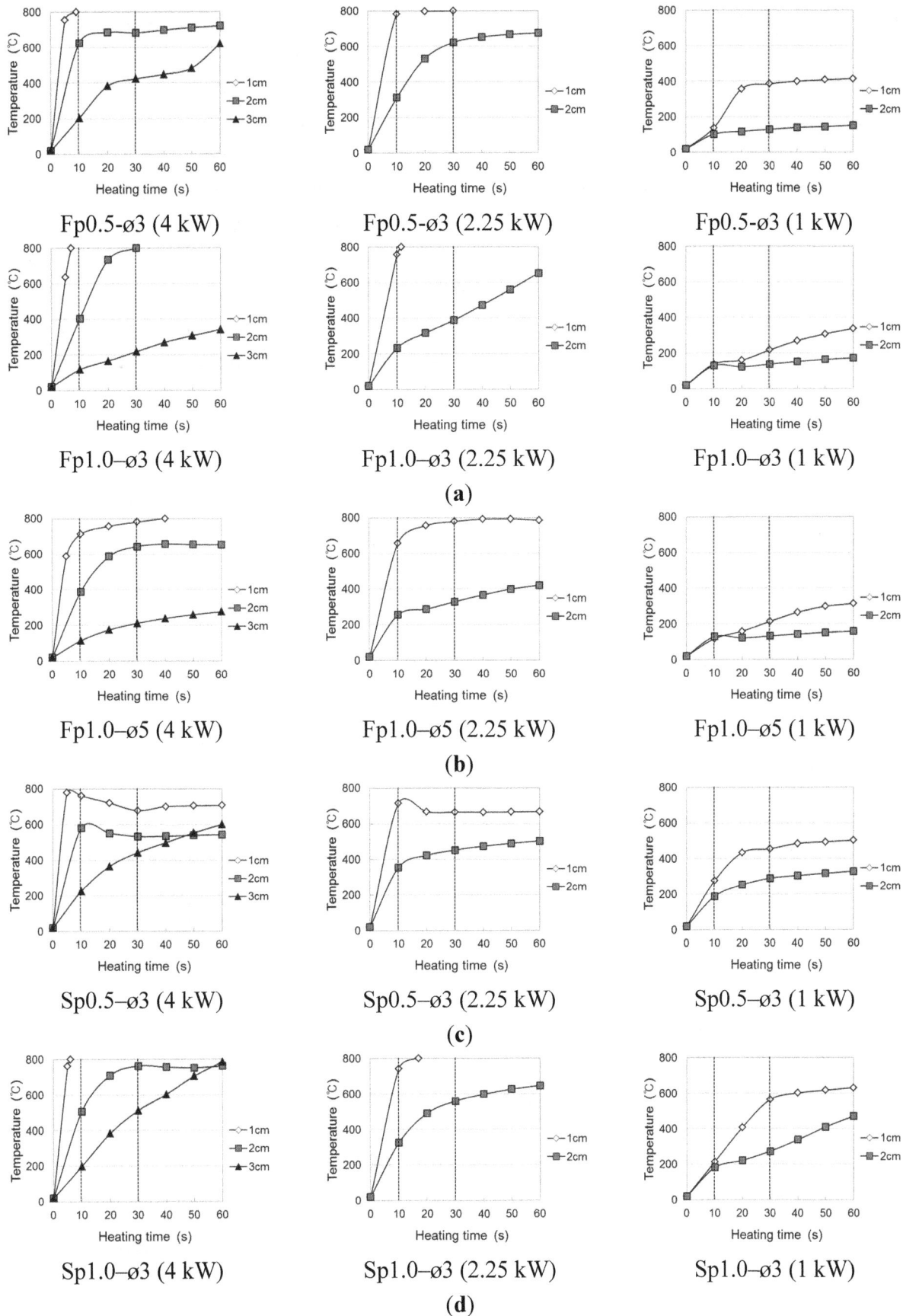

Figure 7. Temperature elevation of conductive resistor (Punching metal). (**a**) Wire punching metal of diameter 3 mm; (**b**) wire punching metal of diameter 5 mm; (**c**) stainless punching metal of thickness 0.5 mm; (**d**) stainless punching metal of thickness 1.0 mm.

At a radio frequency power of 4 kW with a metal mesh as the conductive resistor and an induction heating distance of 1 or 2 cm from the coil to the conductive resistor, the temperature sharply increased with the heating time from 10 to 30 s and thereafter tended to remain constant or slightly decrease. This indicated that the induction heating energy was concentrated in both the wire fabric and stainless fabric (Figure 8; heating distance of 2 cm). Therefore, the function of the circuit that facilitated the induction heating was terminated because the outside, which was heated the fastest, was fractured as it reached the melting point. For this reason, the temperature decreased, while a new circuit was created at the center, where the heat efficiency was relatively low. In particular, in the case of the 10-mesh wire fabric (specimen F10), the highest temperature was the same as that in the case of the stainless fabric, despite the significant decrease in heating area. This was attributed to the low melting point of the zinc plating applied to prevent the wire fabric from corroding, which caused an electrical resistance across the interface of the melted and non-melted parts. Another reason is that the wire fabric may not readily serve as a second circuit for adhesion among the wires because of the melting of the zinc plating. However, the temperature increased gradually when the circuit was not fractured at a distance of 3 cm. By the same principle, when the radio frequency power was 2.25 kW, with a heating distance of 2 or 3 cm, the temperature increased gradually over time. However, when the heating distance was 1 cm, the rates of temperature elevation hardly differed between 4 and 2.25 kW; additionally, as the heating distance gradually increased, the efficiency of the heating by the radio frequency power increased.

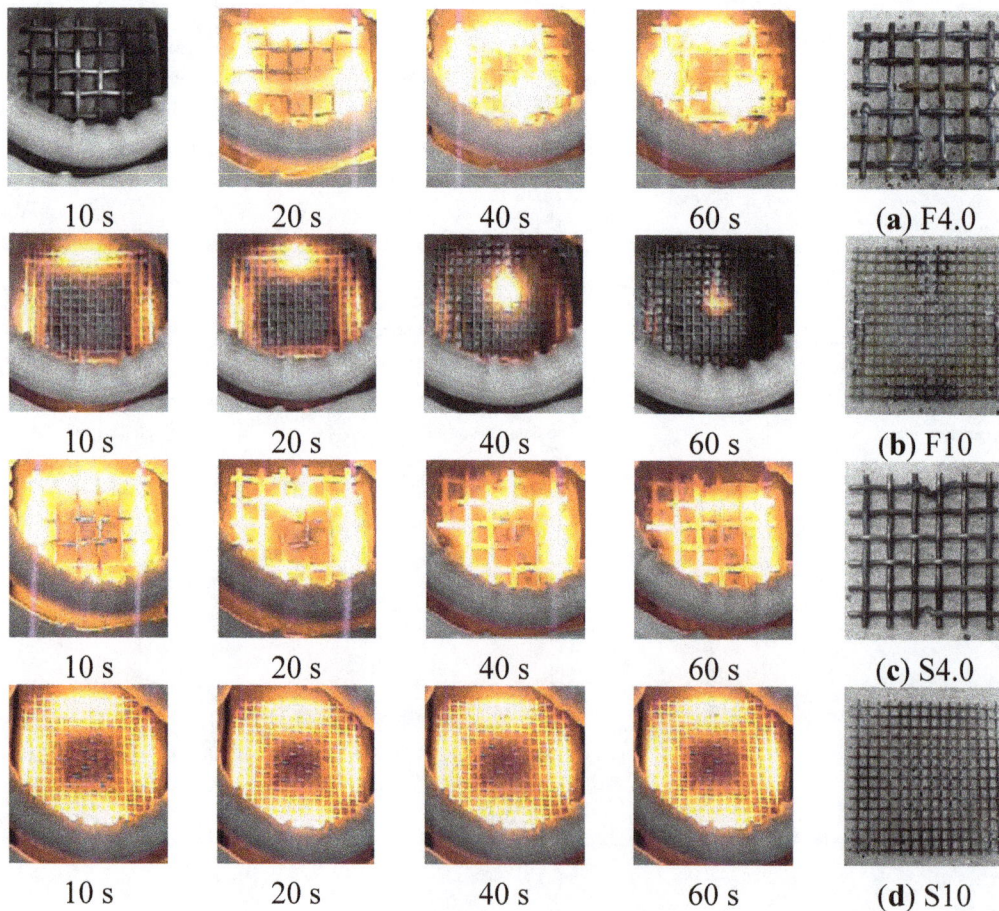

Figure 8. *Cont.*

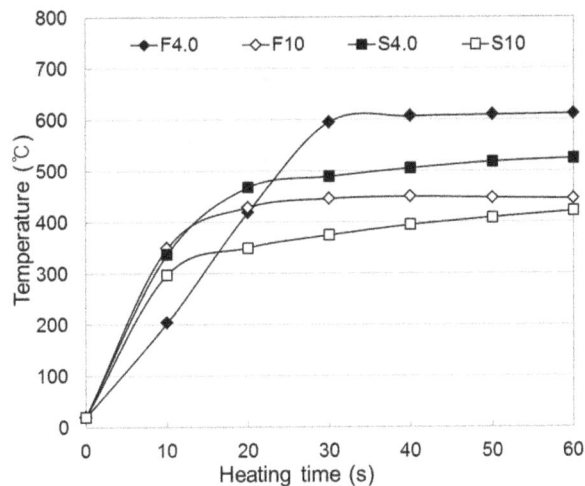

Figure 8. Appearance of heating conductive resistor by induction (metal mesh, 4 kW–2 cm).

The thickness of the conductive resistor affected the temperature elevation characteristics. Temperature increased gradually in the 4-mesh stainless fabric and wire fabric, in which a second circuit was readily created within the cross section by the induction heating as a result of the large cross section of each wire of the metal mesh. Regarding the effect of the material type on the temperature elevation characteristics, the highest temperature was shown by the wire, which, as a magnetic material, exhibited a heat emission reaction by a hysteresis loss through heating by an eddy current. In contrast, the stainless steel showed a more uniform heating tendency at the cementitious joints because of higher thermal conductivity.

In the case of the punching metal (Figure 9), the rate of temperature elevation was exceedingly high compared with that in the case of the metal mesh. Hence, the rate could not be measured because the temperature reached 800 °C within 10 s at the 4-kW heating condition, with the difference increasing with a decrease in the heating distance. This was because the contact point of the punching metal in a wire fabric is large, whereas the contact area of a horizontal line and vertical line is small. Hence, a second eddy current easily occurs as a result of induction heating. Although the resistor thickness had hardly any effect, in case of the 0.5-mm stainless punching metal, the temperature decreased because of a fracture after reaching the highest temperature. As for the difference caused by the punching metal opening rate, the rate of temperature elevation was slightly lower at ø5 mm (opening rate 62.5%) than at ø3 mm (opening rate 32.4%). However, in the case of the 1-kW radio frequency power, the temperature could not exceed 400 °C, where the weakening of the cementitious material was anticipated, even after 60 s of heating.

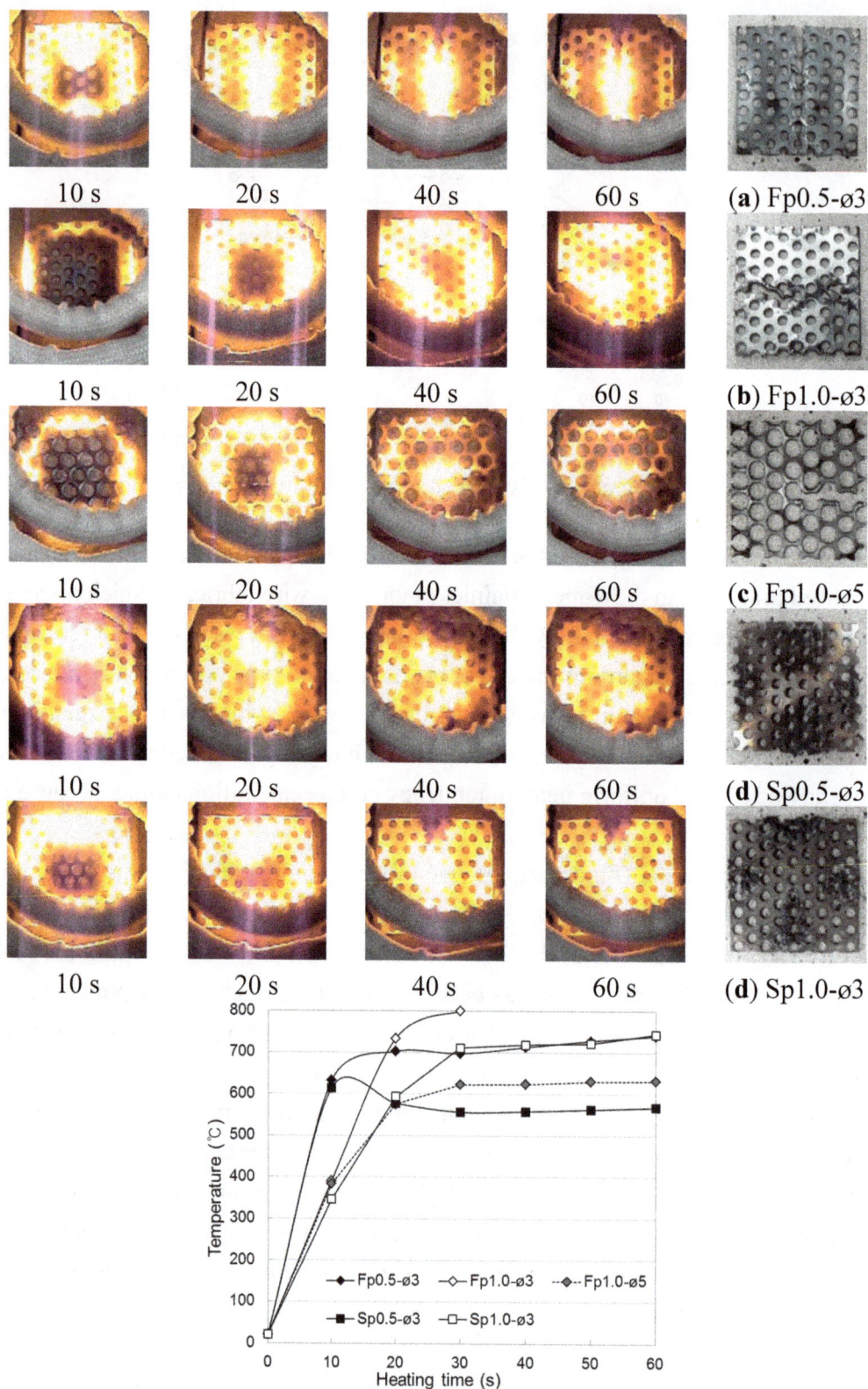

Figure 9. Appearance of heating conductive resistor by induction (punching metal, 4 kW–2 cm).

Therefore, in the cases where the metal mesh and punching metal were used as conductive resistors, 2 cm was judged to be the appropriate distance for the induction coil to allow the conductive resistor to weaken by heating the cementitious joint. If the energy efficiency and temperature elevation characteristics are considered, the appropriate power would be 2.25 kW in the case of a heating distance of 1 cm and 4 kW in the case of a heating distance of 1 to 2 cm.

4. Evaluation of Disassembly Properties of a Cementitious Joint during Induction Heating

4.1. Overview of Experiment

Experiments were performed to analyze the temperature elevation characteristics of the induction heating and the disassembly performance as a function of the type of conductive resistor. Specifically, experiments were carried out under a condition in which a wire mesh, stainless metal mesh, or punching metal was inserted into adhered mortar as a conductive resistor and one in which steel fiber was mixed into the adhered mortar directly. The disassembly performance of the cementitious joint was determined by the effect of selectively heating (by induction heating) the conductive resistor analyzed in the previous section, and the consequent weakening of the cementitious joint. In addition, the following issues were analyzed in the case of the steel fiber: whether a closed circuit formed depending on the degree of fiber distribution inside, and the effect on the weakening of the adhered mortar of varying the composition using fibers with different thicknesses and lengths. Through the analysis of the adhered mortar using a conductive resistor, integrative evaluations of the assembly and disassembly properties were also conducted. Table 2 lists the experimental factors used to evaluate the assembly properties.

Table 2. Experimental factors for evaluation of assembly property.

Specimen Symbol	Conductive Resistor	
	Conductive Resistor Shape	Specification
N	-	-
F4.0	Wire Fabric (zinc plated)	Diameter 1.60 mm × P4.75 mm (4 Mesh)
F10		Diameter 0.55 mm × P1.99 mm (10 Mesh)
S4.0	Stainless fabric (SUS304)	ø1.60 mm × P4.75 mm (4 Mesh)
S10		ø0.53 mm × P2.01 mm (10 Mesh)
Fp0.5–ø3	Steel punching metal	T0.5 mm–ø3 mm × P5 mm
Fp1.0–ø3		T1.0 mm–ø3 mm × P5 mm
Fp1.0–ø5		T1.0 mm–ø5 mm × P6 mm
Sp0.5–ø3	Stainless punching metal	T0.5 mm–ø3 mm × P5 mm
Sp1.0–ø3		T1.0 mm–ø3 mm × P5 mm
SF06–*	Steel fiber 6 mm	*: contents rate (1%, 2%, 3%)
SF13–*	Steel fiber 13 mm	

4.2. Production of Materials and Specimens

The metal mesh and punching metal used as the conductive resistors were the same as those discussed in the previous section. Steel fiber 1 (SF06), as shown in Figure 10, was manufactured by shear-machining a thin steel plate produced by cold rolling into the shape of a fiber with the length of 6 mm and cross section of 0.25 mm per side for the aspect ratio of 24. The dimensions of steel fiber 2 (SF13) were a length of 13 mm and diameter of 0.16 mm for an aspect ratio of 81. The specific gravity (7.85 t/m^3) was the same. Steel fiber 2 was brass-plated to prevent corrosion.

The adhered mortar into which the wire fabric and punching metal were inserted was produced by mixing water (W) and cement (C) at a W/C ratio of 35%, with 10% polymer to prevent the hardware from sinking (Table 3).

(a) **(b)**

Figure 10. Steel fiber. **(a)** Steel fiber SF06 (= 6 mm); **(b)** steel fiber SF13 (= 13 mm).

Table 3. Combined ratio of adhered mortar mixed with steel fiber.

Steel Fiber Volume Mixed Rate (V_f)	Cementitious Binder (kg)					
	Water	Cement	Polymer	Antifoaming Agent	Thickener	Chemical Agent
0.0	0.35	1	0.1	0.001	-	-
1.0, 2.0	0.40	1	0.1	0.001	-	-
3.0	0.40	1	0.1	0.001	0.002 *	0.015 **

* Thickener: 0.5 wt·% of volume; ** Chemical agent: high performance air entrain water-reducing agent 1.5% of cement weight.

Steel fiber-mixed mortar has the potential to separate if the mortar does not have the proper fluidity and viscosity as a binder, because it contains a large proportion of steel fiber. Therefore, through preliminary experimentation, the mortar combination that achieved the best fiber distribution property was determined based on the proper viscosity and fluidity. An ordinary mortar with a W/C ratio of 40% was composed at a 1% to 2% mixture rate of steel fiber, whereas a ball of fiber was created by the sudden deterioration and separation of the material when steel fiber was mixed in at a rate of 3%. The fluidity and plastic viscosity were adjusted using a thickener and a high-performance, air-entrained, and water-reducing agent, and excluding the use of a mineral admixture to minimize the effect on strength. The particle size of the fine aggregates was adjusted to 2.5 mm or less.

4.3. Experimental Method

The induction heating device and methods were the same as those described in the previous section. To analyze the assembly and disassembly properties of a cementitious joint with a conductive resistor, a tensile bond strength test was performed on a specimen at 28-day age before and after induction heating. For each specimen, 40 mm × 40 mm × 10 mm of adhered mortar was piled at the center of a mortar substrate of 100 mm × 100 mm × 20 mm.

Under each condition, conductive resistors were installed in the middle of the adhered mortar. For the test, the adhered mortar surface was made even with the use of sandpaper to ensure that the tensile load would be applied vertically. Subsequently, the test was carried out after the resistor was attached to the device with the use of epoxy resin and left undisturbed for 24 h. The specimen shape and test device are shown in Figure 11.

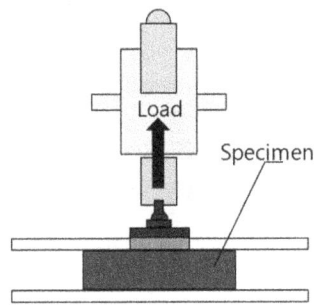

Figure 11. Tensile bond strength test.

Shear bond strength test was performed on a specimen at 28-day age before and after induction heating. The specimen shape was the same as that in the tensile bond strength test; the test device is shown in Figure 12. The shear loading and adhered mortar displacement were measured using a load cell and displacement meter, respectively.

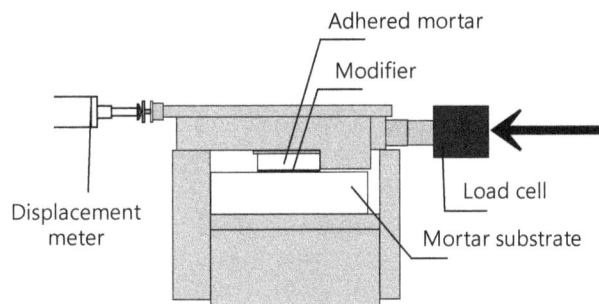

Figure 12. Shear bond strength test.

The pore-size distribution was measured with the use of a mercury intrusion porosimeter (MIP) to assess the weakening of the cementitious joint by induction heating. Vacuum deaeration was carried out for 2 weeks on each specimen in contact with a conductive resistor after it was taken from the specimen and cooled to room temperature after heating. Additionally, the pore-size distributions were obtained using the MIP at low and high pressures.

4.4. Analysis of Assembly Property of Cementitious Joint with Conductive Resistor

The compressive, tensile, and flexural strengths of the conductive mortar with steel fibers at 28-day age are shown in Figures 13–16, respectively. Compared with SF06 (6-mm-long steel fiber), all of the strengths of the 13-mm-long SF13 were superior, and the tensile and flexural strengths increased significantly with an increase in the mixed fiber. Judged on the basis of its strength, SF13 showed the best assembly properties. Hence, SF13 could be used to produce conductive mortar with outstanding assembly properties by achieving a fluidity and material separation resistance, while improving the strength based on the use of admixtures to promote adhesion between the fiber and matrix.

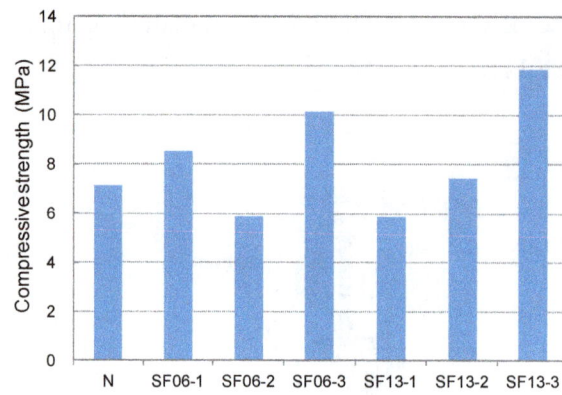

Figure 13. Compressive strength of steel fiber mortar.

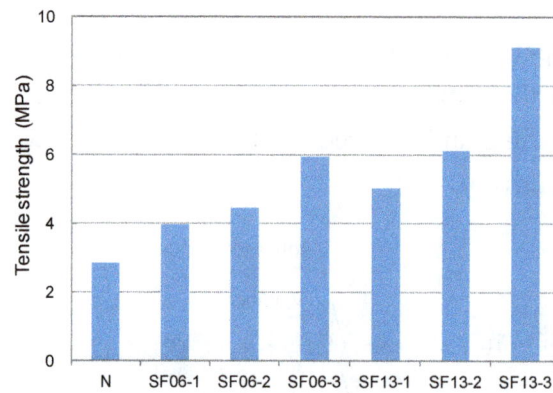

Figure 14. Tensile strength of steel fiber mortar.

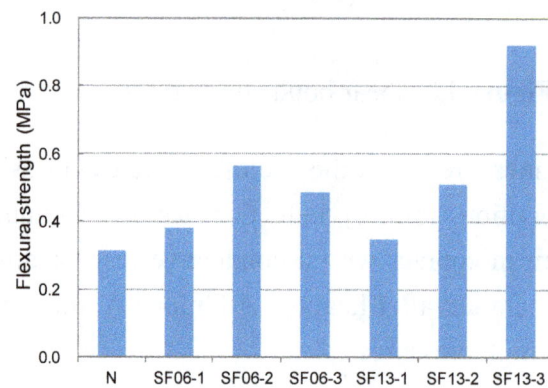

Figure 15. Flexural strength of steel fiber mortar.

Figure 16. Fractual surface of fiber reinforced mortar.

The results of the shear and tensile bond strength tests conducted by inserting various types of conductive resistors are shown in Figure 17. In most cases, the strengths were similar to or higher than those obtained without inserting a conductive resistor. The shear bond strength was relatively low in the case of the punching metal, which had the widest separation area on both sides of the adhered mortar. The bond strength also improved when steel fibers were inserted, as shown in Figure 18a,d. Shear and tensile fractures occurred at the interface of the metal resistor, mostly when fabric and punching metal were used, as shown in Figure 18b,c. Specifically, in the case of the wire fabric, separation often occurred at both the adhered mortar-resistor and mortar-mortar substrate interfaces. However, in the case of the tensile bond strength, the adhered mortar-mortar substrate interfaces separated in all the specimens except for the case of Fp1.0-ø3 mm. Therefore, the assembly properties were satisfactory when a conductive resistor was inserted.

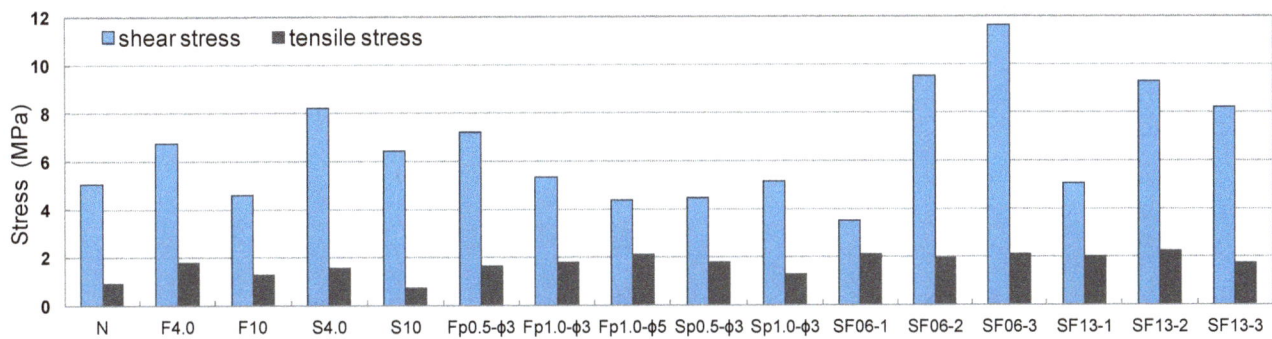

Figure 17. Shear and tensile bond strengths of cementitious joint using a conductive resistor.

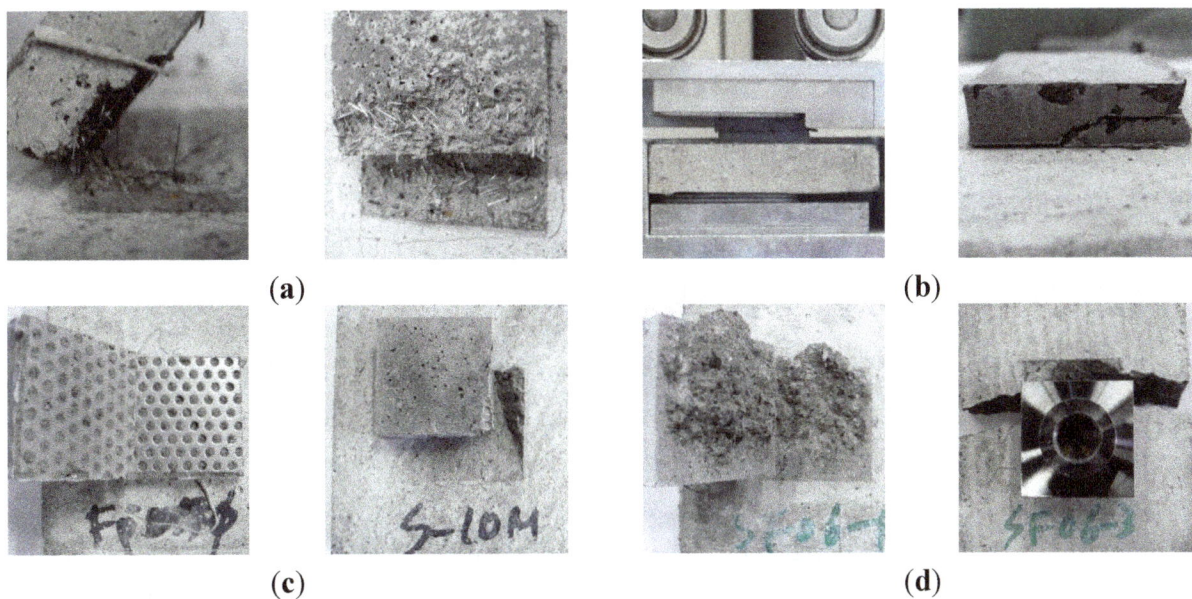

Figure 18. Shapes of bond fracture of cementitious joint using a conductive resistor. (a) Improvement of bond strength by mixing of steel fiber (shear); (b) shear fracture by metal mesh and punching metal; (c) fracture near punching metal (tensile) and metal mesh (shear); (d) exfoliation and fracture of mortar substrate (tensile).

4.5. Evaluation of Disassembly Properties through Induction Heating of a Cementitious Joint with a Conductive Resistor

4.5.1. Changes in Surface Temperature of Mortar Mixed with Steel Fiber through Induction Heating

Figure 19 shows the measurement results for the surface temperature of the adhered mortar during induction heating at a heating distance of 1 cm from the conductive resistor and a frequency power of 2.25 kW. The results show outstanding temperature elevation characteristics in the conductive resistor induction-heating experiment.

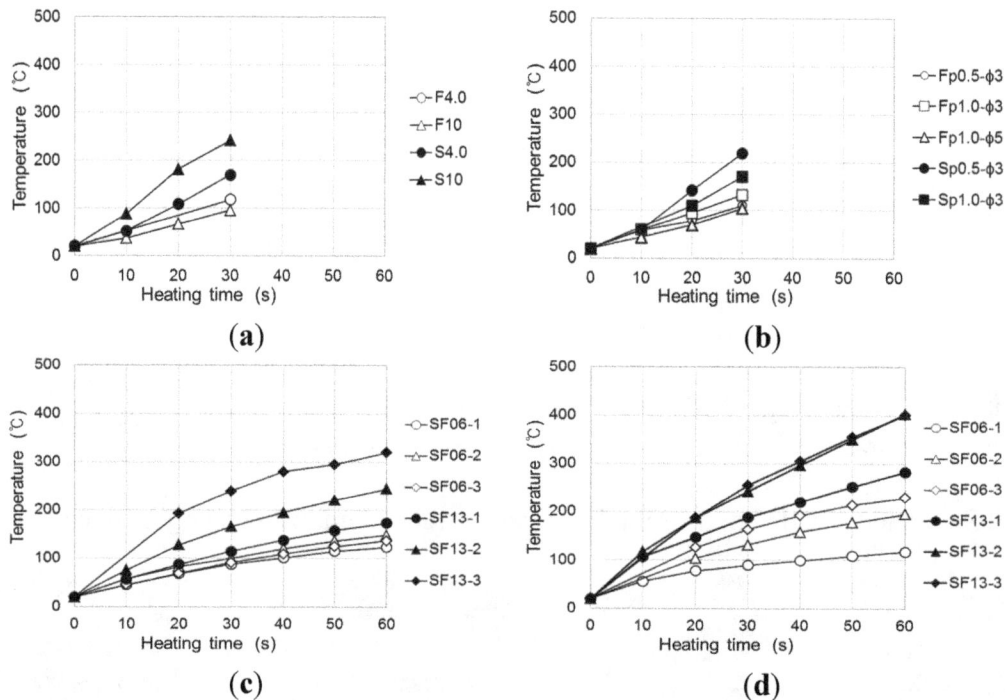

Figure 19. Temperature elevation characteristics of induction heating of adhered mortar to which conductive resistor is mixed. (**a**) Metal mesh (2.25 kW); (**b**) punching metal (2.25 kW); (**c**) steel fiber (2.25 kW); (**d**) steel fiber (4.0 kW).

Consequently, stainless steel (SUS304), in which uniform cracks may occur, showed slightly higher temperature elevation characteristics for the adhered mortar, depending on the type of resistor material, indicating that this material heats adhered mortar efficiently. The values did not significantly differ irrespective of whether metal mesh or punching metal was used. However, in the former case, the 10 mesh, which was evenly distributed within the mortar, showed an even higher efficiency in the heating mortar despite the thinness of the wire. In the case of the steel fiber, although the temperature increased with the duration of the induction heating, it did not depend on the radio frequency power as it did with the metal mesh or punching metal. The temperature elevation characteristics of SF06 (6-mm-long steel fiber) did not differ significantly with the composition at a power condition of 2.25 kW. However, it was significantly affected by the mixed fiber ratio at 4 kW. In contrast, SF13 showed outstanding overall temperature elevation characteristics compared with SF06. In the case of SF13-3, a mortar temperature elevation of ~250 °C could be obtained through 30 s of heating at 2.25 kW. However, in the case of the steel fiber, the linked condition among the fibers, depending on the internal distribution, significantly affected the efficiency of the induction heating. Therefore, the temperature elevation error was slightly larger compared with those for the wire fabric and punching metal.

4.5.2. Changes in Pore Structure of the Cementitious Joint

Figure 20 shows the results of the MIP test for changes in the pore structure due to elevated temperature of a cementitious joint with punching metal. Overall, high peak values appeared near 0.1 μm, and the cumulative amount of pores in the adhered mortar increased the volume by ~10% as a result of induction heating. However, because induction heating only causes intensive heating on the surface of parts in contact with the conductive resistor, this method does not ensure uniform heating throughout the adhered mortar over a short time. Consideration of the size of the porosimeter specimen (a hexahedron of 5 mm) revealed that C–S–H and calcium hydrogen were dehydrated by the high temperature at the surfaces in contact with the conductive resistor. In turn, this also increased the amount of macropores in the 1-μm range to some extent, although the rate of increase in the amount of fine pores in the range of 0.01–0.5 μm by heating below 300 °C was higher. Therefore, the peak value increased, and the number of fine pores decreased when the fine pore diameter increased with increasing dehydration in contrast with totally heated mortar [13].

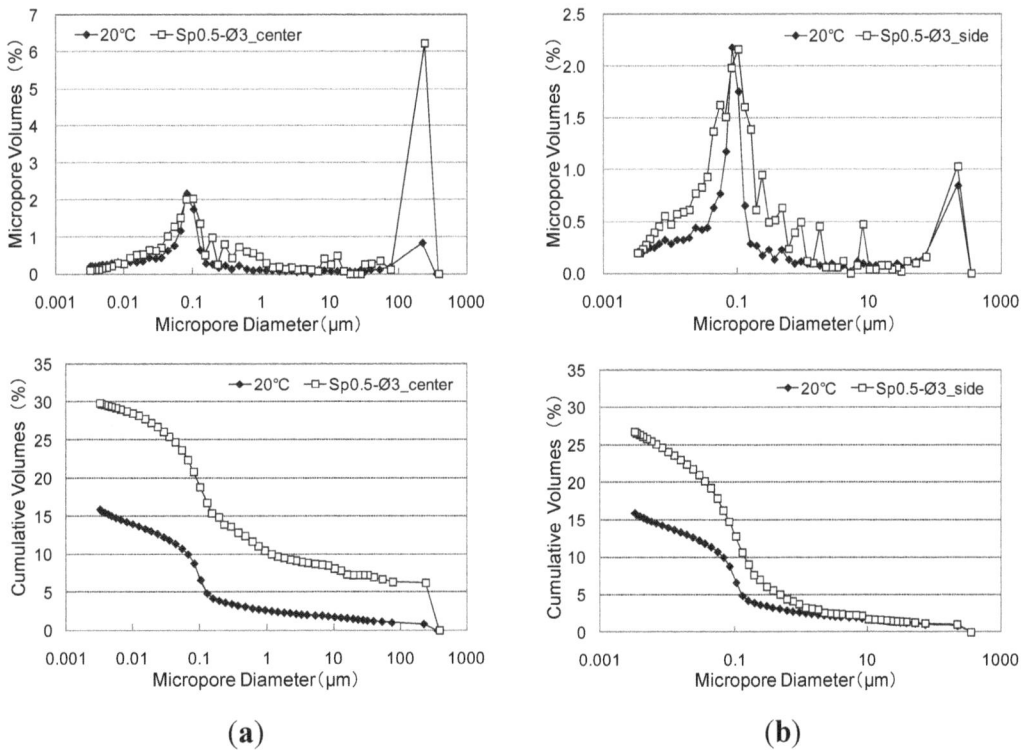

Figure 20. Changes in the pore structure of cementitious joint through induction heating. (**a**) Sp0.5–ø3_2.25 kW (center); (**b**) Sp0.5–ø3_2.25 kW (side).

4.5.3. Evaluation of Disassembly Property through Residual Bond Strength after Induction Heating

Figures 21 and 22 show the residual bond strength after the induction heating of a cementitious joint with a conductive resistor. As a result of induction heating for 30 s with 2.25 kW of power, the bond strength of the joint decreased significantly as a result of the selective heating of the conductive resistor, and the disassembly was the most pronounced when using a punching metal or metal mesh. The surface temperature of the adhered mortar was 100 °C–250 °C, which was slightly low for the cementitious material to be weakened (Figure 19). However, as shown in Figure 23a,b, carbonization and cracks occurred near intensively heated conductive resistors as result of the high-temperature heating. Therefore, in the majority of cases, separation occurred at the interface of the mortar substrate and adhered mortar. Thus, separation often occurred because of local fracturing of the weakened mortar near the punching metal (Figure 23c).

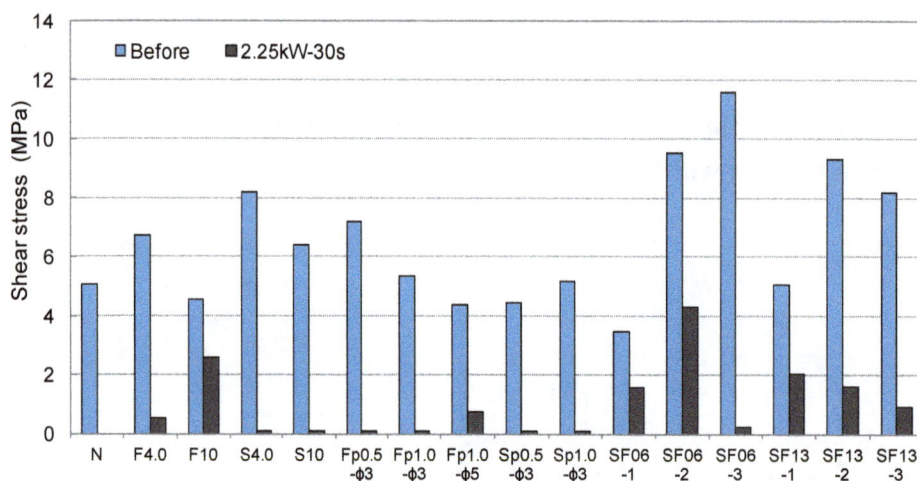

Figure 21. Residual shear bond strength after induction heating of the cementitious joint using a conductive resistor.

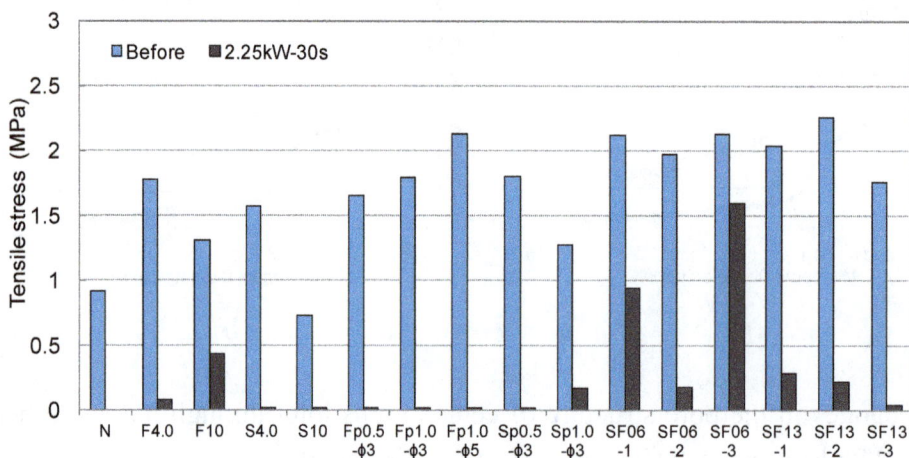

Figure 22. Residual tensile bond strength after induction heating of the cementitous joint using a conductive resistor.

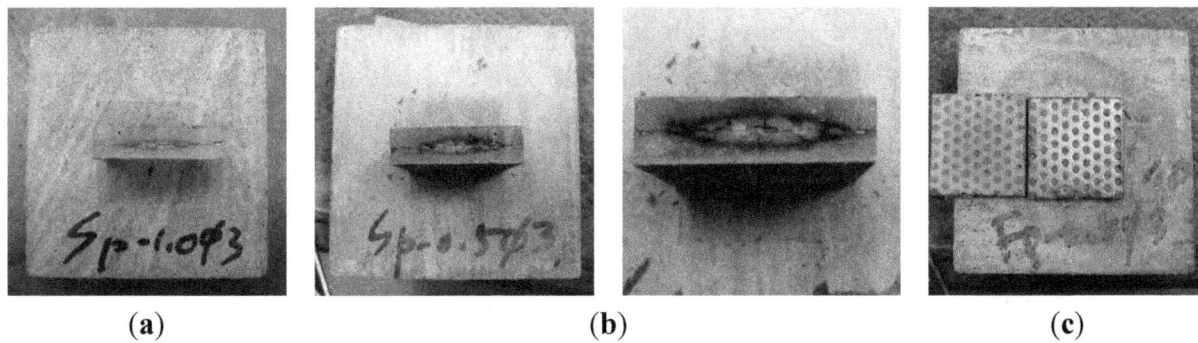

(a)   **(b)**   **(c)**

Figure 23. Heating of conductive resistor and weakening of the cementitious joint through induction heating. **(a)** Sp1.0–ø3 (2.25 kW, 30 s); **(b)** Sp0.5–ø3 (2.25 kW, 40 s); **(c)** Fp1.0–ø3 (2.25 kW, 30 s).

When the metal mesh was used as the conductive resistor, the stainless fabric, which was heated throughout by induction heating, showed an outstanding separation ability compared with the wire fabric. Additionally, separation occurred under no external load immediately after the majority of the specimens were heated, just as when the punching metal was used. In particular, with the punching metal, the temperature elevation characteristics of the metal part itself were similar to those in the metal mesh. However, the punching metal was assumed to undergo thermal expansion. Thus, the adhered part readily weakened because it had a wide area of heat emission. In addition, the steel fibers were assumed to serve as a second circuit when they were distributed and linked to one another inside the matrix, and they demonstrated heat emission characteristics by induction heating. However, in the case of SF06, the bond strength error after heating appeared to be large, and no weakening of the adhesion occurred in some specimens. Hence, the stability of the heating efficiency was assumed to depend on the internal distribution of the fibers. In contrast, SF13 mixed at 2% to 3% showed an outstanding heating efficiency and disassembly. Therefore, the critical mixture rate at which the internal distribution of steel fibers formed a percolation structure was ~2% in the case of SF13. The steel fibers within the joint could be readily separated by the difference in the thermal expansion values of the directly heated steel fiber and mortar and by the weakening of the mortar near the steel fibers.

5. Conclusions

By reviewing the temperature elevation characteristics of conductive resistors, the following conclusions were reached about the assembly properties of cementitious joints and their disassembly via induction heating:

(1) Cementitious joints with conductive resistors showed assembly properties equivalent to or more outstanding than those of ordinary finishing items for joint mortar.

(2) Among the conductive resistors, stainless steel showed more uniform temperature elevation characteristics than wire. Furthermore, punching metal readily caused internal eddy currents and showed the most pronounced disassembly properties by the weakening of cementitious joints.

(3) The temperature elevation and weakening of cementitious joints by induction heating were significantly affected by the radio frequency power from the heating induction coil and the distance of the conductive resistor from the heating induction coil.

(4) Conductive resistors could be selected in many ways, depending on the shape and environment of the joint material and the joint shape in each part of a building. The most appropriate conductive resistors for mortar-weakening by thermal conduction were those with at least 10-mesh wire intervals; no deterioration of the adhesion strength was observed when metal mesh was used. In the case of punching metal, a wide interface presumably formed between the adhered mortars, which thus undermined the assembly performance. Therefore, the use of punching metal with diameters of ø5 mm or greater was deemed preferable for the assembly and disassembly properties. Additionally, when SF13 was mixed at a rate of 2% or more, the disassembly performance was outstanding.

(5) Cementitious joints could be disassembled by manual induction heating at 2.25 kW for 30 s. In this case, 1659 W·h disassembly energy was used and 0.921 kg-CO_2 emission was estimated by using the IH method. Moreover, because such joints have the advantage of being reusable by completely disassembling the materials and parts, induction heating is an appropriate method for disassembling parts that require recycling and the reuse of high-priced materials.

Acknowledgments

This work was supported by the Dong-A University research fund.

Author Contributions

Jaecheol Ahn conducted the experiments and wrote the initial draft and the final manuscript of the manuscript. Takafumi Noguchi and Ryoma Kitagaki designed the project. All authors contributed to the analysis of the data and read the final paper.

Conflicts of Interest

The authors declare no conflict of interest.

References

1. Tsujino, M.; Noguchi, T.; Tamura, M.; Kanematsu, M.; Maruyama, I. Application of Conventionally recycled coarse aggregate to concrete structure by surface modification treatment. *J. Adv. Concr. Technol.* **2007**, *5*, 13–25.

2. Noguchi, T.; Kitagaki, R.; Nagai, H.; Tsujino, M. Completely recyclable concrete of aggregate-recovery type by using microwave heating technology. In Proceedings of the 2nd International RILEM Conference on Progress of Recycling in the Built Environment, Sao Paulo, Brazil, 2–4 December 2009; pp. 333–343.

3. Kobayashi, K.; Mamiya, T.; Inoue, T. Issues and evaluation of environmental load of building demolition waste disposal: Clarificaiton of issues a quantitive influence evaluation based on field survey. *J. Environ. Eng.* **2004**, *582*, 115–121.

4. Choi, J.C.; Park, H.J.; Kim, B.M. The influence of induction heating on the microstructure of A356 for semi-solid forging. *J. Mater. Process. Technol.* **1999**, *87*, 46–52.

5. Favennec, Y.; Labbé, V.; Bay, F. Induction heating processes optimization a general optimal control approach. *J. Comput. Phys.* **2003**, *187*, 68–94.

6. Liu, Q.; Schlangen, E.; García, A.; Ven, M. Induction heating of electrically conductive porous asphalt concrete. *Constr. Build. Mater.* **2010**, *24*, 1207–1213.

7. Liu, Q.; Schlangen, E.; Ven, M.; Bochove, G.; Montfort, J. Evaluation of the induction healing effect of porous asphalt concrete through four point bending fatigue test. *Constr. Build. Mater.* **2012**, *29*, 403–409.

8. Liu, Q.; García, A.; Schlangen, E.; Ven, M. Induction healing of asphalt mastic and porous asphalt concrete. *Constr. Build. Mater.* **2011**, *25*, 3746–3752.

9. Schneider, U. *Verhalten von Beton bei hohen Temperaturen-Behaviour of Concrete at High Temperatures, Deutscher Ausschuss für Stahlbeton, Heft 337*; Vertrieb Durch Verlag von Wilhelm Ernst & Sohn: Berlin, Germany, 1982; pp. 13–61.

10. Hirokazu, S.; Kazumi, K.; Kouichi, H.; Yasuo, K. Development of recovering technology of high quality aggregate from demolished concrete with heat and rubbing method. *Proc. Jpn. Concr. Inst.* **2000**, *22*, 1093–1110.

11. García, A.; Schlangen, E.; van de Ven, M.; van Vliet, D. Induction heating of mastic containing conductive fibers and fillers. *Mater. Struct.* **2011**, *44*, 499–508.

12. Rudolf, R.; Mitschang, P.; Neitzel, M. Induction heating of continuous carbon fibre reinforced thermoplastics. *Compos. Pt. A-Appl. Sci. Manuf.* **2000**, *31*, 1191–1202.

13. Ahn, J. Microwave dielectric heating to disassemble a modified cementitious joint. *Mater. Struct.* **2013**, *46*, 2077–2090.

9

Shape Effect of Electrochemical Chloride Extraction in Structural Reinforced Concrete Elements Using a New Cement-Based Anodic System

Jesús Carmona, Miguel-Ángel Climent, Carlos Antón, Guillem de Vera and Pedro Garcés *

Civil Engineering Department, Universidad de Alicante, Ctra. San Vicente s/n,
San Vicente del Raspeig 03690, Spain; E-Mails: jcarmona@ua.es (J.C.);
ma.climent@ua.es (M.-A.C.); c.anton@ua.es (C.A.); guillem.vera@ua.es (G.V.)

* Author to whom correspondence should be addressed; E-Mail: Pedro.Garces@ua.es;

Academic Editor: Maryam Tabrizian

Abstract: This article shows the research carried out by the authors focused on how the shape of structural reinforced concrete elements treated with electrochemical chloride extraction can affect the efficiency of this process. Assuming the current use of different anode systems, the present study considers the comparison of results between conventional anodes based on Ti-RuO$_2$ wire mesh and a cement-based anodic system such as a paste of graphite-cement. Reinforced concrete elements of a meter length were molded to serve as laboratory specimens, to closely represent authentic structural supports, with circular and rectangular sections. Results confirm almost equal performances for both types of anode systems when electrochemical chloride extraction is applied to isotropic structural elements. In the case of anisotropic ones, such as rectangular sections with no uniformly distributed rebar, differences in electrical flow density were detected during the treatment. Those differences were more extreme for Ti-RuO$_2$ mesh anode system. This particular shape effect is evidenced by obtaining the efficiencies of electrochemical chloride extraction in different points of specimens.

Keywords: electrochemical chloride extraction; cement-based anodic system; shape effect

1. Introduction

Electrochemical chloride extraction (ECE) is nowadays considered an appropriate technique as a means of extending the service life of reinforced concrete structures. This method allows reducing the corrosion rate of steel rebar caused by chloride ions (Cl^-). It basically consists of a migration of Cl^- ions from the reinforcement outward to the structure surface, thereby preventing its corrosive effect on steel. The removal of chloride ions is produced by applying a DC current between an anode in contact with an electrolyte, which is located on the outer surface of the concrete structure being treated, and its own rebar playing the role of cathode in this electrolytic process. Current density is in the range of 0.5–5.0 A/m^2 relative to the exposed concrete surface, and the application time lasts a few weeks. This technique has been studied and developed since the 1970s [1–3], and thereafter the effectiveness of ECE has been solidly founded on a wide and rigorous series of research works [4–7]. An essential milestone in the development of this application was the registration of a patent named "Norcure" by Vennesland and Opsahl. [8], where the anode system was based on a Ti-RuO$_2$ mesh. Afterwards, new studies were carried out aimed at improving some application conditions in order to simplify procedures and reduce costs, especially in the anode system. For this purpose, the use of multifunctional materials was considered, in particular cement-based conductive materials. Concrete, mortars and pastes are poor electricity conducting cement-based materials. However, their conductivity can be substantially enhanced by the addition of conductive materials such as carbon fibers or graphite powder. This procedure has been applied during recent years in order to provide new physical and chemical properties to cement-based materials [9–15]. One application has been the production of anodic systems for electrochemical treatments, such as cathodic protection [16–19]. Also, recent studies have focused on the development of cement-based anodic systems for ECE applications. The authors of this paper as well as others have recently carried out an investigation line using a conductive cement paste as anode for ECE treatments [20,21]. These studies have sufficiently proved that the efficiency of ECE when it is applied with an anode composed by a paste of graphite powder and Portland cement is similar to the one obtained with a reference anode (Ti-RuO$_2$ mesh). The present study is a continuation thereof. Once the efficiency of this technique was demonstrated, the new objective was to check the characteristics of the electric transmission through this kind of anode. The shape and density of the electric flow caused by the direct current from the cathode (reinforced steel bars) to the anode were already cited by Tritthart [4]. Also, the electric flow of ECE was treated in an indirect way regarding the ECE capability to extract Cl^-, according to the rebar arrangement of the treated element by Hope et al. [22] and Ihekwaba et al. [23]. More recently, the relationship between the rebar arrangement of structural elements and the ECE efficiency was studied by Garcés et al. [24]. All previously cited researchers have verified different particularities of this effect, but only in a qualitative way. In consequence, the objective of the present study consists in the evaluation of the electric flow structure produced during ECE applications with different types of anode systems in a quantitative way. It was based on checking the different efficiencies of ECE among core samplings extracted from several points of the different shaped specimens after the application of this electrochemical process.

2. Experimental Program and Materials

2.1. Case Studies Carried out

The research was carried out by applying ECE to four specimens of reinforced concrete, with the followings features:

Case study 1: application of ECE to a cylindrical specimen with Ti-RuO$_2$ mesh anode.

Case study 2: application of ECE to a cylindrical specimen with graphite-cement paste anode.

Case study 3: application of ECE to a rectangular section specimen with Ti-RuO$_2$ mesh anode. ECE efficiency is obtained for a core sample extracted on the center of the biggest face of the specimen.

Case study 4: application of ECE to a rectangular section specimen with Ti-RuO$_2$ mesh anode. ECE efficiency is obtained for a core sample extracted on the cover zone of the rebar, just over the steel.

Case study 5: application of ECE to a rectangular section specimen with graphite-cement paste anode. ECE efficiency is obtained for a core sample extracted on the center of the biggest face of the specimen.

Case study 6: application of ECE to a rectangular section specimen with graphite-cement paste anode. ECE efficiency is obtained for a core sample extracted on the cover zone of the rebar, just over the steel.

To obtain the initial content of Cl$^-$, one specimen of each shape was also prepared with the same procedures and materials, but without ECE application. Both were called the reference specimens.

2.2. Materials Used

2.2.1. Concrete

In the present research, the electric flow density produced during ECE application in real structural elements is the main consideration, and the way to assess this density is to obtain the resultant ECE efficiency. For this approach, shape and size of the laboratory specimens are crucial. Therefore, these elements must be bigger than usual. They were molded as structural support-shaped, 1 m length, three of them in cylindrical shape with circular section 200 mm diameter, and another three as prisms, with rectangular section 200 × 300 mm. The cylindrical ones were reinforced with 6 longitudinal rebar 8 mm diameter, hexagonally arranged, including 3 stirrups 6 mm diameter uniformly distributed. Regarding the prismatic-shaped, the reinforcement was composed of 4 rebar 16 mm diameter located in the corners and 4 stirrups 8 mm diameter. In all cases, concrete cover was 40 mm thick. Concrete had a composition as is shown in Table 1. Water-cement ratio (w/c) was 0.6, higher than usual in construction. The reason is to obtain a porous concrete, in order to make easier the ionic transport across the concrete mass, evidencing in a clearer way the shape effect of the structural element during ECE application. In order to simulate a serious Cl$^-$ contamination, 3.3% NaCl was added to the mixing water. Thus, the concrete contained 2% of Cl$^-$ relative to cement mass.

Table 1. Dosage of the concrete for laboratory specimens.

Material	Dosage
Portland cement	350 kg/m³
w/c Ratio	0.6
Distilled water	210 kg/m³
Limestone aggregate 4/6	466 kg/m³
Limestone aggregate 6/12	679 kg/m³
Limestone sand	630 kg/m³
NaCl	3.3% (2% Cl⁻ relative to cement mass)

Both water/cement ratio and maximum content of Cl⁻ parameters were beyond the acceptable ranges in Spanish Code on Structural Concrete [25].

With the above-mentioned conditions, concrete reached a compression strength of 17.8 N/mm² (UNE EN 12390-3:2009), a porosity of 17.0% (UNE 83980:2014) and a bulk density of 2.16 T/m³ (UNE EN 12390-7:2009).

2.2.2. Anodic Systems

Reference anodic system consists of a mesh of braided wire 1 mm thick of titanium with ruthenium oxide. This wire is braided in diamond shapes 33 mm per 12 mm of diagonal length. The mesh was firmly attached covering the whole vertical surface of the specimens, between two layers of absorbent synthetic fabric to keep moisture. The electrical resistance of this mesh per unit length (1 m length × 1.2 m width) is 0.041 Ω/m. The mesh is connected to the positive pole of the electric source by a copper wire.

On the other hand, the conductive graphite-cement paste (GCP) is a homogeneous mixture of graphite powder, Portland cement and distilled water as is shown in Table 2. This product is applied by spraying with a compressed air gun on the vertical specimen surfaces forming a 2 mm thick overlay. The high water-to-solid ratio (w/s = 0.8) meets the fluidity requirements of the spraying application system. Dosage and thickness of GCP is adopted as a consequence of the good performances shown by this anode system in recent researches by the same authors [20,21].

Table 2. Dosage of graphite-cement paste for 5 kg of paste.

Material	Dosage (for 5 kg of paste)
Portland cement	1.389 kg
Graphite powder	1.389 kg
Distilled water	2.222 kg

2.2.3. Electrolyte and Way of Moistening

The electrolyte for all the treatments was tap water, which was provided by a system able to maintain constant humidity during ECE process. That system consisted of a water pump and several pipes with droppers, assembled around the surface of each specimen. This constant drip irrigation system assures the presence of the electrolyte during the electrochemical procedure.

2.2.4. Electric Power Source

The power source provided a direct current density, which was set in the range of 2–5 A/m^2 relative to the exposed surface of the specimen. ECE treatments were designed with an electric charge density of 5×10^6 C/m^2, also relative to the exposed surface of the specimen. The voltage is therefore consequence of the constant current density and the resistance of the system. Thus, this voltage tends to raise as the ECE process progresses, since the resistivity of the specimens increases. In order to avoid an excessive increase, voltage must be monitored and controlled throughout the treatment. To control the voltage evolution and any circumstances during the whole research, specimens in treatment were remotely controlled by a WIFI IP camera. The European Standard CEN/TS 14038-2 proposes not exceeding the level of 40 V, for safety reasons as well as to prevent damages in the anode [26].

2.3. Assembly of Specimens

2.3.1. Specimens with Ti-RuO$_2$ Mesh Anode System

One of each of the cylindrical and prismatic specimens was equipped with a conventional anode composed of a Ti-RuO$_2$ mesh, as described in Section 2.2.2. The system consisted of two absorbent polymeric layers housing the Ti-RuO$_2$ mesh between them. Those three layers were successively and firmly set up covering the whole exposed concrete surface of specimens. In this way, the specimens were able to retain moisture and thus, to maintain the electrolyte action during ECE process (see Figure 1). Specimens assembled as mentioned were placed into a recipient with water in order to immerse the pump of drip irrigation system. As for electrical connection, a protrusion was made in an extreme of the mesh to connect with the positive pole of the current source, closing the anodic circuit.

Figure 1. Anodic system composed by a Ti-RuO$_2$ mesh embedded between two absorbent polymeric layers, wrapping concrete specimen.

2.3.2. Specimens with Graphite-Cement Paste (GCP) Anode System

As far as the rest of specimens (one of each type), a layer of GCP was applied by spraying on all vertical surfaces thereof. The spray was made with a compressed air gun, as a shotcrete system, forming an overlay around 2 mm thick. Then, specimens were moist-cured for 10 days in a curing chamber of relative humidity >95%. Finally, the paste coating was covered by the same polymeric layer mentioned above, and the same irrigation system was also installed to ensure the presence of the electrolyte. Anodic connections were made through two interconnected graphite felt fabric strips, 4 cm wide, firmly attached to the GCP layer, and united in turn with the positive pole of the source.

All specimens closed its cathodic circuit by connecting a rebar to the negative pole of the current source, assuming that all rebar are interconnected by the stirrups.

2.4. ECE Applications

2.4.1. Specimens with Ti-RuO$_2$ Mesh Anode System

ECE was applied with a current density of 5 A/m^2, and charge density was 5×10^6 C/m^2. Wetting remained constant. Voltage, which was continuously monitored, never reached 40 V.

2.4.2. Specimens with Graphite-Cement Paste (GCP) Anode System

Voltage increasing was higher than in Study 1. The voltage evolution is perhaps the most notable difference between both studies, which implies higher increase of resistivity with GCP anode than with a conventional Ti-RuO$_2$ mesh anode. In order to prevent values above 40 V, two means were used, namely:

- a reduction in current density, but never under 1 A/m^2; and
- the inclusion of pauses along the treatment.

The only effect of this change is the increase of the process time to achieve the same charge density of 5×10^6 C/m^2. ECE efficiency is directly related with electric charge density, while the times of application or the current density does not affect thereto, as was stated by Elsener et al. [27] and Polder et al. [28], and more recently by Sánchez de Rojas et al. [11]. Therefore, the comparison of ECE efficiencies among the different specimens is perfectly plausible because charge density was 5×10^6 C/m^2 for all studies.

2.5. Extraction of Core Samples

In order to work out the profiles of chloride content before and after ECE applications, different core samples were extracted from the specimens. All extracted cores were cylindrical 95 mm diameter and 40 mm depth, reaching just to the rebar location (see Figure 2).

From those cores, concrete dust samples were obtained by grinding each concrete core using a grinder device as recommended by RILEM TC 178-TMC [29]. This device is able to drill and pulverize accurately 2 mm of concrete each time with only turning the grinding crown 360°. Since the cores were 40 mm depth (concrete cover was 40 mm thick), 20 samples of concrete dust are obtained from each core, each sample belonging to every 2 mm thick layer. The subsequent chemical analysis will yield the chloride concentration at each depth.

Before carrying out the six studies, it was necessary to know the initial chloride content in the concrete specimens. To this end, a cross section was extracted, and the content of Cl$^-$ of both reference specimens was analyzed; one of each of the cylindrical and prismatic shape. These results were the initial chloride content profiles for the different studies carried out with each type of specimens.

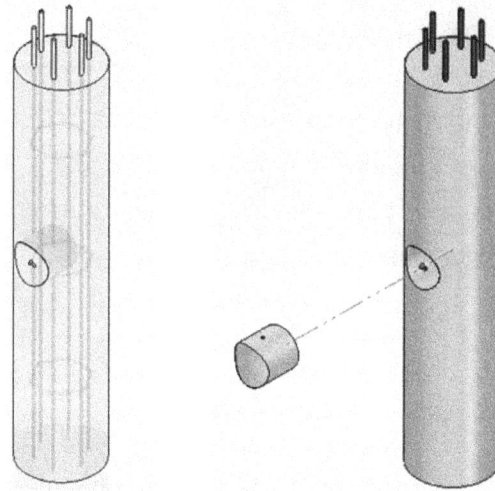

Figure 2. Extraction of core samples of cylindrical reinforced concrete specimens.

Regarding the cylindrical ones, the extraction point was the central zone of the specimen, although the extraction efficiency should be independent of the sampling point. Indeed, in specimens of circular section with a regular rebar distribution, anode system is always equidistant from the cathode. The isotropy of this system implies that the electric field established during ECE treatment takes a uniform, or uniformly distributed, configuration (see Figure 3).

Nevertheless, geometry and heterogeneous rebar arrangements of rectangular section elements may produce differences in the electric flow created between the anode and cathode systems during ECE, as was mentioned in the Section 1. In 1998, Tritthart proposed his theory about electric flow produced during ECE process [4]. In the present study, this electric flow has been calculated through a standard finite element method. Results are shown in Figure 3.

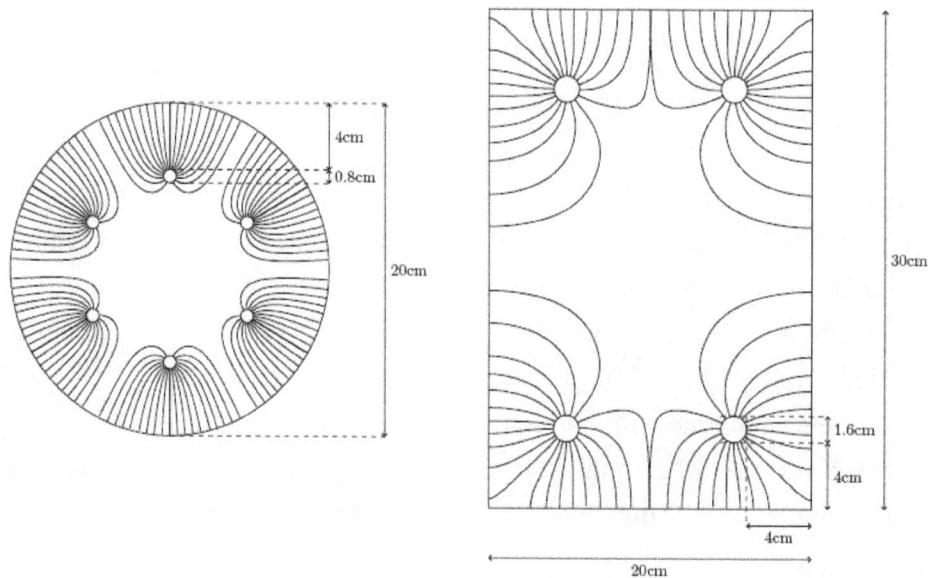

Figure 3. Different electric flow schemes during ECE application in an isotropic vertical element (circular section and regularly reinforced) and in an anisotropic vertical specimen (rectangular section and conventionally reinforced with rebar in the corners). Current streamlines, calculated using a standard finite element method, are shown.

This electric flow configuration suggested that ECE efficiency must be different depending on the shape of the treated structural element. For the isotropic ones, it is evident that ECE efficiency is equal in whatever point of the element. On the contrary, ECE efficiency must vary along the same horizontal plane of an anisotropic treated element. This research tries to confirm and quantify the differences, and to evaluate the influence of the type of anode system in this topic. For this purpose, two different core samples were extracted from the rectangular shaped specimens (see Figure 4).

    i. Number 1 on the center of the biggest face of the specimen shown in Figure 4.

    ii. Number 2 in the same horizontal plane, but on the concrete cover zone of the rebar (Figure 4).

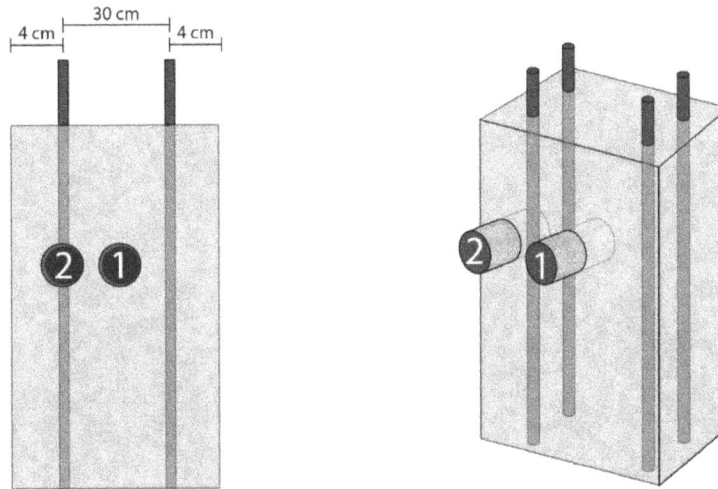

Figure 4. Extraction of two different core samples on prismatic reinforced concrete specimens. Point 1 is on the center of the biggest face of the specimen, and point 2 on the concrete cover, over the steel bar.

2.6. Chloride Analysis

The determination of acid-soluble contents of the concrete samples was performed by potentiometric titration, according to the method stated by Climent *et al.* [30,31]. All Cl⁻ concentrations of concrete in this paper refer to acid-soluble chlorides, and are expressed as percentages, relative to cement mass. The amount of Cl⁻ contained in each sample, before and after ECE, allows plotting the different Cl⁻ profiles as shown in the figures of the Results and Discussion Section. The integration of the evolution of those profiles by the action of ECE represents the efficiency of the ECE treatments.

3. Results and Discussion

In this section, the set of results obtained during the related experiences is exposed. The objective is to test the effect of electrical flux lines produced during the application of ECE due to the anisotropy of structural elements, and how the use of different anode system could affect this consequence. To this end, the efficiency of ECE in different cases is obtained. On the one hand, considering the type of anode system (Ti-RuO$_2$ or GCP), and, on the other hand, taking into account the different shape of treated specimens.

3.1. Study 1. Circular Section with Ti-RuO₂ Mesh Based Anode System

Current density was 5 A/m², and charge density 5×10^6 C/m². Wetting remained constant. Cl⁻ profiles before and after ECE treatment are represented in Figure 5. Also, the profile of differences between initial and final Cl⁻ content is plotted, showing the efficiency of ECE. The average of this parameter was 83.06%. Contents of Cl⁻ are always expressed as percentage relative to cement mass in concrete.

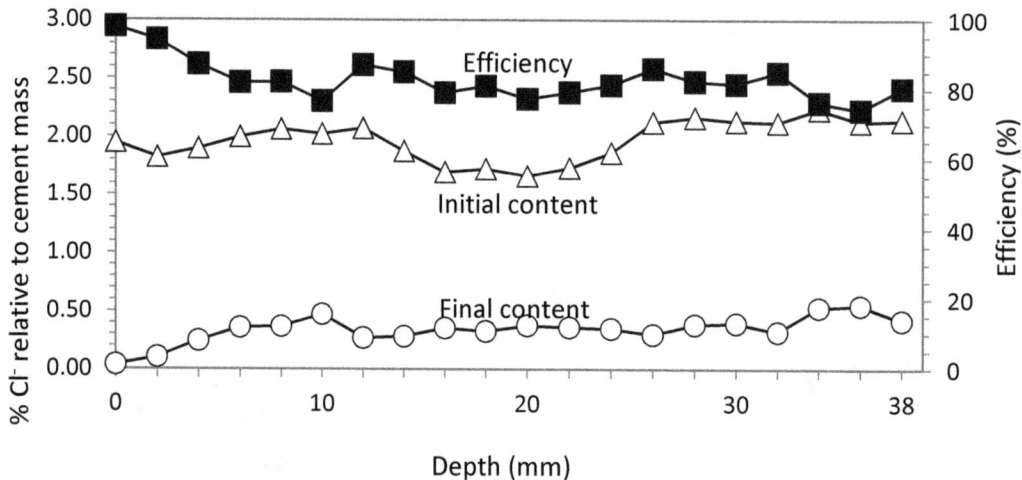

Figure 5. Study 1. Circular section specimen. ECE with Ti-RuO₂ mesh anodic system. Current density 5 A/m². Current charge density 5×10^6 C/m². Contents of Cl⁻ before and after ECE and differences in percentage relative to cement mass, which conform the efficiency profile.

Voltage evolution during the process remains almost constant during the first five days, evolving upwards to achieve an increase of only 12 V at the end of treatment, and always below 40 V.

3.2. Study 2. Circular Section with GCP Based Anode System

In this second study, ECE is applied with the same electrical parameters as in Study 1, *i.e.*, current density 5 A/m² and charge density 5×10^6 C/m². The anode system consists of a sprayed graphite-cement paste, as described in Section 2.3.2.

Figure 6 represents the Cl⁻ concentration profiles obtained before and after ECE application with a graphite-cement paste (GCP) anodic system, wetting constant. Extraction average of Cl⁻ was 82.60%.

In Study 2, the feeding voltage increase was higher than in Study 1. Pauses during the treatment were included to control the feeding voltage so that it remained below 40 V. Every time the voltage was near 40 V, a pause was applied. These pauses consisted of a current interruption of 24 h. When the current was restarted, a reduction of the feeding voltage in the range of 25%–30% of the last recorded value was found. For this study, three pauses were needed.

Efficiency results are very similar for ECE applied with both anode systems, as can be seen in Figure 7.

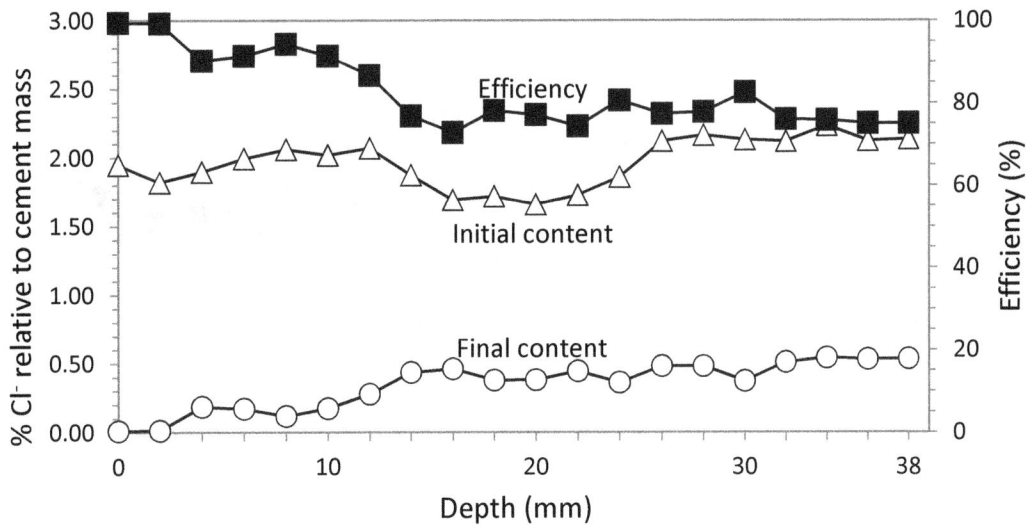

Figure 6. Study 2. Circular section specimens. ECE with GCP anodic system. Current density 5 A/m^2. Current charge density 5×10^6 C/m^2. Contents of Cl$^-$ before and after ECE and differences in percentage relative to cement mass, which conform the efficiency profile.

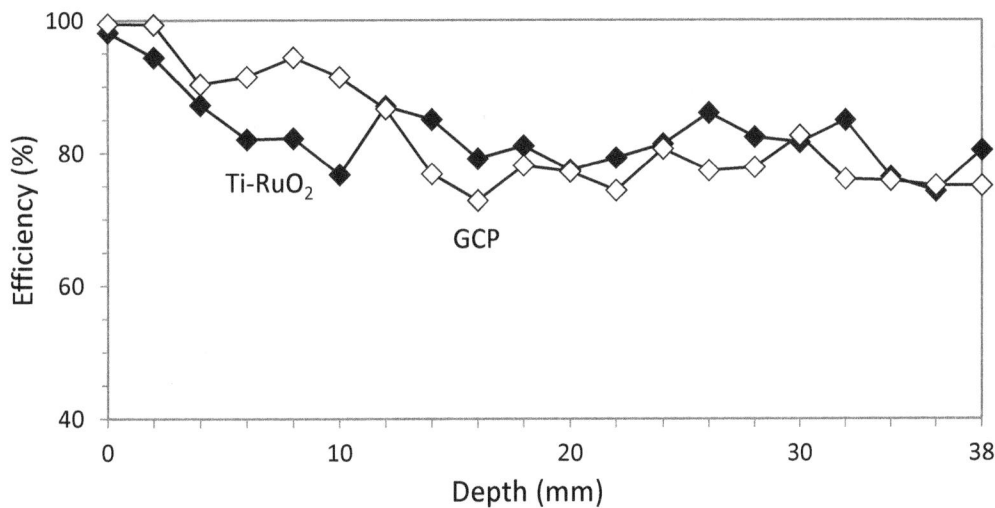

Figure 7. Comparative efficiency results for studies 1 and 2. Circular section specimens. ECE efficiency average with Ti-RuO$_2$ mesh was 82.79% against 82.60% with GCP anodic system.

3.3. Rectangular Section with Ti-RuO$_2$ Mesh Based Anode System

3.3.1. Study 3. Core Sample in the Center of the Biggest Face

As was said in Section 2.4, reinforced concrete specimens support type are also used for the second part of the laboratory tests, but in this case of rectangular section. Results of ECE efficiency for the core sample in the center of the biggest face (location 1 in Figure 3) are shown in Figure 8. Efficiency average for Study 3 was 52.52%.

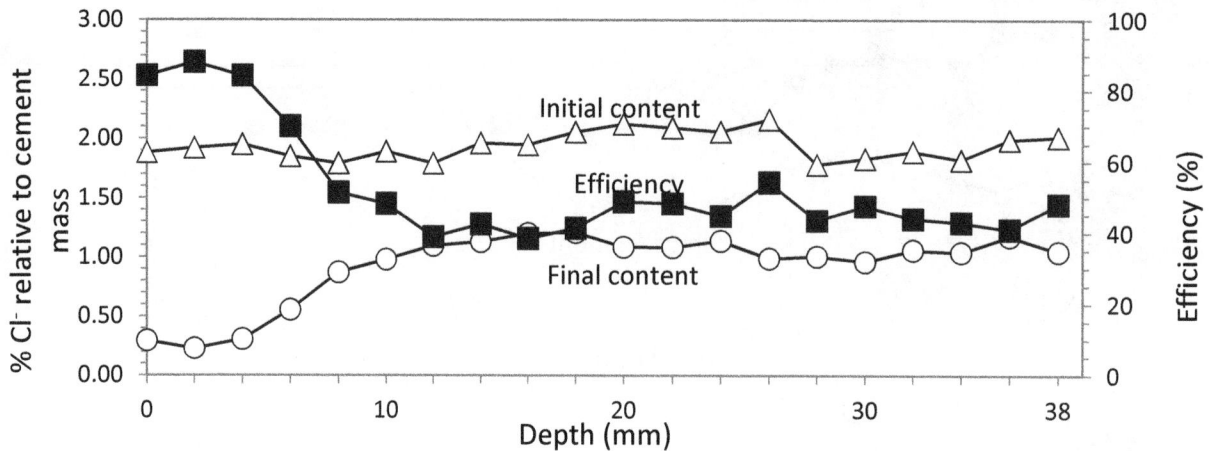

Figure 8. Study 3. Rectangular section specimens. ECE with Ti-RuO₂ mesh anodic system. Variable current density between 3–5 A/m². Current charge density 5×10^6 C/m². Contents of Cl⁻ before and after ECE and differences in percentage relative to cement mass, which conform the efficiency profile. Core sample extracted in the center of the biggest face.

In this case, the voltage evolution was faster than in circular section specimens. Consequently, around the middle of the treatment, the voltage was near 40 V. At that time, current density range was reduced from 5 to 3 A/m² and the voltage dropped to 25 V. As is known, the only effect of this change is the increase of the process time to achieve the same charge density of 5×10^6 C/m².

3.3.2. Study 4. Core Sample in the Concrete Cover Zone

Nevertheless, when the core sample was extracted on the same horizontal plane, but on the concrete cover over the rebar (location 2 in Figure 3), chloride analysis shows an ECE efficiency almost double (see Figure 9). The average efficiency reached in this case 85.22%. Electric parameters (current density = 3–5 A/m² and charge density = 5×10^6 C/m²) were the same as in Study 3.

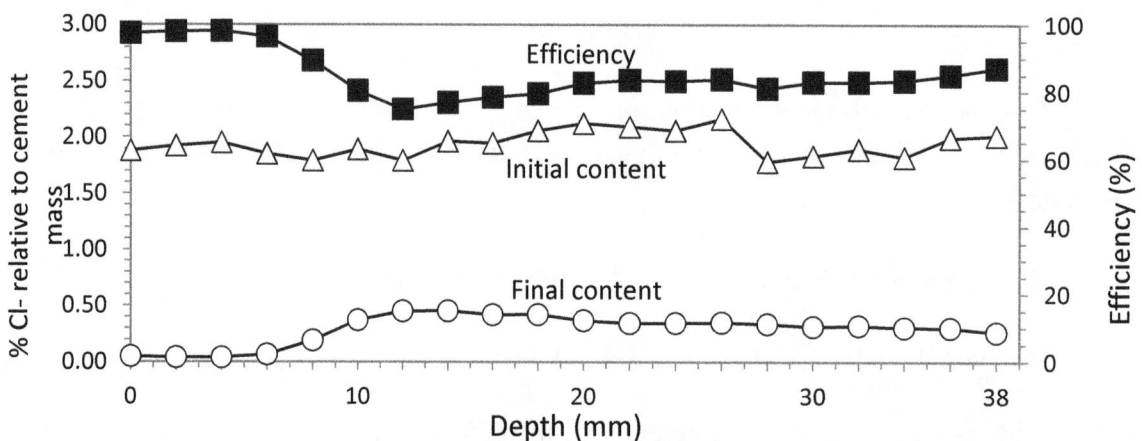

Figure 9. Study 4. Rectangular section specimens. ECE with Ti-RuO₂ mesh anodic system. Variable current density between 3–5 A/m². Current charge density 5×10^6 C/m². Contents of Cl⁻ before and after ECE and differences in percentage relative to cement mass, which conform the efficiency profile. Core sample extracted of the concrete cover up to the vertical rebar.

3.4. Rectangular Section with GCP Based Anode System

3.4.1. Study 5. Core Sample in the Center of the Biggest Face

This time, ECE was applied to rectangular section specimens using a GCP anode system as described before. The most significant difference consisted of a higher resistivity, manifested by a stronger increase of voltage during ECE process. Consequently, to maintain voltage below 40 V it was necessary to reduce current density to a range between 2–3 A/m^2. Charge density was also 5×10^6 C/m^2 (see Figure 10).

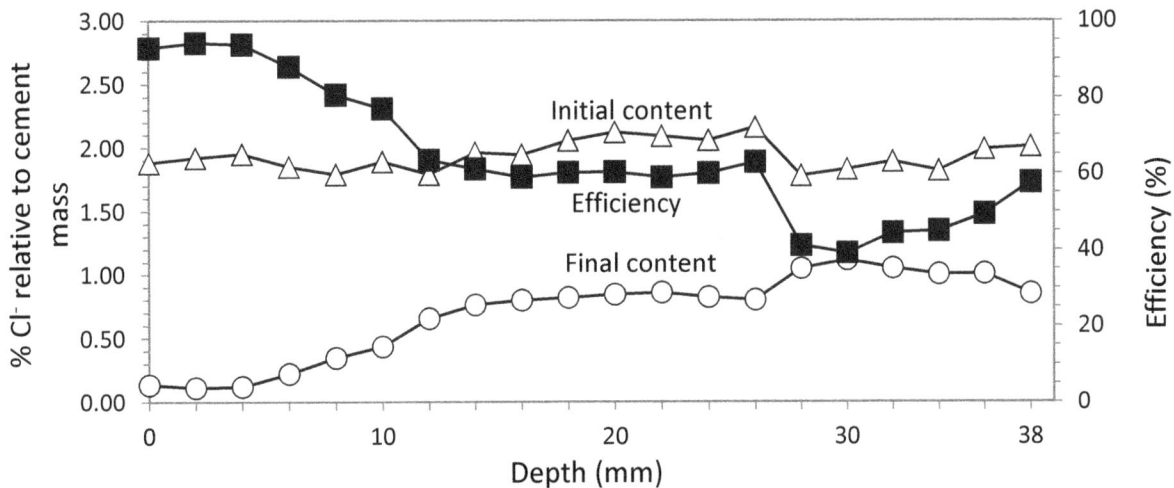

Figure 10. Study 5. Rectangular section specimens. ECE with GCP based anodic system. Variable current density between 2–3 A/m^2. Current charge density 5×10^6 C/m^2. Contents of Cl$^-$ before and after ECE and differences in percentage relative to cement mass, which conform the efficiency profile. Core sample extracted in the center of the biggest face.

The average efficiency was in this case 64.38%, a better value than in Study 3 (Ti-RuO$_2$ mesh anode system—sample in the center of the biggest face).

3.4.2. Study 6. Core Sample in the Concrete Cover Zone

In this last study, the core sample was extracted from the concrete cover zone, just over the rebar location. The values of the electric parameters were the same of Study 5: 2–3 A/m^2 and 5×10^6 C/m^2. Initial and final Cl$^-$ contents and efficiency are shown in Figure 11. The average efficiency in this case was 76.15%.

The different efficiencies obtained in studies 3, 4, 5 and 6 are shown in Figure 12, in order to compare them easier.

To summarize, results of efficiency averages of the six referred to studies are shown in Table 3.

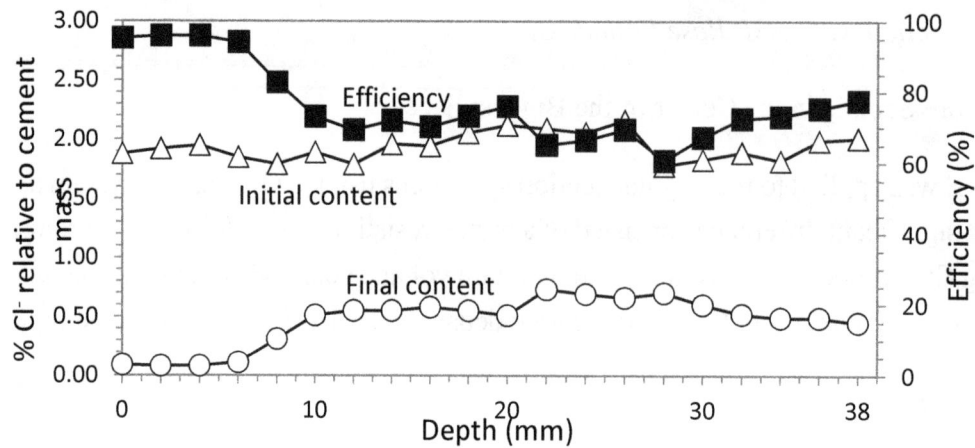

Figure 11. Study 6. Rectangular section specimens. ECE with GCP based anodic system. Variable current density between 2–3 A/m^2. Current charge density 5×10^6 C/m^2. Contents of Cl$^-$ before and after ECE and differences in percentage relative to cement mass, which conform the efficiency profile. Core sample extracted of the concrete cover up to the rebar.

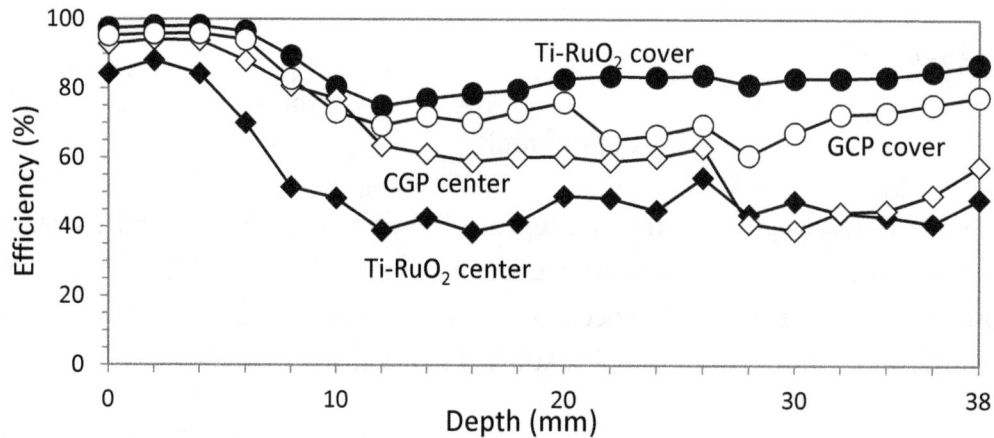

Figure 12. Comparative efficiency results for studies 3, 4, 5 and 6. Rectangular section specimens. ECE efficiency average with Ti-RuO2 mesh was 52.52% in the vertical axis of the largest side and 85.22% in the concrete cover, whereas for GCP, results were 64.38% and 76.15% for the same locations.

Table 3. Summary of obtained efficiencies in the different studies carried out applying ECE to vertical shape specimens of reinforced concrete.

Study	Horizontal section shape of specimen	Anode system	Core sample location	Average efficiency (%)
1	Circular section	Ti-RuO$_2$ mesh		82.79
2	Circular section	GCP		82.60
3	Rectangular section	Ti-RuO$_2$ mesh	Center of the biggest face	52.52
4	Rectangular section	Ti-RuO$_2$ mesh	Concrete cover over rebar	85.22
5	Rectangular section	GCP	Center of the biggest face	64.38
6	Rectangular section	GCP	Concrete cover over rebar	76.15

As for circular section specimens, ECE efficiencies show practically equal performance for both anode systems along the different depths and only 0.2% less efficiency average in GCP. The higher resistivity of GCP is also evidenced, but it is possible to pass the same electric charge density only by reducing current density or including some pauses in the treatment. This result confirms the feasibility of using a GCP anode system for ECE applications, as was also stated in previous research [20,21]. The high level of ECE efficiency, in either anode system and either sampling point, shows the uniformity and high density of electric flow lines produced between a regular cathode (rebars uniformly distributed) and an equidistant anodic surface, as was exposed in Section 2.5 and Figure 2.

Regarding specimens with rectangular sections and no uniformly distributed rebars (anisotropic element), ECE efficiency obtained in different points of the same horizontal section denoted the expected differences in the electric flow density at those positions, confirming the set out hypothesis. Indeed, the studies carried out with Ti-RuO$_2$ mesh anode system gave significant difference of ECE efficiencies: 52.52% for the center of the biggest face *versus* 85.22% for the concrete cover over the rebar (62% more efficiency in the cover zone than in the center zone). This is consistent with the resultant electric flow density produced during ECE application to an anisotropic structural element, such as the prismatic specimens of this research (see Section 2.5 and Figure 2).

The same effect was detected in GCP anode system, but the difference of the electric flow density for the same horizontal section are significantly lower than in Ti-RuO$_2$ mesh anode system (18% more efficiency in the cover zone than in the center of the biggest face). This effect would probably be related to the different shape of electric flow lines produced during ECE due to the fact that the contact between the Ti-RuO$_2$ and the concrete surface is not perfect at all positions. As it is a metal mesh with certain rigidity and elasticity, the absolute adherence to the concrete surface is difficult to achieve. To confirm this question, subsequent studies are advisable. In any case, the fact that GCP is an overlay and its location as anode is the same structural element surface implies undoubtedly an advantage over other systems. Besides, efficiency averages among both extracted cores of each specimen after ECE application are also very similar, 68.87% for Ti-RuO$_2$ mesh anode system and 70.26% for GCP anode system; 1.4% higher for GCP. All these qualities, coupled with the fact that the cost of the GCP materials is 8–10 times lower than that of the Ti-RuO$_2$ mesh anode system, strengthen the option of GCP as anode system for ECE treatments.

4. Conclusions

The use in this research of specimens with a size close to natural scale allows determining the effect of electric flow density produced during ECE application to different shaped structural elements. Firstly, the measurement of the resultant percentage of removed Cl$^-$ by the ECE action confirms the feasibility of achieving a similar ECE efficiency with a cement-based anode system such as GCP (a sprayed graphite-cement paste as overlay) to that obtained with the conventional Ti-RuO$_2$ mesh anode system, whatever shape and arrangement of reinforced concrete elements to be treated.

As for the peculiar electric flow configuration produced by ECE application on anisotropic structural elements, GCP anode system shows similar effects to those of Ti-RuO$_2$ mesh anode system, but with less extreme differences. This particular shape effect in the electric flow lines is deduced considering the different efficiencies of the treatments, which are significantly more marked with the conventional

Ti-RuO$_2$ mesh anode system. Further studies could be carried out to find out the reason of the obtained differences in ECE efficiency between both types of anodes.

It has also been noted that the shape and density of the electric flow as were calculated through a standard finite element method (see Figure 2) is directly related with the efficiency of ECE treatment. It is therefore probable to find the same effect in the rest of electrochemical applications to improve the service life of reinforced concrete structures, such as cathodic protection and realkalization. Thus, the results of this research suggest the advisability to move towards the design of isotropic structural elements for sustainability considerations.

Acknowledgments

This research was funded by the Spanish Ministerio de Economía y Competitividad (and formerly by the Spanish Ministerio de Ciencia e Innovación) and ERDF (European Regional Development Fund) through projects BIA2010-20548 and MAT2009-10866, and also through the project PROMETEO/2013/035 of Generalitat Valenciana (Spain).

Author Contributions

This paper includes a part of the results of the PhD thesis research by Jesús Carmona under the supervision of Pedro Garcés and Miguel Ángel Climent. Carlos Antón participated in the experimental research, and Guillem de Vera performed the finite element modelization of the ECE current streamlines. All five authors contributed to the analysis and conclusions, and revised the paper.

Conflicts of Interest

The authors declare no conflict of interest.

References

1. Lankard, D.R.; Slater, J.E.; Hedden, W.A.; Niesz, D.E. *Neutralization of Chloride in Concrete*; Report No. FHWA-RD-76-60; Federal Highway Administration: Washington, DC, USA, 1975; pp. 1–143.

2. Slater, J.E.; Lankard, D.R.; Moreland, P.L. *Electrochemical Removal of Chlorides from Concrete Bridge Decks*; Transportation Research Record No. 604; Transportation Research Board: Washington, DC, USA, 1976; pp. 6–15.

3. Morrison, G.L.; Virmani, Y.P.; Stratton, F.W.; Gilliland, W.J. *Chloride Removal and Monomer Impregnation of Bridge Deck Concrete by Electro-Osmosis*; Report No. FHWA-Ks-RD-74-1; Kansas Department of Transportation, Federal Highway Administration (USA): Washington, DC, USA, 1976; p. 38.

4. Tritthart, J. *Electrochemical Chloride Removal: An Overview and Scientific Aspects*; American Ceramic Society: Westerville, OH, USA, 1998; pp. 401–441.

5. Mietz, J. *Electrochemical Rehabilitation Methods for Reinforced Concrete Structures. A State of the Art Report*; Book No 709; European Federation of Corrosion by the Institute of Materials: Graz, Austria, 1998.

6. Andrade, C.; Castellote, M.; Alonso, C. An overview of electrochemical realkalisation and chloride extraction. In Proceedings of the 2nd Int. RILEM/CSIRO/ACRA Conference on Rehabilitation of Structures, Melbourne, Australia, 21–23 September 1998; Ho, D.W.S., Godson, I., Collins, F., Eds.; pp. 1–12.

7. Bertolini, L.; Elsener, B.; Pedeferri, P.; Polder, R. *Corrosion of Steel in Concrete*; WYLEY-VCH Verlag GmbH & Co. KGaA: Veinheim, Germany, 2008; pp. 345–374.

8. Vennesland, Ø.; Opsahl, O.A. Removal of Chlorides from Concrete. U.S.A & U.K. Patent 4832803, May 1989.

9. Chung, D.D.L. Electrically conductive cement-based materials. *Adv. Cem. Res.* **2004**, *16*, 167–176.

10. Climent, M.A.; Sánchez de Rojas, M.J.; De Vera, G.; Garcés, P. Effect of type of anodic arrangements on the efficiency of electrochemical chloride removal from concrete. *ACI Mater. J.* **2006**, *103*, 243–250.

11. Sánchez de Rojas, M.J.; Garcés, P.; Climent, M.A. Electrochemical extraction of chlorides from reinforced concrete: Variables affecting treatment efficiency. *Mater. Constr.* **2006**, *56*, 17–26.

12. Garcés, P.; Andión, L.G.; Varga, I.; Catalá, G.; Zornoza, E. Corrosion of steel reinforcement in structural concrete with carbon material addition. *Corros. Sci.* **2007**, *49*, 2557–2566.

13. Ivorra, S.; Garcés, P.; Catalá, G.; Andión, L.G.; Zornoza, E. Effect of silica fume particle size on mechanical properties of short carbon fiber reinforced concrete. *Mater. Des.* **2010**, *31*, 1553–1558.

14. Garcés, P.; Zornoza, E.; Alcocel, E.G.; Galao, Ó.; Andión, L.G. Mechanical properties and corrosion of CAC mortars with carbon fibers. *Constr. Build. Mater.* **2012**, *34*, 91–96.

15. Garcés, P.; Fraile, J.; Vilaplana-Ortego, V.; Cazorla, D.; Alcocel, E.G.; Andión, L.G. Effects of carbon fibres on the mechanical properties and corrosion levels of reinforced Portland cement mortars. *Cem. Concr. Res.* **2005**, *35*, 324–331.

16. Fu, X.; Chung, V. Carbon fiber reinforced mortar as an electrical contact material for cathodic protection. *Cem. Concr. Res.* **1995**, *25*, 689–694.

17. Hou, J.; Chung, D.D.L. Cathodic protection of steel reinforced concrete facilitated by using carbon fiber reinforced mortar or concrete. *Cem. Concr. Res.* **1997**, *27*, 649–656.

18. DePeuter, F.; Lazzari, L. New conductive overlay for CP in concrete: Results of long term testing. Corrosion/93, paper n° 325. In Proceedings of the NACE Conference, Houston, TX, USA, 1993.

19. Bertolini, L.; Bolzoni, F.; Pastore, T.; Pedeferri, P. Effectiveness of a conductive cementitious mortar anode for cathodic protection of steel in concrete. *Cem. Concr. Res.* **2004**, *34*, 681–694.

20. Pérez, A.; Climent, M.A.; Garcés, P. Electrochemical extraction of chlorides from reinforced concrete using a conductive cement paste as the anode. *Corros. Sci.* **2010**, *52*, 1576–1581.

21. Cañón, A.; Garcés, P.; Climent, M.A.; Carmona, J.; Zornoza, E. Feasibility of electrochemical chloride extraction from structural reinforced concrete using a sprayed conductive graphite powder-cement as anode. *Corros. Sci.* **2013**, *77*, 128–134.

22. Hope, B.B.; Ihekwaba, V.; Hansson, C.M. Influence of Multiple Rebar Mats on Electrochemical Removal of Chloride from concrete. *Mater. Sci. Forum* **1995**, *192–194*, 993–890.

23. Ihekwaba, N.M.; Hope, N.M.; Hansson, C.M. Structural shape effect on rehabilitation of vertical concrete structures by ECE technique. *Cem. Concr. Res.* **1996**, *26*, 165–175.

24. Garcés, P.; Sánchez de Rojas, M.J.; Climent, M.A. Effect of the reinforcement bar arrangement on the efficiency of electrochemical chloride removal technique applied to the reinforced concrete structures. *Corros. Sci.* **2006**, *48*, 531–545.

25. Comisión Permanente del Hormigón, Ministerio de Fomento. *Instrucción de Hormigón Estructural EHE-08 (Spanish Code on Structural Concrete EHE-08)*; Comisión Permanente del Hormigón, Ministerio de Fomento: Madrid, Spain, 2008. (Only available in Spanish)

26. BSI. *Electrochemical Re-Alkalization and Chloride Extraction Treatments for Reinforced Concrete—Part 2: Chloride Extraction*; CEN/TS 14038-2:2011; BSI: London, UK, 2011.

27. Elsener, B.; Molina, M.M.; Böhni, H. The Electrochemical Removal of Chlorides from Reinforced Concrete. *Corros. Sci.* **1993**, *35*, 1563–1570.

28. Polder, R.B.; Walker, R.; Page, C.L. Electrochemical chloride removal tests of concrete cores from a coastal structure. In Proceedings of the International Conference on Corrosion and Corrosion Protection of Steel in Concrete, University of Sheffield, Sheffield, UK, 24–28 July 1994.

29. Vennesland, Ø.; Climent, M.A.; Andrade, C. Recommendation of RILEM TC 178-TMC: Testing and modeling chloride penetration in concrete. Methods for obtaining dust samples by means of grinding concrete in order to determine the chloride concentration profile. *Mater. Struct.* **2013**, *46*, 337–344.

30. Climent, M.A.; Viqueira, E.; de Vera, G.; López, M.M. Analysis of acid-soluble chloride in cement, mortar and concrete by potentiometric titration without filtration steps. *Cem. Concr. Res.* **1999**, *29*, 893–898.

31. Climent, M.A.; de Vera, G.; Viqueira, E.; López, M.M. Generalization of the possibility of eliminating the filtration step in the determination of acid-soluble chloride content in cement and concrete by potentiometric titration. *Cem. Concr. Res.* **2004**, *34*, 2291–2295.

10

The Effects of Different Fine Recycled Concrete Aggregates on the Properties of Mortar

Cheng-Chih Fan [1], Ran Huang [2], Howard Hwang [3] and Sao-Jeng Chao [4],*

[1] Institute of Materials Engineering, National Taiwan Ocean University, No. 2 Pei-Ning Road, Keelung 20224, Taiwan; E-Mail: chengchihfan@gmail.com

[2] Department of Harbor and River Engineering, National Taiwan Ocean University, No. 2 Pei-Ning Road, Keelung 20224, Taiwan; E-Mail: ranhuang@ntou.edu.tw

[3] Graduate Institute of Architecture and Sustainable Planning, National Ilan University, No.1, Sec. 1, Shen-Lung Road, I-Lan 26047, Taiwan; E-Mail: hmhwang@niu.edu.tw

[4] Department of Civil Engineering, National Ilan University, No.1, Sec. 1, Shen-Lung Road, I-Lan 26047, Taiwan

* Author to whom correspondence should be addressed; E-Mail: chao@niu.edu.tw;

Academic Editor: Maryam Tabrizian

Abstract: The practical use of recycled concrete aggregate produced by crushing concrete waste reduces the consumption of natural aggregate and the amount of concrete waste that ends up in landfills. This study investigated two methods used in the production of fine recycled concrete aggregate: (1) a method that produces fine as well as coarse aggregate, and (2) a method that produces only fine aggregate. Mortar specimens were tested using a variety of mix proportions to determine how the characteristics of fine recycled concrete aggregate affect the physical and mechanical properties of the resulting mortars. Our results demonstrate the superiority of mortar produced using aggregate produced using the second of the two methods. Nonetheless, far more energy is required to render concrete into fine aggregate than is required to produce coarse as well as fine aggregate simultaneously. Thus, the performance benefits of using only fine recycled concrete aggregate must be balanced against the increased impact on the environment.

Keywords: fine recycled concrete aggregate; recycled concrete aggregate; recycled aggregate; compressive strength; ultrasonic pulse velocity; mortar

1. Introduction

The practical use of recycled concrete aggregate produced by crushing concrete waste reduces the consumption of natural aggregate as well as the amount of concrete waste that ends up in landfills. The crushing of concrete waste produces coarse recycled concrete aggregate (CRCA) and fine recycled concrete aggregate (FRCA), as defined by particle size. A number of studies [1–11] have shown that CRCA can be used as a replacement for coarse natural aggregate in structural concrete; however, relatively little research has been conducted on the application of FRCA in structural concrete [11–15].

Several studies [11–19] have used laboratory crushers for the crushing concrete of waste to produce FRCA. Test results of these materials reveal that the FRCA produced using this crushing process leaves a large amount of cement paste attached to the surface of FRCA, which can have a detrimental effect on the material properties. Thus, researchers have shifted their attention to the influence of production process on the properties of the resulting FRCA.

Lee [20] investigated two discrete crushing processes using a jaw crusher and an impact crusher to obtain two types of FRCA: RF-A and RF-B, with specific gravity values of 2.39 and 2.28 and water absorption of 6.59% and 10.35%, respectively. Various quantities of fine natural aggregate (FNA) were then replaced with the two types of FRCA, whereupon the resulting mortars were tested. The mortar in which FNA was replaced entirely by RF-A presented higher density and greater compressive strength than did the samples made entirely with RF-B. These results also indicate that the water absorption of FRCA influences the properties of the mortar, particularly at higher replacement ratios.

Sim and Park [21] applied advanced recycling methods to the production of FRCA with a specific gravity of 2.28 and water absorption of 6.45%. They replaced various proportions of FNA with FRCA and tested the resulting concrete specimens. Compressive strength was shown to decline with an increase in the replacement ratio of FRCA. When the replacement ratio reached 100%, the compressive strength of the mortar at 28 days was approximately 33% lower than that of the original samples and all specimens with over 60% FNA replacement presented a significant drop in compressive strength.

Florea and Brouwers [22] investigated the influence of concrete crushing method on the particle size distribution and density of recycled concrete aggregate (RCA). As a standard, they adopted concrete with compressive strength of 60 MPa at 91 days, to which they applied three methods for the crushing of concrete: (1) RC-1 refers to RCA from concrete waste that was passed through a jaw crusher just once before being screened; (2) RC-2 refers to RCA that passed through a jaw crusher ten times before being screened; and (3) RC-3 refers to RCA produced from three consecutive crushing processes using the Smart Crusher SC 1, designed specifically for concrete waste. RC-3 presented the optimal particle size distribution, between 125 μm and 200 μm, with a density of 2.50 g/cm^3, which increased with the size of the particles. RC-3 particles between 2 mm and 4 mm in size had a density of 2.61 g/cm^3. They concluded that optimizing the crushing method could enhance the quality of the resulting RCA.

Ulsen et al. [23] produced a variety of FRCAs by crushing recycled aggregates smaller than 19 mm using a jaw crusher in conjunction with a vertical shaft impact (VSI) crusher at various rotational speeds: (1) CDW-sand refers to FRCA produced using a jaw crusher prior to screening; (2) VSI-55 refers to FRCA produced using a jaw crusher followed by a VSI crusher at 55 m/s prior to screening; (3) VSI-65 refers to FRCA produced using a jaw crusher followed by a VSI crusher at 65 m/s prior to screening; and (4) VSI-75 refers to FRCA produced using a jaw crusher followed by a VSI crusher at 75 m/s prior to screening. Their results indicate that the rotational speed of the VSI crusher had no effect on the particle shape or particle size distribution of the FRCA; however, it did affect water absorption and porosity. The water absorption of CDW-sand, VSI-55, VSI-65, and VSI-75 were 12%, 9%, 8.1%, and 7%, respectively, whereas the porosity percentages were 11.9%, 6.9%, 5%, and 6%, respectively.

Song and Ryou [24] introduced a washing stage to the production of FRCA using a combination of chemical and physical processes. The washing process had the following effects: water absorption dropped from 5.8% to 1.92%; the ratio of absolute volume increased from 62.3% to 65.1%; and impurity content dropped from 0.46 to 0.18%. Clearly, this washing process can enhance the physical properties of the resulting FRCA.

Koshiro and Ichise [25] employed a heat grinder system for the processing of concrete waste from a demolished building, which resulted in FRCA with density of 2.57 g/cm^3 and water absorption of 2.52%. Their results demonstrate the efficacy of heat grinder systems in the production of high-quality FRCA suitable for the structure of new buildings.

Clearly, the methods used in the processing of concrete waste influence the quality of the resulting FRCA. Most previous studies have obtained FRCA produced under laboratory conditions or using the methods typically employed in large-scale recycling facilities, in which CRCA and FRCA are produced simultaneously. In this study, we obtained FRCA from a recycling facility in Yilan, Taiwan, which using a crushing process that produces only fine recycled concrete aggregate. We then prepared and tested specimens using a variety of mix proportions to determine the influence of production process and FRCA proportion on the properties of the resulting mortar, which is a constituent of concrete.

2. Experimental

2.1. Materials

This study employed Type I Portland cement and FNA comprising clay slate and river sand, which was processed in a gravel plant. Figure 1 presents the particle size distribution of the FNA and Table 1 lists the basic physical properties.

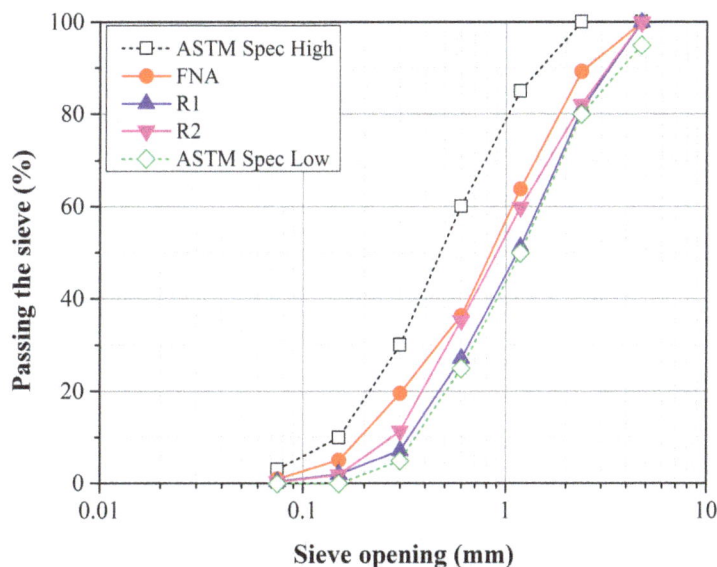

Figure 1. Particle size distribution curves of FNA, R1 and R2 (ASTM C33-13 [26]).

Table 1. Physical properties of FNA, R1 and R2.

Physical properties	FNA	R1	R2
Saturated surface dry density (kg/m^3) (ASTM C128-12 [27])	2653	2347	2404
Water absorption (%) (ASTM C128-12 [27])	1.3	8.9	6.6
Fineness modulus (ASTM C136-14 [28])	2.9	3.3	3.1

Figure 2 illustrates two processes commonly used for the production of fine recycled concrete aggregate. The first process produces coarse and fine aggregates simultaneously by crushing waste concrete with a large jaw crusher and then separating the resulting aggregate using a vibrating screen. Aggregate larger than 19 mm in diameter is sent through two cone crushers, after which the product is separated using a vibrating screen. Aggregate between 4.75 mm and 19 mm is transported to a coarse aggregate stockpile area. Aggregate smaller than 4.75 mm is sent to a roller sand washer to wash away sludge smaller than 150 μm. The resulting product is fine aggregate (ranging in size from 150 μm to 4.75 mm), which is stored in a fine aggregate stockpile area. The FRCA produced by this process is denoted as R1, the properties or which are presented in Figure 1 and Table 1.

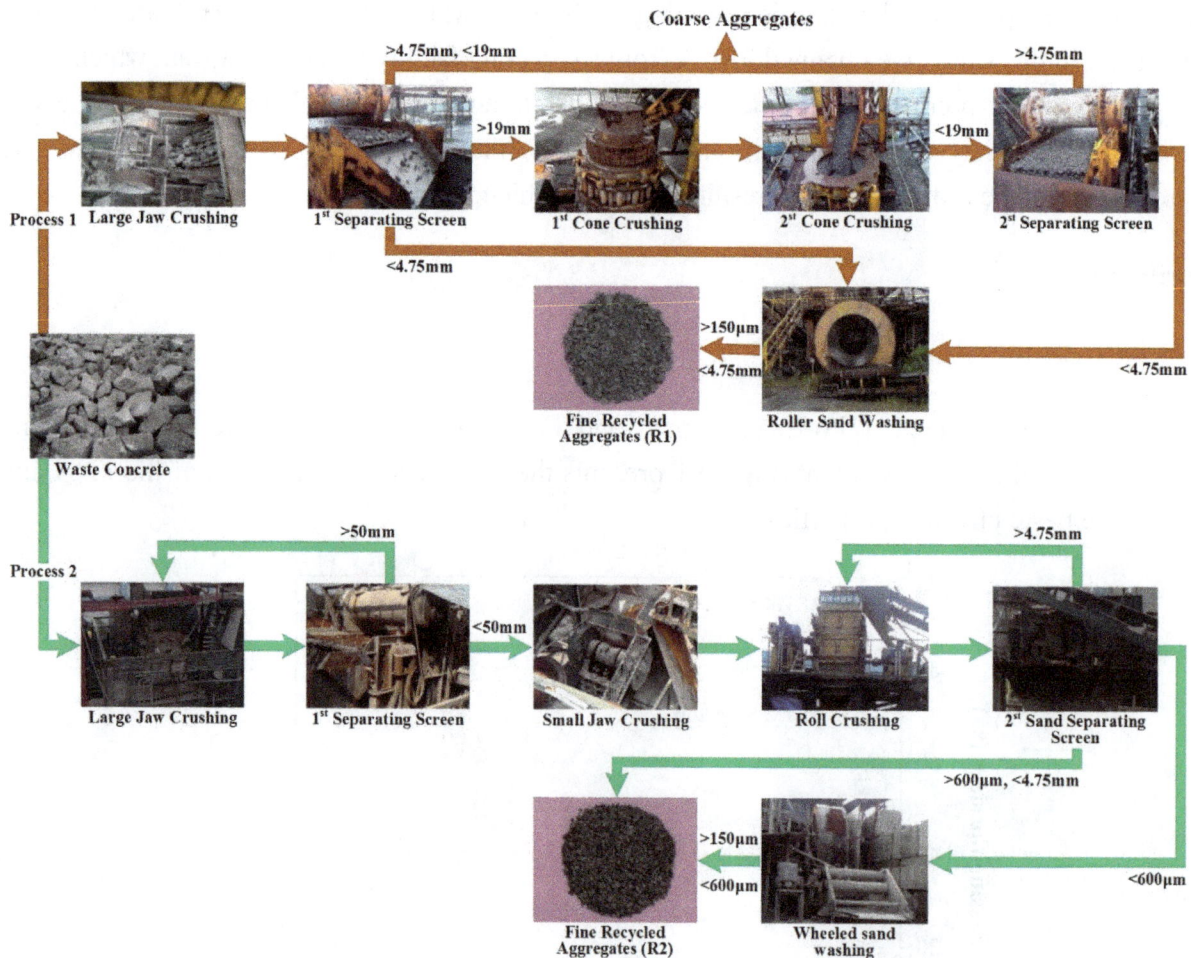

Figure 2. Processes used in the production of fine recycled concrete aggregates.

The other method employed in this study produces only fine aggregate using multiple processes. Concrete waste first undergoes crushing in a large jaw crusher, whereupon the resulting aggregate is separated using a vibrating screen. Fragments exceeding 50 mm are sent back to the large jaw crusher to undergo repeated crushing, and this cycle is repeated until all of the aggregate is less than 50 mm. The resulting material is then crushed using a small jaw crusher and a roll crusher and separated using a vibrating screen. Aggregate larger than 4.75 mm is returned to the roll crusher until it is all smaller than 4.75 mm. Aggregate between 600 μm and 4.75 mm is sent to a fine aggregate stockpile; whereas aggregate smaller than 600 μm is sent to a wheeled sand washer for the removal of sludge smaller than 150 μm. The remaining aggregate between 150 μm and 600 μm is sent to the fine aggregate stockpile. The second type of FRCA produced using the above-mentioned process is denoted as R2, the basic properties of which are also presented in Figure 1 and Table 1.

2.2. Mix Proportions

Table 2 presents the mix proportions adopted for the mortar in this study. The water/cement ratios were set at 0.35 and 0.55; however, the mortar from the former mix displayed poor flowability; thus a superplasticizer (0.5% of the weight of the cement) was added to increase flowability. The replacement level of FNA by R1 and R2 were set at volume fraction of 0%, 25%, 50%, and 100%. As shown in Table 2, the weight ratio for cement and FNA in the control groups is 1:2. Because the densities of R1 and R2 are different from that of FNA, the weights of R1 and R2 in other groups are different from that of FNA used in the control groups, as shown in Table 2.

Table 2. Mix proportions of mortar specimens.

Mix notation	W/C ratio	FRCA content (%)	Mix proportions (kg/m³)					
			Water (kg)	Cement (kg)	FNA (kg)	R1 (kg)	R2 (kg)	Superplasticizer (kg)
AControl	0.35	0	245	700	1400			4
A25R1	0.35	25	245	700	1050	310		4
A50R1	0.35	50	245	700	700	619		4
A100R1	0.35	100	245	700		1239		4
A25R2	0.35	25	245	700	1050		317	4
A50R2	0.35	50	245	700	700		634	4
A100R2	0.35	100	245	700			1269	4
BControl	0.55	0	340	620	1240			-
B25R1	0.55	25	340	620	930	274		-
B50R1	0.55	50	340	620	620	548		-
B100R1	0.55	100	340	620		1097		-
B25R2	0.55	25	340	620	930		281	-
B50R2	0.55	50	340	620	620		562	-
B100R2	0.55	100	340	620			1124	-

2.3. Fabrication of Specimens

The FRCA used in this study presents a higher water absorption than does FNA. We therefore applied pre-wetting to R1 and R2 sample for 24 h. We then adjusted the measurement of surface moisture (ASTM C70-13 [29]) prior to mixing in order to achieve saturated-surface-dry (SSD) conditions using the water compensation method. Mixing was performed according to set proportions and the resulting mortar for each mixture was used to produce the following: six cylindrical specimens with a diameter of 100 mm and height of 50 mm, three cylindrical specimens with a diameter of 100 mm and height of 200 mm, four $285 \times 25 \times 25$ mm prismatic specimens, and nine $50 \times 50 \times 50$ mm cubic specimens. After casting, the specimens were covered with plastic sheeting to prevent evaporation. The samples stood in the laboratory for 24 h before being de-molded and held in saturated lime water for curing at a mean temperature of 23 ± 2 °C until the time of testing.

2.4. Testing

Various tests were performed to characterize the attributes of the mortar, including flow tests, absorption tests, density tests, drying shrinkage tests, compressive strength tests, and ultrasonic pulse velocity (UPV) tests.

Flow testing was performed in accordance with ASTM C1437-13 [30]. Freshly mixed mortar was poured into a flow mold with a bottom diameter of 100 mm (D_0) on a flow table. The flow mold was then lifted off and the flow table was vibrated. The mean diameter (D_a) of the resulting flow was calculated using results obtained after conducting the test four times, as follows:

$$\text{Flow (\%)} = (D_a - D_0)/D_0 \times 100 \text{ (\%)} \tag{1}$$

Density testing was performed using cylindrical specimens with a diameter of 100 mm and height of 50 mm in accordance with ASTM C642-13 [31]. The specimens were weighed at 28 days in SSD condition (W_s). The specimens were weighed while suspended in boiling water 5 h (W_a) and again after being removed from the water that had cooled to 25 °C (W_b). The density of the specimens was calculated using the following formula.

$$\text{Density (kg/m}^3\text{)} = [W_s/(W_b - W_a)] \times 1{,}000 \text{ (kg/m}^3\text{)} \tag{2}$$

Absorption testing was performed using cylindrical specimens with a diameter of 100 mm and height of 50 mm, in accordance with ASTM C642-13 [31]. Specimens at 28 days were first placed in an oven at 105 ± 5 °C and dried until achieving constant weight (W_d). They were then soaked in water to achieve the SSD condition before being weighted (W_s). The water absorption was calculated using the following formula:

$$\text{Absorption (\%)} = (W_s - W_d)/W_d \times 100 \text{ (\%)} \tag{3}$$

Drying shrinkage was tested on $285 \times 25 \times 25$ mm prismatic specimens in accordance with ASTM C596-09 [32]. We first measured the initial length (L_i), which is the difference between the comparator reading of the specimen and the reference bar at 3 days. The specimens were then placed in a chamber under relative humidity of $50\% \pm 4\%$ at a temperature of 23 ± 2 °C for curing. The length of the specimens was then measured at 7 days, 14 days, 21 days, and 28 days (L_x). The extent of drying shrinkage was calculated as follows:

$$\text{Drying shrinkage} = (L_i - L_x)/G \tag{4}$$

where G is the nominal effective length, 250 mm.

Compressive strength was tested on $50 \times 50 \times 50$ mm cube specimens in accordance with ASTM C109-13 [33]. Specimens were retrieved, dried, and tested at 7 days, 14 days, and 28 days to gauge compressive strength.

The UPV test was performed on cylindrical specimens with a diameter of 100 mm and height of 200 mm in accordance with ASTM C597-09 [34]. The measurement device used in this test was the Pundit Plus, manufactured by CNS Farnell Limited. Converters were placed at both ends of the specimens at 28 days, and the ultrasonic frequency was set at 54 kHz. Wave velocities were measured twice and then averaged to obtain the UPV value.

3. Results and Discussion

3.1. Properties of Fine Recycled Concrete Aggregates

Figure 1 presents the particle size distribution curves for FNA, R1, and R2, as well as the distribution range deemed acceptable in ASTM C33-13 [26]. The particle size distribution curves of all samples fell within the acceptable range. R1 was produced in a single stage crushing process and R2 was produced using multi-stage crushing. Thus, R2 contains a larger quantity of finer particles, which places the particle size distribution curve of R2 above that of R1.

Figure 3 illustrates the appearance of FNA, R1, and R2. Clearly, particles in R1 are rougher in shape and more angular than those in R2 as well as more grayish in color. The difference in shape can be attributed to the repeated crushing and lack of coarse aggregate in the production of R2, such that it contains a higher percentage of fine aggregate. With regard to color, both materials were produced from concrete waste and thus had cement paste adhered to the larger fragments, as shown in Figure 4. We can therefore assume that the difference in color is due to a higher percentage of cement paste in R1 (more grayish in color) than that found in R2.

Figure 3. Comparison of appearances and particle size distributions of FNA, R1 and R2.

Figure 4. Microscopic observation of fine recycled concrete aggregates.

The attributes in Table 1 show that FRCAs (R1 and R2) have a lower density and a higher water absorption than does FNA. In addition, R1 has lower density and higher water absorption than does R2. The porosity of cement paste in FRCAs is higher than that of FNA; therefore, FRCAs the density is lower, and water absorption is higher [13–15,18–20,22,23,35]. As mentioned above, R1 contains a larger amount of cement paste than does R2 and therefore has lower density and higher water absorption. For the same reason, the fineness modulus of R1 exceeds that of R2, as shown in Table 1. These findings clearly demonstrate how the production process can influence the basic physical properties of FRCA.

3.2. Flowability

In accordance with the formulas in ASTM C1437-13 [30], we employed a flow table with a diameter 254 mm, such that the maximum flow value would be 154% (*i.e.*, (254 − 100)/100 × 100%). In addition, all of the mortar mixtures had a water/cement ratio of 0.55. Regardless of the amount of FRCA in the freshly mixed mortars, they all presented good flow values exceeding 154%.

Figure 5 presents the flow values of freshly mixed mortar with a water/cement ratio of 0.35. All of the mortar mixtures containing FRCA presented lower flow values than did the control groups. Furthermore, the flow values decreased with an increase in replacement ratio; *i.e.*, the proportion of FNA substitution with FRCA. With the same replacement ratio, samples that included R1 presented lower flow values than did those that included R2. This can be attributed to the fact that R1 fragments have a rougher surface texture and greater angularity, which increases the friction among the particles. It should be noted that all of the flow values obtained from mixtures with a water/cement ratio of 0.55 presented flow values exceeding 154% (data not shown).

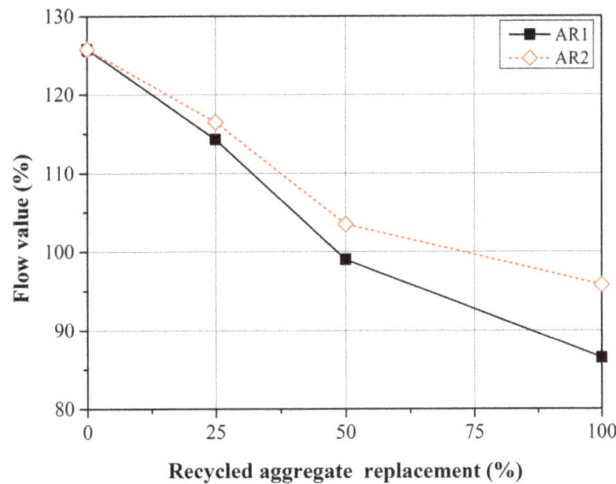

Figure 5. Flow values of mortar with W/C = 0.35.

3.3. Density

Figure 6 displays the mean density values from three mortar specimens at 28 days. All of the specimens containing FRCA had lower densities than did the control groups. In addition, the density of specimens containing FRCA decreased with an increase in replacement ratio. These findings are in agreement with those obtained in previous studies [14,16,18]. With the same replacement ratio, R1 samples present lower density values than do R2 samples. As shown in Table 1, the fact that the density of FRCAs is lower than that of FNA means that the density values of mixtures prepared using FRCA will also be lower. Increasing the replacement ratio also means that a higher proportion of FRCA leads to a reduction in the density of the mortar. The density of R1 is lower than that of R2; therefore with the same replacement ratio, the density of the mortars produced using R1 will also be lower than those produced using R2.

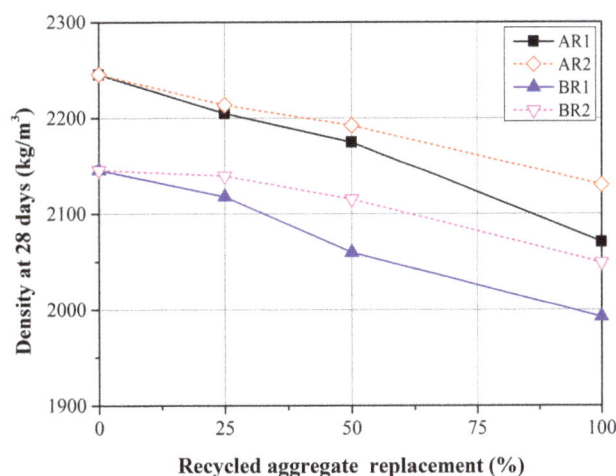

Figure 6. Density of mortars with R1 and R2 replacement at 28 days.

3.4. Absorption

Figure 7 presents the mean water absorption from three mortar specimens at 28 days. The water absorption of the mortar mixtures with FRCA exceeded that of the control groups. Furthermore, the absorption increased with the replacement ratio. With the same replacement ratio, the absorption of mortar mixtures produced using R1 exceeded those produced using R2. This can attributed to the fact that the water absorption of FRCA is higher than that of FNA. These findings are in agreement with those of a previous study [13].

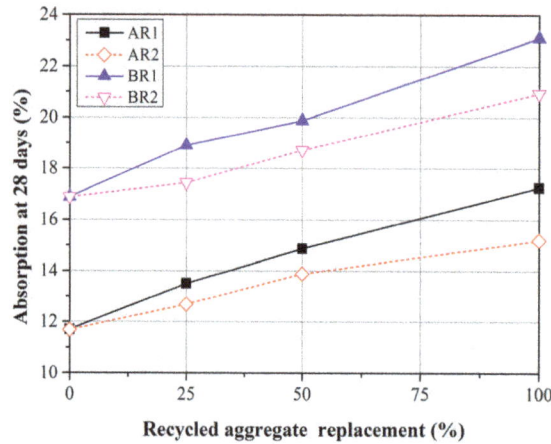

Figure 7. Absorption of mortars with R1 and R2 replacement at 28 days.

3.5. Drying Shrinkage

Figure 8 present the mean drying shrinkage from four mortar specimens with water/cement ratios of 0.35 and 0.55, respectively, at 7 days, 14 days, 21 days, and 28 days. All of the specimens containing FRCA present higher drying shrinkage than did the control groups. This can be explained by the fact that the higher porosity of FRCA enables water to evaporate more rapidly. Increasing the replacement ratio also led to an increase in the drying shrinkage. With the same replacement ratio, the mortar containing R1 presented higher drying shrinkage than do the samples with R2. This can be attributed to the fact that the porosity of R1 exceeds that of R2.

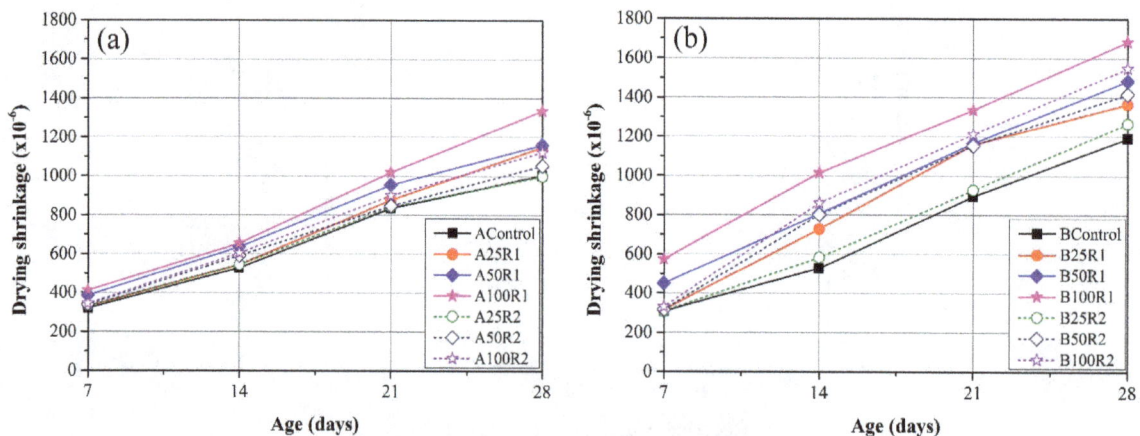

Figure 8. Development of drying shrinkage of mortars with (**a**) W/C = 0.35, and (**b**) W/C = 0.55.

3.6. Compressive Strength

The compressive strength of mortar specimens was tested at 7 days, 14 days, and 28 days. Figure 9 illustrates the development of compressive strength in specimens with water/cement ratios of 0.35 and

0.55, respectively. These values are the mean obtained from three specimens of each mortar mixture. The compressive strength of the specimens containing FRCA is lower than that in the control groups. This can be attributed to the fact that FRCA contains cement paste, which has greater porosity and therefore less compressive strength. Furthermore, compressive strength was shown to decrease with the replacement ratio of FRCA. These findings are in agreement with those obtained in previous studies [16,21]. In specimens containing the same proportions of R1 and R2, the R1 specimens presented less compressive strength due to the higher proportion of cement paste in the R1.

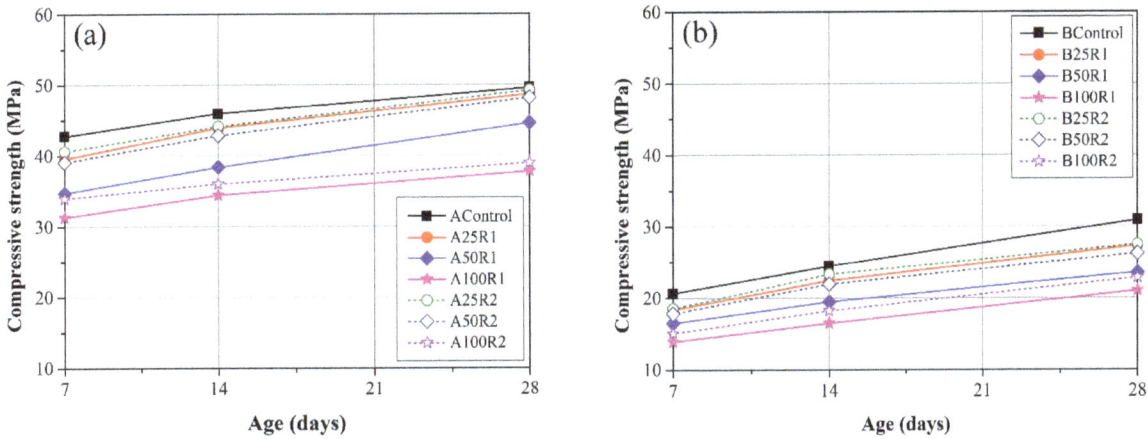

Figure 9. Development of compressive strength in mortars with (**a**) W/C = 0.35, and (**b**) W/C = 0.55.

3.7. Ultrasonic Pulse Velocity (UPV)

Figure 10 presents the mean UPV values from three mortar specimens at 28 days. All of the specimens containing FRCA presented lower UPV values than did the control groups and UPV values decreased with an increase in replacement ratio. These findings are in agreement with those obtained in [16]. Mortar containing R1 had lower UPV values than did the mortar containing R2. The porosity of mortars containing FRCA exceeded that of mortars containing FNA. Porosity inhibits the conduction of ultrasonic pulses, such that a higher FRCA content would lead to greater porosity, which would translate into lower UPV values.

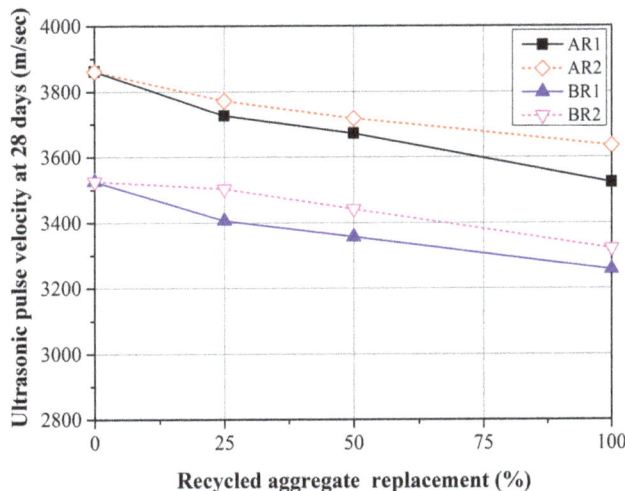

Figure 10. UPV values of mortars with R1 and R2 replacement at 28 days.

Figure 11 presents regression analysis of the UPV values and compressive strength of mortars at 28 days. The resulting regression formula with a R2 value of 0.9295 is $y = 0.0578x - 169.68$, where x and y denote the UPV value and compressive strength, respectively. Regression analysis results indicate that compressive strength increases with an increase in UPV value.

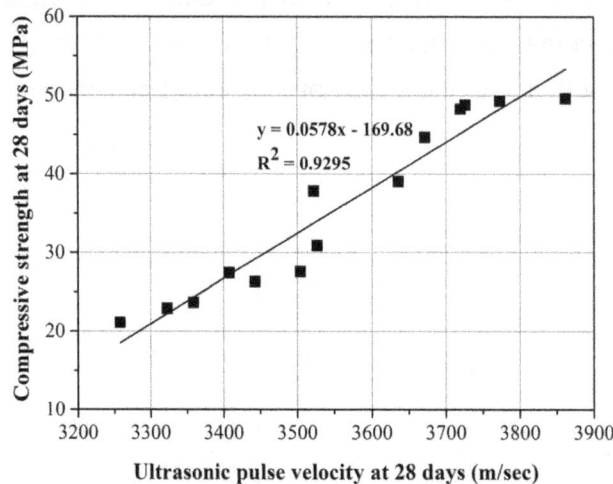

Figure 11. Correlation between compressive strength and UPV values of mortars at 28 days.

4. Conclusions

This study investigated the use of two types of fine recycled concrete aggregate (R1 and R2) obtained using different production processes at a recycling facility in Yilan, Taiwan. R1 was produced using a process that simultaneously produces coarse as well as fine aggregate by crushing concrete waste with a large jaw crusher. R2 was produced using a process that produces only fine aggregate through repeated crushing of concrete waste. We then tested mortar specimens in which various proportions of FRCA were substituted for FNA. This led to the following conclusions:

1. R2 has lower porosity, higher density, and lower water absorption than does R1, all of which are indicators of the superior quality of R2. This also demonstrates that the crushing process can significantly influence the quality of the resulting FRCA.

2. In all of the mortars containing FRCA, an increase in the replacement ratio led to a reduction in flow values, density, compressive strength, and UPV values. This is a clear demonstration that the replacement ratio is an important factor influencing the physical and mechanical properties of the resulting mortar.

3. When comparing mortars containing R1 or R2 at the same replacement ratio, the mortar containing R1 presented a lower flow value, lower density, higher absorption, higher drying shrinkage, lower compressive strength, and lower UPV values. This demonstrates that mortars containing R1 cannot match the physical or mechanical properties of R2, and further demonstrates the importance of the crushing process used in the production of FRCA.

4. Our results demonstrate the superiority of mortars produced using aggregate processed from recycled concrete via multistage crushing to obtain only FRCA. Nonetheless, the performance benefits of using only fine recycled concrete aggregate must be balanced against the additional energy requirements and subsequent impact on the environment.

Author Contributions

Cheng-Chih Fan, Howard Hwang and Sao-Jeng Chao designed experiments, analyzed the data and wrote the manuscript. Ran Huang supervised the project.

Conflicts of Interest

The authors declare no conflict of interest.

References

1. Fonseca, N.; de Brito, J.; Evangelista, L. The influence of curing conditions on the mechanical performance of concrete made with recycled concrete waste. *Cement Concrete Compos.* **2011**, *33*, 637–643.
2. Choi, W.-C.; Yun, H.-D. Compressive behavior of reinforced concrete columns with recycled aggregate under uniaxial loading. *Eng. Struct.* **2012**, *41*, 285–293.
3. Ismail, S.; Ramli, M. Engineering properties of treated recycled concrete aggregate (RCA) for structural applications. *Constr. Build. Mater.* **2013**, *44*, 464–476.
4. Ma, H.; Xue, J.; Zhang, X.; Luo, D. Seismic performance of steel-reinforced recycled concrete columns under low cyclic loads. *Constr. Build. Mater.* **2013**, *48*, 229–237.
5. Huda, S.B.; Alam, M.S. Mechanical behavior of three generations of 100% repeated recycled coarse aggregate concrete. *Constr. Build. Mater.* **2014**, *65*, 574–582.
6. Pedro, D.; de Brito, J.; Evangelista, L. Influence of the use of recycled concrete aggregates from different sources on structural concrete. *Constr. Build. Mater.* **2014**, *71*, 141–151.
7. Soares, D.; de Brito, J.; Ferreira, J.; Pacheco, J. Use of coarse recycled aggregates from precast concrete rejects: Mechanical and durability performance. *Constr. Build. Mater.* **2014**, *71*, 263–272.
8. Thomas, C.; Setién, J.; Polanco, J.A.; Lombillo, I.; Cimentada, A. Fatigue limit of recycled aggregate concrete. *Constr. Build. Mater.* **2014**, *52*, 146–154.
9. Brand, A.S.; Roesler, J.R.; Salas, A. Initial moisture and mixing effects on higher quality recycled coarse aggregate concrete. *Constr. Build. Mater.* **2015**, *79*, 83–89.
10. Huda, S.B.; Shahria Alam, M. Mechanical and freeze-thaw durability properties of recycled aggregate concrete made with recycled coarse aggregate. *J. Mater. Civ. Eng.* **2015**, 04015003, doi:10.1061/(ASCE)MT.1943-5533.0001237.
11. Manzi, S.; Mazzotti, C.; Bignozzi, M.C. Short and long-term behavior of structural concrete with recycled concrete aggregate. *Cement Concrete Compos.* **2013**, *37*, 312–318.
12. Evangelista, L.; de Brito, J. Mechanical behaviour of concrete made with fine recycled concrete aggregates. *Cement Concrete Compos.* **2007**, *29*, 397–401.
13. Evangelista, L.; de Brito, J. Durability performance of concrete made with fine recycled concrete aggregates. *Cement Concrete Compos.* **2010**, *32*, 9–14.
14. Pereira, P.; Evangelista, L.; de Brito, J. The effect of superplasticisers on the workability and compressive strength of concrete made with fine recycled concrete aggregates. *Constr. Build. Mater.* **2012**, *28*, 722–729.

15. Pereira, P.; Evangelista, L.; de Brito, J. The effect of superplasticizers on the mechanical performance of concrete made with fine recycled concrete aggregates. *Cement Concrete Compos.* **2012**, *34*, 1044–1052.

16. Khatib, J.M. Properties of concrete incorporating fine recycled aggregate. *Cement Concrete Res.* **2005**, *35*, 763–769.

17. Shui, Z.; Xuan, D.; Wan, H.; Cao, B. Rehydration reactivity of recycled mortar from concrete waste experienced to thermal treatment. *Constr. Build. Mater.* **2008**, *22*, 1723–1729.

18. Neno, C.; de Brito, J.; Veiga, R. Using fine recycled concrete aggregate for mortar production. *Mater. Res.* **2013**, doi:10.1590/S1516-14392013005000164.

19. Khoshkenari, A.G.; Shafigh, P.; Moghimi, M.; Mahmud, H.B. The role of 0–2 mm fine recycled concrete aggregate on the compressive and splitting tensile strengths of recycled concrete aggregate concrete. *Mater. Design* **2014**, *64*, 345–354.

20. Lee, S.T. Influence of recycled fine aggregates on the resistance of mortars to magnesium sulfate attack. *Waste Manag.* **2009**, *29*, 2385–2391.

21. Sim, J.; Park, C. Compressive strength and resistance to chloride ion penetration and carbonation of recycled aggregate concrete with varying amount of fly ash and fine recycled aggregate. *Waste Manag.* **2011**, *31*, 2352–2360.

22. Florea, M.V.A.; Brouwers, H.J.H. Properties of various size fractions of crushed concrete related to process conditions and re-use. *Cement Concrete Res.* **2013**, *52*, 11–21.

23. Ulsen, C.; Kahn, H.; Hawlitschek, G.; Masini, E.A.; Angulo, S.C.; John, V.M. Production of recycled sand from construction and demolition waste. *Constr. Build. Mater.* **2013**, *40*, 1168–1173.

24. Song, I.H.; Ryou, J.S. Hybrid techniques for quality improvement of recycled fine aggregate. *Constr. Build. Mater.* **2014**, *72*, 56–64.

25. Koshiro, Y.; Ichise, K. Application of entire concrete waste reuse model to produce recycled aggregate class H. *Constr. Build. Mater.* **2014**, *67*, 308–314.

26. ASTM International. *Standard Specification for Concrete Aggregates*; ASTM C33M-13; ASTM International: West Conshohocken, PA, USA, 2013.

27. ASTM International. *Standard Test Method for Density, Relative Density (Specific Gravity), and Absorption of Fine Aggregate*; ASTM C128-12; ASTM International: West Conshohocken, PA, USA, 2012.

28. ASTM International. *Standard Test Method for Sieve Analysis of Fine and Coarse Aggregates*; ASTM C136M-14; ASTM International: West Conshohocken, PA, USA, 2014.

29. ASTM International. *Standard Test Method for Surface Moisture in Fine Aggregate*; ASTM C70-13; ASTM International: West Conshohocken, PA, USA, 2013.

30. ASTM International. *Standard Test Method for Flow of Hydraulic Cement Mortar*; ASTM C1437-13; ASTM International: West Conshohocken, PA, USA, 2013.

31. ASTM International. *Standard Test Method for Density, Absorption, and Voids in Hardened Concrete*; ASTM C642-13; ASTM International: West Conshohocken, PA, USA, 2013.

32. ASTM International. *Standard Test Method for Drying Shrinkage of Mortar Containing Hydraulic Cement*; ASTM C596-09; ASTM International: West Conshohocken, PA, USA, 2009.

33. ASTM International. *Standard Test Method for Compressive Strength of Hydraulic Cement Mortars (Using 2-in. or [50-mm] Cube Specimens)*; ASTM C109M-13; ASTM International: West Conshohocken, PA, USA, 2013.

34. ASTM International. *Standard Test Method for Pulse Velocity Through Concrete*; ASTM C597-09; ASTM International: West Conshohocken, PA, USA, 2009.

35. Kou, S.C.; Poon, C.S. Properties of self-compacting concrete prepared with coarse and fine recycled concrete aggregates. *Cement Concrete Compos.* **2009**, *31*, 622–627.

11

How Properties of Kenaf Fibers from Burkina Faso Contribute to the Reinforcement of Earth Blocks

Younoussa Millogo [1,2], Jean-Emmanuel Aubert [3,†], Erwan Hamard [4,†] and Jean-Claude Morel [5,*]

[1] Unité de Formation et de Recherche en Sciences et Techniques (UFR/ST), Université Polytechnique de Bobo-Dioulasso, 01 BP 1091 Bobo 01, Burkina Faso; E-Mail: millogokadi@gmail.com

[2] Laboratoire de Chimie Moléculaire et de Matériaux (LCMM), UFR/Sciences Exactes et Appliquées, Université de Ouagadougou, 03 BP 7021 Ouagadougou 03, Burkina Faso

[3] Université de Toulouse, UPS, INSA, LMDC (Laboratoire Matériaux et Durabilité des Constructions), 135 avenue de Rangueil, F-31077 Toulouse cedex 4, France; E-Mail: jean-emmanuel.aubert@univ-tlse3.fr

[4] Institut Français des Sciences et Technologies des Transports, de l'Aménagement et des Réseaux, Département Matériaux, GPEM, route de Bouaye, 44344 Bouguenais, CS4, France; E-Mail: erwan.hamard@ifsttar.fr

[5] Ecole Nationale des Travaux Publics de l'Etat, Université de Lyon CNRS-LTDS, UMR 5513, LGCB, 3 rue Maurice Audin, Vaulx-en-Velin cedex, F-69120, France

[†] These authors contributed equally to this work.

[*] Author to whom correspondence should be addressed; E-Mail: morel@entpe.fr;

Academic Editor: Geminiano Mancusi

Abstract: Physicochemical characteristics of Hibiscus cannabinus (kenaf) fibers from Burkina Faso were studied using X-ray diffraction (XRD), infrared spectroscopy, thermal gravimetric analysis (TGA), chemical analysis and video microscopy. Kenaf fibers (3 cm long) were used to reinforce earth blocks, and the mechanical properties of reinforced blocks, with fiber contents ranging from 0.2 to 0.8 wt%, were investigated. The fibers were mainly composed of cellulose type I (70.4 wt%), hemicelluloses (18.9 wt%) and lignin (3 wt%) and were characterized by high tensile strength (1 ± 0.25 GPa) and Young's modulus (136 ± 25 GPa), linked to their high cellulose content. The incorporation of short fibers of kenaf reduced the propagation of cracks in the blocks, through the good adherence of fibers to the clay matrix, and therefore improved their mechanical properties.

Fiber incorporation was particularly beneficial for the bending strength of earth blocks because it reinforces these blocks after the failure of soil matrix observed for unreinforced blocks. Blocks reinforced with such fibers had a ductile tensile behavior that made them better building materials for masonry structures than unreinforced blocks.

Keywords: kenaf fibers; Earth blocks; physicochemical characteristics; mechanical properties; Burkina Faso

---

1. Introduction

Hibiscus cannabinus, or kenaf, is a plant of the Malvaceae family that grows in tropical and sub-tropical areas. It is an annual herbaceous plant (or, rarely, a short-lived perennial) growing to 1.5 to 3.5 m tall, with a woody base. The stems are 1 cm to 2 cm in diameter and often but not always branched. The leaves are 10–15 cm long and variable in shape, with leaves near the base of the stems being deeply lobed with 3–7 lobes, while leaves near the top of the stem are weakly lobed. In Burkina Faso, kenaf leaves are used to prepare sauces. The flowers are 8–15 cm diameter, white, yellow or purple; when white or yellow, the center is still dark purple. The fruit is a 2-cm-diameter capsule containing several seeds. Because of their high mechanical strength, kenaf fibers are usually used in West Africa to manufacture ropes and sacks. In Burkina Faso, some local populations also use the fibers to make masks for traditional ceremonies.

In West Africa, before the introduction of industrial materials such as steel and concrete, adobes were traditionally stabilized or reinforced by locally available organic matter such as plant fibers, plant decoctions and cow dung. Many studies have dealt with the physical and mechanical characteristics of adobes stabilized or reinforced with natural fibers [1–10] but little attention has been paid to how the physicochemical characteristics of the incorporated fibers affect the physical and mechanical properties of the adobes elaborated. Moreover, although the chemical compositions of kenaf fibers from some countries such as Malaysia, Thailand, India, China, the southern United States, Mexico and Korea have been reported in the literature, this is not the case for kenaf fibers from Burkina Faso [11–13] and it is well known that the chemical composition of these fibers is mainly linked to the climate, plant species and type of soil.

The aim of this work is to study the influence of the physicochemical and mechanical characteristics of kenaf fibers from Burkina Faso on the mechanical behavior of adobes reinforced with these fibers. The final main objective is to valorize the fibers of the Kenaf plant in the building materials sector as it is easily produced in Burkina Faso, where it is abundant and cheap.

2. Materials and Testing Procedures

2.1. Raw Materials

The kenaf plant fibers were extracted near Bobo-Dioulasso (village of Farakoba) in the west of Burkina Faso. The extraction was done by hand, by pulling the fibers from the plant stalk. The Kenaf plant and extracted fibers are presented in Figure 1. The raw material used to manufacture adobes was a clayey soil from Rochechinard, a site located in the Isère valley (France). Its geotechnical characteristics have been published in a previous paper [14]. This raw material is composed of kaolinite (45 wt%), quartz (23 wt%), illite (14 wt%), goethite (7 wt%) and calcite (4 wt%) [15].

(a)   (b)

Figure 1. Pictures of kenaf plant (**a**) and extracted fibers (**b**).

2.2. Procedures

2.2.1. Physicochemical, Mineralogical and Mechanical Characteristics of Fibers

The fibers were considered as circular and their diameter was measured using a caliper with a precision of 0.01 mm. The natural humidity of the fibers was determined by drying fresh fibers in an oven at 105 °C for 24 h. The water absorption (W) was calculated on fibers soaked drinking water for 24 h [16].

To assess the mineralogical composition of the fibers, X-ray diffraction, thermal gravimetric analysis and Fourier transform infrared (FTIR) spectroscopy were performed on a crushed sample (size < 80 μm).

The thermal gravimetric analysis of the sample was performed at a constant heating rate of 10 °C/min. The X-ray diffraction apparatus used was a Siemens D5000 power X-ray Diffractometer equipped with a monochromator using a Kα ($\lambda = 1.789$ Å) cobalt anticathode. The thermal gravimetric curve of the fibers was obtained with a Netzsch SATA 449 F3 Jupiter apparatus and the infrared spectrum was obtained with a Nicolet 510FT-IR spectrometer operating in the range 4000−400 cm^{-1}. A JEOL 6380 LV scanning electron microscope (JEOL, Croissy Sur Seine, France) equipped with a backscattered electron (BSE) detector was used for SEM observations on the fibers and the study of the morphology of the fibers was completed using an area video microscope, Keyence VH-5911 (Keyence, Osaka, Japan).

The fibers for chemical analysis were reduced to powder in a miller. The experimental technique used was the Van Soest procedure using four extracted detergents as NDS (Neutral Detergent Soluble), NDF (Neutral Detergent Fiber), ADF (Acid Detergent Fiber) and ADL (Acid Detergent Lignin) in order to quantify the amount of cellulose, hemicelluloses, lignin and ash of the fibers [17,18].

Cellulose is a linear polymer of β-(1-4)-D-glucopyranode. It exists in five types (I, II, III, IV and V). Type I is a native cellulose and has the best physical and mechanical properties (Young's modulus of 150 GPa). Cellulose is a crystalline polysaccharide. Hemicellulose is composed of different types of saccharides such as xylose, mannose and glucose. It is strongly bound to the cellulose fibrils by hydrogen bonds [16]. Lignins are amorphous polymers formed by aromatic units such as guaiacycle, syringyl and phenylpropane. It acts as a cementing agent, binding the cellulose fibers together.

The useful mechanical characteristics of the fibers are limited to their tensile behavior, which is classical for fibers (no shear and no compressive strength). The tensile behavior has already been described elsewhere [15].

2.2.2. Preparation and Mechanical Characterization of Earth Blocks

Earth blocks were prepared using the technique of Pressed Adobe Blocks (PABs) [14]. The soil used for the manufacture of PABs was sieved to obtain particles with dimensions less than 5 mm. Then the prepared soil sample was mixed with 30-mm-long pieces of fiber for up to 0.8 wt% in relation to the dry weight of the soil. The average water/dry soil ratio used to manufacture the reinforced samples was 20 wt%. The soil composite was mixed for 15 min until it became a homogeneous paste. The different pastes were introduced into the rectangular prism mold ($295 \times 140 \times 100$ mm^3) of a GEO 50 manual press to produce the pressed soil blocks. The pressure applied during compaction was approximately 2 MPa. The as-molded PABs were dried in the laboratory at room temperature (average 22 °C), with a humidity of 60% for an average period of 3 weeks until the block weight became constant. The unconfined compression test was performed on samples standing on their smallest faces in order to reduce the friction in the central part of the sample (Figure 2). It was very important for the test to be carried out in unconfined conditions in order to validate the assumption that the stress tensor was constant in the central part of the sample. This assumption enabled the stress to be calculated from the force measured by the press sensor. The details of the procedure are given elsewhere [19–21].

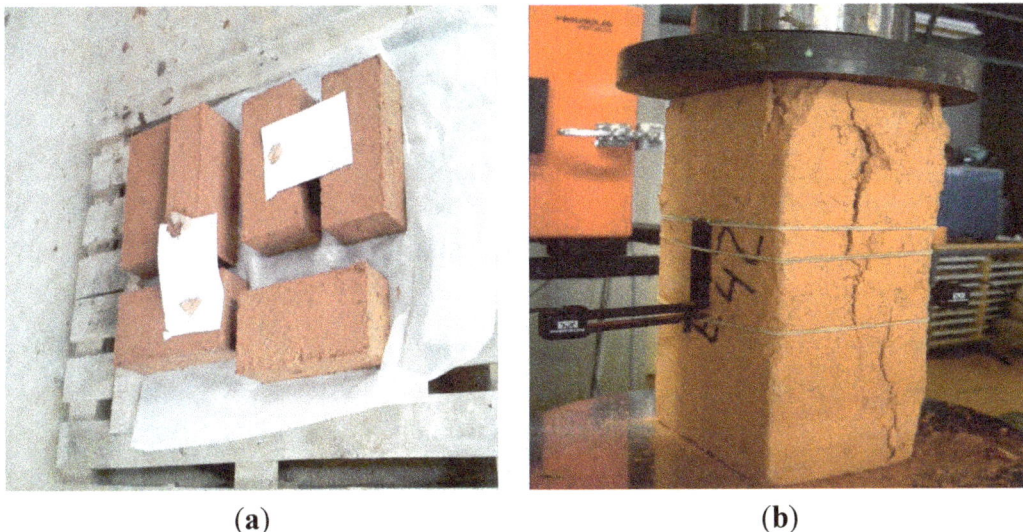

(a)   (b)

Figure 2. Compression test samples before test (**a**), sample after test with an extensometer in the central part of the sample measuring the strain (**b**).

Extensometers placed in the central part of the sample measured the strain necessary for the calculation of Young's modulus. It is now well known that earth materials have a very small elastic domain and that plasticity begins before 20% of the compressive strength [22,23]. As it was difficult to assess small strains at the beginning of the test, it was decided to work in cycles to measure the elasticity modulus. The procedure is not a standard one but details are given elsewhere [19–23].

The three-point bending strength tests were carried out using an INSTRON hydraulic press on the PAB specimens. The test was conducted at a constant speed of 0.01 mm·s^{-1}. The load sensor used had a capacity of 50 kN. The mechanical tests have been carried out following the procedures given in previous papers [24,25].

3. Results and Discussion

3.1. Physicochemical, Mineralogical and Mechanical Characteristics of Fibers

The X-ray diffraction pattern of crushed fibers is presented in Figure 3.

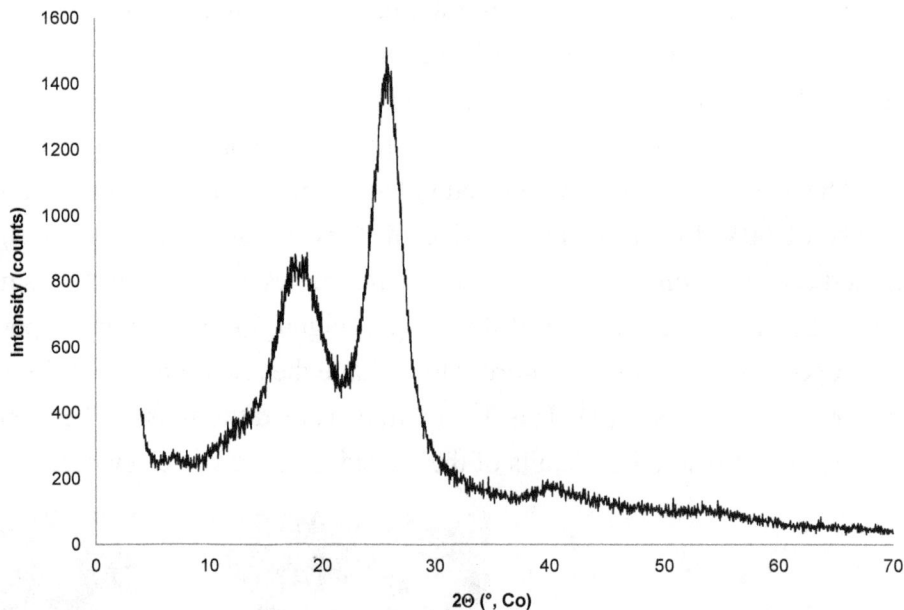

Figure 3. X-ray diffraction pattern of the fibers.

This pattern shows the presence of crystallized native cellulose with broad peaks located at 17.68°, 18.68°, 25.8°, 39.84° and 40.74°. The peaks at 17.68° and 18.68° are individualized in the X-ray diffraction pattern. This result indicates a significant amount of cellulose I in the studied fibers and is the same as those obtained with kenaf fibers from Malaysia and Korea [11,12]. The main peak at 39.84° in 2θ is not a doublet; it proves that the cellulose contained in the above fibers is cellulose I and this confirms its high Young's modulus. In order to study the crystallinity of the cellulose contained in the fibers, the crystalline index of cellulose (I_c) was evaluated using Segal's method with the equation: $I_c = (I_{002}-I_{am})/I_{002} \times 100$ where I_{002} is the intensity of the XRD peak at $2\theta = 25.8°$ and I_{am} the peak at $2\theta = 18.68°$ [12]. The value obtained was 43. This low value is in the range of values reported in another paper on kenaf fibers [12] and is linked to the presence of amorphous compounds such as hemicelluloses and lignin in the raw fibers, which decrease the crystalline index.

To complete the mineralogical composition, and especially the characterization of amorphous compounds, the crushed fibers were analyzed by thermal gravimetric analysis (TGA) and Fourier transform infrared (FTIR) spectroscopy. Figure 4 shows the TGA curves for the sample.

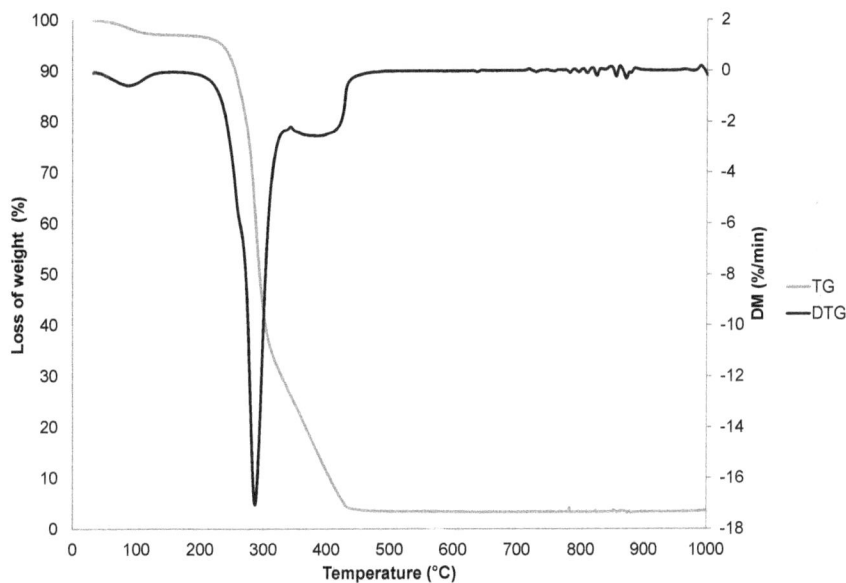

Figure 4. TG and DTG curves of the fibers.

The thermal characterization of the fibers during their sintering was carried out as described in the works of El-Shekeil *et al.* [26] and Morān *et al.* [27]. The first mass loss step around 100 °C was due to water loss from the fibers. The small mass loss step around 280 °C is a shoulder expressing the decomposition of thermally unstable compounds such as non-structural hemicelluloses and concerns the depolymerization of these hemicelluloses. This peak could also express the degradation of lignin. The major decomposition peak around 300 °C was attributed to cellulose decomposition. This large mass loss indicated that cellulose was a major compound of the fibers. The non-negligible peak around 400 °C is probably attributable to the structural hemicelluloses, lignin and cellulose which were pyrolyzed at this temperature. The attribution of this peak to lignin decomposition is supported by the fact that the lignin macromolecule is heat resistant. Lignin is thermally stable because, in its molecular structure, there is the possibility to form hydrogen bonds, which are known to stabilize the molecules. The thermal phenomenon could also be due to oxidative degradation of the charred residue.

The FTIR spectrum of the fibers is presented in Figure 5.

The interpretations of this spectrum were supported by the works of Ivanova *et al.* [28], Morān *et al.* [27] and Shin *et al.* [11]. The broad band around 3400 cm^{-1} corresponds to the stretching of the O-H bond of cellulose molecules. This band also characterizes the stretching of the O-H bond of water absorbed by fibers. The spectrum exhibits C-H stretching around 2900 cm^{-1}. The peak at 1732 cm^{-1} is associated with C=O stretching of the acetyl group in hemicelluloses. This peak can be attributed to the p-coumeric acids of lignin or hemicelluloses. It could also express the presence of pectin in the fibers. The presence of proteins in the fibers is proved by the 1633 cm^{-1} and 1596 cm^{-1} peaks, which characterize the presence of primary and secondary amides. The peak at 1242 cm^{-1} is due to the C-O stretching of the aryl group in lignin. In the frequency range 800–1200 cm^{-1}, valence vibrations of C-O, C-C and ring structures and external deformational vibrations of CH$_2$, COH, CCO and CCH groups of cellulose are visible.

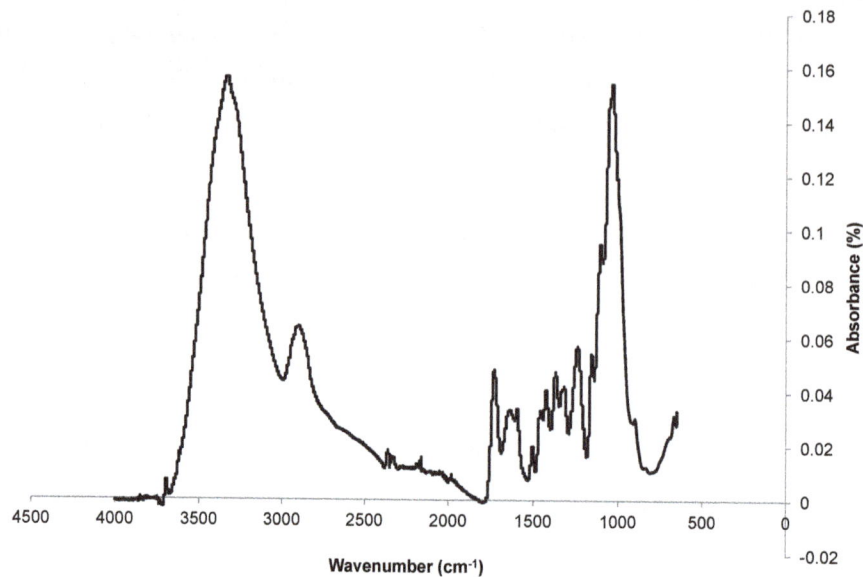

Figure 5. Infrared spectrum of the fibers.

The video-microscopy image and the SEM micrograph of the fibers are presented in Figure 6.

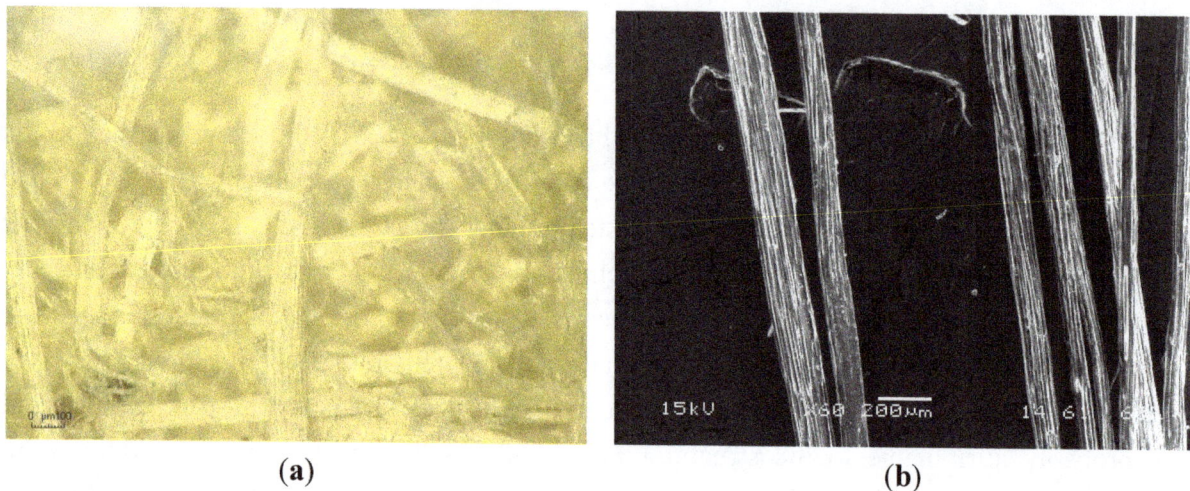

(a)   (b)

Figure 6. Video microscope image (**a**) ×175 and SEM micrograph (**b**) ×60 of the fibers.

The observation of kenaf fibers reveals that they are arranged in beams with a rough surface that would be favorable for adherence with the earth in reinforced adobes. The fiber surfaces are covered by veins and they are surrounded by dark residues which could be pieces of the bark of the plant. The mean estimated diameter obtained with 20 isolated fibers was 60 μm. This value is within the range obtained by measurement with a micrometer. The fibers observed by SEM show a streaked structure with grooves. The isolated fibers are oriented in a single direction. The fiber surfaces are covered with dispersed impurities. According to the work of Jonoobi *et al.* on kenaf fibers [12], these impurities are hemicelluloses, lignin, pectin and waxy substances.

The chemical composition of the studied fibers and other kenaf fibers is presented in Table 1.

Table 1. Chemical composition of various kenaf fibers.

Reference	Cellulose (wt%)	Hemicelluloses (wt%)	Lignin (wt%)	Ash (wt%)
Present Study	70	19	3	1.3
[29]	53 ± 4	18 ± 1.4	8 ± 1.2	-
[13]	45–57	21.5	8–13	-
[12]	58 ± 1	22 ± 1	17.5 ± 1.3	2.4 ± 0.4
[11]	60.8	19.2	14.7	-
[30]	31–39	21.5	15–19	-

The cellulose, hemicelluloses and lignin contents diverge for different kenaf fibers. The cellulose content of the fibers studied here is greater than that of the other kenaf fibers. This disparity may be due to the climate, soil nature and the species of the plant. The hemicelluloses content is in the same range of values, whereas the lignin content is smaller than in any other study.

Table 2 presents the chemical composition of some other natural fibers for comparison with the kenaf fibers.

Table 2. Chemical composition of the fibers studied here compared to other types of fibers [9,13,28,29].

Reference	Cellulose (wt%)	Hemicelluloses (wt%)	Lignin (wt%)
Fibers studied here	70	19	3
Flax	71	19–20.6	2.2
Hemp	70–74	18–22.4	3.7–5.7
Jute	61–71	14–20	12–13
Ramie	68–76	13–17	0.6–0.7
Sisal	63–64	12.0	10–14
Banana	63–64	10	5
Cotton	85–90	5.7	-
Coir	32–43	0.15–0.25	40–45
Cereal straw	38–45	15–31	12–20
OPEFB * fibers	59	2.1	25
Groundnut shell	36	19	30
Bagasse	40–46	24.5–29	12.5–20
Rice husk	31	24	14
Coconut coir	47.7	25.9	17.8

Note: * Oil Palm Empty Fruit Bunch.

The discussion is focused on the cellulose content, which has the most influence on the mechanical characteristics of fibers because of its high tensile strength. The cellulose content of the fibers shown above is almost the same for flax, hemp (a variety of *cannabinus*), jute and ramie. The cellulose content of kenaf fibers is greater than that of the fibers of banana, cereal straw, oil palm empty fruit bunch, sisal, groundnut shell, rice husk, bagasse and coconut coir but cotton plant fibers are richer in cellulose than kenaf [30]. The fact that kenaf fibers have better mechanical properties than sisal, coir, coconut and oil palm empty fruit bunch is explained by their cellulose content.

The physical and mechanical properties of the fibers studied in the present work (diameter, natural humidity, specific weight, water absorption, specific weight, elasticity modulus and tensile strength) are given in Table 3. The diameters and the specific weights of kenaf fibers were in the same range as values for sisal, coconut and Lechuguilla fibers [2,8] but were smaller than those reported for straw fibers and oil palm empty fruit bunch fibers [5,9]. The natural humidity of the fibers was lower but their water absorption was higher than that of sisal, coconut and Lechuguilla fibers [2,8]. Straw fibers presented 500%–600% higher water absorption than kenaf fibers [5].

The tensile strength had a mean value of 1 GPa, with a high standard deviation (around 0.25 GPa) depending on the natural variability of the fibers. The tensile strength of kenaf fibers was higher than the tensile strength of sisal, lechuguilla, coconut, coir and oil palm empty fruit bunch fibers [2,8,9,13]. The higher cellulose content of the kenaf from Burkina Faso (Table 2) may be the reason why the mechanical characteristics of the fibers are higher than those of fibers of other origins. It should not be forgotten that, in the context of local materials, it is always necessary to test the materials to assess their performance [19], because of the variability of their composition depending on the soil, climate and plant species.

Table 3. Physical and mechanical properties of Kenaf fibers.

Properties	Results
Diameter (mm)	0.13
Natural humidity: H (%)	6.1
Water absorption: W (%)	307
Specific weight: γ (g/cm^3)	1.04
Elasticity modulus (MPa)	136 ± 25
Tensile strength: σ (GPa)	1 ± 0.25

The elasticity modulus of kenaf was close to that of native cellulose (type I) and cellulose I could, therefore, be a major component of the fibers studied. The tensile strength was approximately twice that of steel with half the stiffness. These results reveal that kenaf fibers are 4 times more deformable than steel. This deformability is favorable when the fibers are used to reinforce soil for the production of PAB, which is a material with a stiffness approximately 200 times less than steel [14]. This point will be explored in the next section.

3.2. Mechanical Properties of Earth Blocks

To highlight the behavior of PABs within a masonry structure, a stress-strain study was carried out using extensometers in unconfined compression tests with three cycles of loading and unloading (Figure 7). As observed, compressive strength increased with fibers additions. This result is due to the fact that fiber incorporation in the blocks prevents the propagation of cracks through their good adherence to the clay matrix because of their rough surface.

In a previous paper [15], two fiber lengths were tested. It was shown that the 3 cm length was more effective. Therefore only 3-cm-long fibers are considered here. The compressive strength increased to an optimum value near 0.4 wt% and then decreased. This is a classical result in soils reinforced by fiber [30]. At this optimum, the compressive strength increase was 16%. This increase in the compressive strength is small but, in general, the compressive strength of earth blocks is sufficient for buildings a few stories high. What is important is to check that the addition of fibers does not decrease the compressive strength (this is discussed in detail elsewhere [15]) as the addition of fibers may weaken the clay matrix by increasing the void content [30].

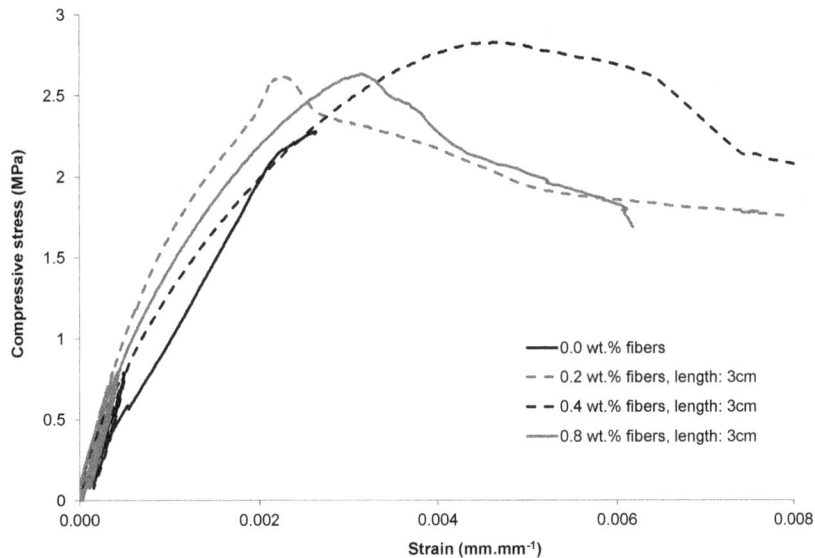

Figure 7. Compressive stress-strain behavior of fiber-reinforced PABs.

Fibers are added to earth for two main reasons: to limit shrinkage cracks during the drying process of the material after manufacture of the blocks, and to improve their tensile ductility. That is why it is necessary to conduct both bending and tensile tests. Figure 8 presents the flexural behavior of fiber-reinforced PABs.

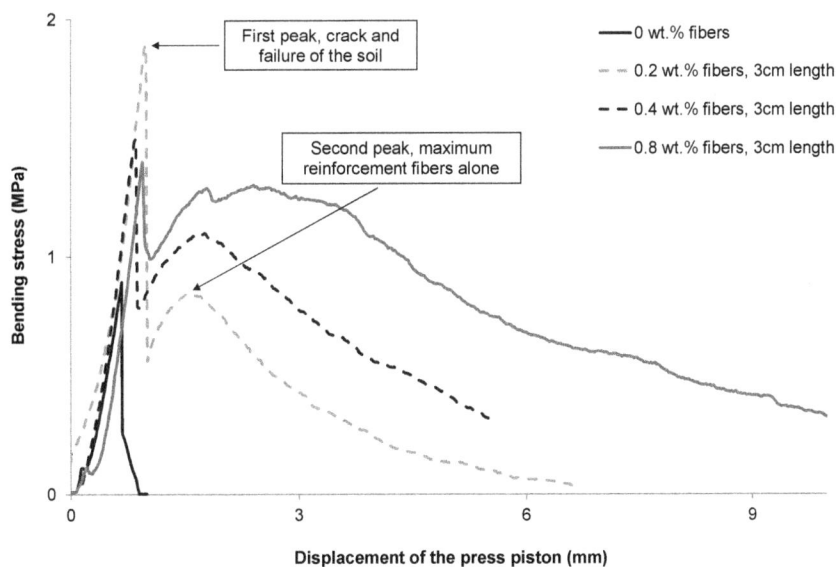

Figure 8. Flexural stress-train behavior of fiber-reinforced PABs, the span length is 28 cm.

As it is shown in Figure 8, fibers incorporation increased the tensile strength of the PABs. This result is due to the good adherence of fibers with clay matrix and mainly their high tensile strength linked to their high content of cellulose (70 wt%), known for its high tensile strength (average 500 MPa). The importance of fiber incorporation is that they reinforce the blocks after the failure of the soil matrix, increasing the ductility of PABs.

The tensile behavior of unreinforced PABs shows only one stress peak, whereas those of PABs reinforced by fibers show two stress peaks [4]. The first stress peak is attributed to the cracking of the clay matrix of the PABs for all samples (reinforced and unreinforced). From the first peak, the residual stress is due to the fibers crossing the cracks that appear with the first peak. This is the new ductile behavior provided solely by the fibers crossing the crack. The area beneath the curves gives the ductility gained by adding fibers. This area is approximately proportional to the fiber content. It can be seen that, whether the fiber content is smaller than or equal to 0.8 wt%, the embedded length of the fibers crossing the crack is enough to mobilize most of the fibers' tensile strength.

However, adding more fibers than 0.8 wt% may weaken the soil matrix and the friction mobilized by the embedded length of the fibers crossing the cracks may decrease, becoming too small to anchor the fibers. The fibers would be pulled out and their tensile strength would not be mobilized.

The second peak is specific to reinforced PABs and gives the maximum average strength of the fibers. To explain this phenomenon, two types of fibers crossing the crack must be considered separately. Let L_{me} being the minimum embedded length necessary to mobilize enough friction to reach the tensile strength of the fibers and let L_e be the embedded length of a fiber (the smallest length from either side of the crack).

In the Figure 8, for a given displacement the following can be set:

Case $L_e < L_{me}$:   (a) fibers with decreasing tensile stress. The fiber is pulled out and its embedded length is decreasing.
(b) Fibers with an increasing tensile stress; the embedded length of these fibers is greater than the fibers in case (a).

Case $L_e > L_{me}$:   (c) Broken fibers; the reinforcement is zero.
(d) Fibers with an increasing tensile stress, the embedded length of these fibers is smaller than the fibers in case (c).

As the fibers are randomly distributed, the balance between all four types of fibers results in a decrease in strength after the second peak. In tensile tests the unreinforced PABs are brittle, whereas those reinforced with fibers have more ductile behavior, and this is very important for masonries, which usually show brittle behaviour. More ductile behavior of the blocks and mortar makes the masonry more stable against earthquakes and differential settlements, limiting crack growth.

In conclusion, the increased ductility of reinforced samples is linked to fibers that hold the cracks together after the failure of the clay matrix. The tensile strength of the fibers crossing the cracks is mobilized thanks to their embedded length, where a bond is created essentially between cellulose molecules (negative charge of O-H bonds) and, for example, flocculated cations such as Fe^{3+}, Ca^{2+}, and Mg^{2+} in the soil.

4. Conclusions

The physicochemical and mechanical characteristics of kenaf fibers from Burkina Faso and the mechanical properties of PABs reinforced with these fibers were investigated in this work. The conclusions drawn can be summarized as follows:

1) Kenaf fibers have a tensile strength higher than 750 MPa with a high elasticity modulus (136 ± 25 GPa) compared to other natural fibers.

2) Kenaf fibers contain large amounts (70 wt%) of cellulose type I ($I_c = 40$), with hemicelluloses (19 wt%) and lignin (3 wt%).

3) The incorporation of kenaf fibers in PABs improves mainly the ductility in tension of the blocks thanks to the high mechanical strength of fibers and their strong adherence to the clay matrix.

4) The incorporation of fibers is particularly beneficial as far as the bending strength of earth blocks is concerned.

5) In order to obtain optimum mechanical behaviour, an amount smaller than 0.8 wt% of short fibers (30 mm) is recommended.

Acknowledgments

This article is dedicated to Karfa Traoré, a pathfinder in the study of clay materials in Burkina Faso who passed away on 27 August 2012 in Ouagadougou.

Author Contributions

Younoussa Millogo and Jean-Emmanuel Aubert carried out the experiments and wrote the paper; Jean-Claude Morel also contributed to the writing of the paper; Erwan Hamard performed the experiments, especially the mechanical tests (compressive and tensile strengths) under the mentorship of Jean-Claude Morel. All the authors participated in the final corrections of the paper.

Conflicts of Interest

The authors declare no conflict of interest.

References

1. Binici, H.; Aksogan O.; Shah, T. Investigation of fibre reinforced mud bricks as a building material. *Constr. Build. Mater.* **2005**, *19*, 313–318.

2. Ghavami, K.; Toledo Filho, R.D.; Barbosa, N.P. Behaviour of composite soil reinforced with natural fibres. *Cem. Concr. Compos.* **1999**, *21*, 39–48.

3. Toledo Filho, R.D.; Ghavami, K.; England, G.L.; Scrivener, K. Development of vegetable fibre-mortar composites of improved durability. *Cem. Concr. Compos.* **2003**, *25*, 185–196.

4. Mesbah, A.; Morel, J.C.; Walker, P.; Ghavami, K. Development of a direct tensile test for compacted soil blocks reinforced with natural fibers. *J. Mater. Civil Eng.* **2004**, *16*, 95–98.

5. Bouhicha, M.; Aouissi, F.; Kenai, S. Performance of composite soil reinforced with barley straw. *Cem. Concr. Compos.* **2005**, *27*, 617–621.

6. Kumar, A.; Walia, S.B.; Mohan, J. Compressive strength of fiber reinforced highly compressible clay. *Constr. Build. Mater.* **2006**, *20*, 1063–1068.

7. Yetgin, S.; Cavdar, O.; Cavdar, A. The effects of the fiber contents on the mechanic properties of the adobes. *Constr. Build. Mater.* **2008**, *22*, 222–227.

8. Juárez, C.; Guevara, B.; Durán-Herrera, A. Mechanical properties of natural fibers reinforced sustainable masonry. *Constr. Build. Mater.* **2010**, *24*, 1536–1541.

9. Ismail, S.; Yaacob, Z. Properties of laterite bricks reinforced with oil palm empty fruit bunch fibres. *Pertanika J. Sci. Technol.* **2011**, *19*, 33–43.

10. Quagliarini, E.; Lenci, S. The influence of natural stabilizers and natural fibres on the mechanical properties of ancient Roman adobe bricks. *J. Cult. Herit.* **2010**, *11*, 309–314.

11. Shin, H.K.; Jeun, J.P.; Kim, H.B.; Kang, P.H. Isolation of cellulose fibers from kenaf using electron beam. *Radiat. Phys. Chem.* **2012**, *81*, 936–940.

12. Jonoobi, M.; Harun, J.; Tahir, P.M.; Shakeri, A.; SaifulAzry, S.; Makinejad, M.D. Physicochemical characterization of pulp and nanofibers from kenaf stem. *Mater. Lett.* **2011**, *65*, 1098–1100.

13. Akil, H.M.; Omar, M.F.; Mazuki, A.A.M.; Safiee, S.; Ishak, Z.A.M.; Abu Bakar, A. Kenaf fiber reinforced composites: A review. *Mater. Des.* **2011**, *32*, 4107–4121.

14. Kouakou, H.; Morel, J.C. Strength and elasto-plastic properties of non-industrial building materials manufactured with clay as a natural binder. *Appl. Clay Sci.* **2009**, *44*, 27–34.

15. Millogo, Y.; Morel, J.C.; Aubert, J.E.; Ghavami, K. Experimental analysis of pressed adobe blocks reinforced with *Hibiscus cannabinus* fibers. *Constr. Build. Mater.* **2014**, *52*, 71–78.

16. Toledo Filho, R.D. Natural Fibre Reinforced Mortar Composites: Experimental, Characterisation. Ph.D. Thesis, DEC-PUC/Imperial College, London, UK, 1997, p. 472.

17. Van Soest, P.D.; Wine, R.H. The use of detergents in the analysis of fibrous feed II. A rapid method for determination of fiber and lignin. *J. Assoc. Off. Agric. Chem.* **1967**, *50*, 50–55.

18. Fulgencio, S.C.; Jaime, C.; Juan, G.R. Determination of hemicelluloses, cellulose, and lignin contents of dietary fibre and crude fiber of several seed hulls. *Data Comp.* **1983**, *177*, 200–202.

19. Ciblac, T.; Morel, J.C. *Sustainable Masonry: Stability and Behavior of Structures*; Wiley: Hoboken, NJ, USA, 2014; p. 280.

20. Aubert, J.E.; Maillard, P.; Morel, J.C.; Al Rafii, M. Towards a simple compressive strength test for earth bricks? *Mater. Struct.* **2015**, doi:10.1617/s11527-015-0601-y.

21. Aubert, J.E.; Fabbri, A.; Morel, J.C.; Maillard, P. A soil block with a compressive strength higher than 45 MPa! *Constr. Build. Mater.* **2013**, *47*, 366–369.

22. Olivier, M.; Mesbah, A. Modèle de comportement pour sols compactés. In Proceedings of the First International Conference on unsaturated soils, Paris, France, 6–8 September 1995.

23. Bui, Q.B.; Morel, J.C.; Hans, S.; Walker, P. Effect of moisture content on the mechanical characteristics of rammed earth. *Constr. Build. Mater.* **2014**, *54*, 163–169.

24. Morel, J.C.; Pkla, A. A model to measure compressive strength of Compressed Earth Blocks with the "3 points bending test". *Constr. Build. Mater.* **2002**, *16*, 303–310.

25. Morel, J.C.; Pkla, A.; di Benedetto, H. Interprétation en compression ou traction de l'essai de flexion en trois points. *Rev. Fr. Génie Civil* **2003**, *7*, 221–237.

26. El-Shekeil, Y.A.; Sapuan, S.M.; Abdan, K.; Zainudin, E.S. Influence of fiber content on the mechanical and thermal properties of Kenaf fiber reinforced thermoplastic polyurethane composites. *Mater. Des.* **2012**, *40*, 299–303.

27. Morān, J.I.; Alvarez, V.A.; Cyras, V.P.; Vāsquez, A. Extraction of cellulose and preparation of nanocellulose. *Cellulose* **2008**, *15*, 149–159.

28. Ivanova, N.V.; Korolenko, E.A.; Korolik, E.V.; Zhbankov, R.G. IR spectrum of cellulose. *J. Appl. Spectrosc.* **1989**, *51*, 847–851.

29. Godin, B.; Ghysel, F.; Agneessens, R.; Schmit, T.; Gofflot, S.; Lamaudière, S.; Sinnaeve, G.; Goffart, J.P.; Gerin, P.A.; Stilmant, D.; *et al.* Détermination de la cellulose, des hemicelluloses, de la lignine et des cendres dans diverses cultures lignocellulosiques dédiées à la production de bioéthanol de deuxième génération. *Biotechnol. Agron. Soc. Environ.* **2010**, *14*, 549–560.

30. Morel, J.C.; Gourc, J.P. Behavior of sand reinforced with mesh elements. *Geosynth. Int.* **1997**, *4*, 481–508.

12

A Novel Schiff Base of 3-acetyl-4-hydroxy-6-methyl-(2H)pyran-2-one and 2,2'-(ethylenedioxy)diethylamine as Potential Corrosion Inhibitor for Mild Steel in Acidic Medium

**Jonnie N. Asegbeloyin [1], Paul M. Ejikeme [1], Lukman O. Olasunkanmi [2,3],
Abolanle S. Adekunle [2,3] and Eno E. Ebenso [2,*]**

[1] Department of Pure and Industrial Chemistry, University of Nigeria, Nsukka 40001, Enugu State, Nigeria; E-Mails: niyi.asegbeloyin@unn.edu.ng (J.N.A.); paul.ejikeme@unn.edu.ng (P.M.E.)

[2] Material Science Innovation and Modelling (MaSIM) Research Focus Area, Faculty of Agriculture, Science and Technology, North-West University (Mafikeng Campus) Private Bag X2046, Mmabatho 2735, South Africa; E-Mails: waleolasunkanmi@gmail.com (L.O.O.); sadekpreto@gmail.com (A.S.A.)

[3] Department of Chemistry, Faculty of Science, Obafemi Awolowo University, Ile-Ife 220005, Nigeria

*   Author to whom correspondence should be addressed; E-Mail: Eno.Ebenso@nwu.ac.za;

Academic Editor: Jorge de Brito

Abstract: The corrosion inhibition activity of a newly synthesized Schiff base (SB) from 3-acetyl-4-hydroxy-6-methyl-(2H)-pyran-2-one and 2,2'-(ethylenedioxy)diethylamine was investigated on the corrosion of mild steel in 1 M HCl solution using potentiodynamic polarization and electrochemical impedance spectroscopic techniques. Ultraviolet-visible (UV-vis) and Raman spectroscopic techniques were used to study the chemical interactions between SB and mild steel surface. SB was found to be a relatively good inhibitor of mild steel corrosion in 1 M HCl. The inhibition efficiency increases with increase in concentration of SB. The inhibition activity of SB was ascribed to its adsorption onto mild steel surface, through physisorption and chemisorption, and described by the Langmuir adsorption model. Quantum chemical calculations indicated the presence of atomic sites with potential nucleophilic and electrophilic characteristics with which SB can establish electronic interactions with the charged mild steel surface.

Keywords: Schiff base; electrochemical techniques; mild steel; adsorption; quantum chemical calculations

1. Introduction

Mild steel is used in many industrial and structural applications due to its good mechanical strength and relatively low cost [1,2]. Acidic solutions commonly used in many industrial activities, including the steelmaking finishing process, constitute unfriendly corrosive media for mild steel [3]. The use of organic corrosion inhibitors has been identified as one of the most economical ways of reducing corrosion rate and protecting steel-made industrial facilities against corrosion [3,4]. The ability of Schiff base ligands to form stable complexes closely packed in the coordination sphere of metal ion introduces another class of compounds for corrosion inhibition [5]. Schiff bases are adsorbed on metal surfaces due to the presence of >C=N– groups [6]. This adsorption behavior leads to spontaneous formation of a monolayer covering the metal surface, consequently acting as effective corrosion inhibitor.

3-acetyl-4-hydroxy-6-methyl-(2H) pyran-2-one, commonly referred to as dehydroacetic acid, and its derivatives have been of research interest because of their interesting coordination chemistry, pharmaceutical importance and biological activities [7–11]. DNA binding and antibacterial screening of dehydroacetic acid complexes of Ru(II) and Ru(III) containing PPh$_3$/AsPh$_3$ have been recently reported by Chitrapriya *et al.* [12]. Dehydroacetic acid is well known for its fungicidal [13], herbicidal and antimicrobial activities [10]. It is also widely used in food technology, as a vitamin C stabilizer and as a preservative in food products like fish sausages [14]. Therefore, dehydroacetic acid and possibly its Schiff bases, are non-toxic and eco-friendly. The presence of O and N heteroatoms as well as >C=N– and >C=O functional groups in dehydroacetic acid/2,2'-(ethylenedioxy)diethylamine Schiff base may facilitate electronic interactions with a mild steel surface, leading to adsorption on the steel surface and consequently, inhibition of steel corrosion.

The present work is in furtherance of the continual search for eco-friendly, easy to synthesize and effective corrosion inhibitors. In this work, a novel Schiff base (SB) was synthesized based on condensation of 3-acetyl-4-hydroxy-6-methyl-(2H)pyran-2-one and 2,2'-(ethylenedioxy)diethylamine and investigated for its corrosion inhibition activities using electrochemical methods, spectroscopic techniques and quantum chemical calculations.

2. Results and Discussion

2.1. Synthesis of SB

The results of IR, NMR and elemental analyses confirmed successful synthesis of SB with the chemical structure (SB) shown under the experimental Section 3.3. The percentage yield and melting point of the whitish compound (C$_{22}$H$_{28}$N$_2$O$_8$) synthesized were 85% and 155 °C, respectively. The infrared, proton and carbon-13 NMR spectroscopic results of the product are given in Table 1. Detail assignments of these spectroscopic data have been described somewhere else [15].

Table 1. IR, ^{1}HNMR, ^{13}CNMR spectroscopy data and elemental analysis of SB.

IR (KBr cm^{-1})	1665 (C=N), 3454 (OH), 1703 (C=O),1254 (C-O), 1358 (C-N)	
^{1}H NMR (CDCl$_3$, δ, ppm)	2.11 (s, 3H, -C=C-CH$_3$); 2.50 (s, 3H, N=C-CH$_3$); 5.62 (s, 1H, CH$_3$=C-H); 3.27δ (s, 4H); 3.70δ (m, 6H); 13.80δ (s, 1H, enolic OH)	
^{13}C NMR (CDCl$_3$, δ, ppm)	18.27, 19.58 (-CH$_3$), 183.66 (O-$\underline{C}$=O); 176.03 ($\underline{C}$-OH); 162.47 (H$_3$C-$\underline{C}$-O); 107.49 (C=N); 68.52 (-CH$_2$-O); 44.12 (CH$_2$ –N)	
Elemental analysis	Experimental	C=58.32; H=6.24; N=6.25
	Calculated	C=58.92; H=5.98; N=6.44

2.2. Electrochemical Measurements

2.2.1. Open Circuit Potential (OCP)

The results of OCP measurements on mild steel corrosion in 1 M HCl with and without various concentrations of SB are as shown in Figure 1. A relatively steady potential ($\Delta E = \pm 1$ mV) was reached at about 1000 s of immersion in all cases. There was initial decrease in potential for mild steel corrosion in 1 M HCl without the inhibitor (the blank system). This was attributed to dissolution of air oxide film on the mild steel surface [16,17]. A later increase in potential for this system after 50 s may be due to the formation of insoluble iron (III) oxide [16] on the mild steel surface leading to a more passive state of the steel. Initial increase in potential was observed in the presence of inhibitors followed by a sharp decrease after 75 s.

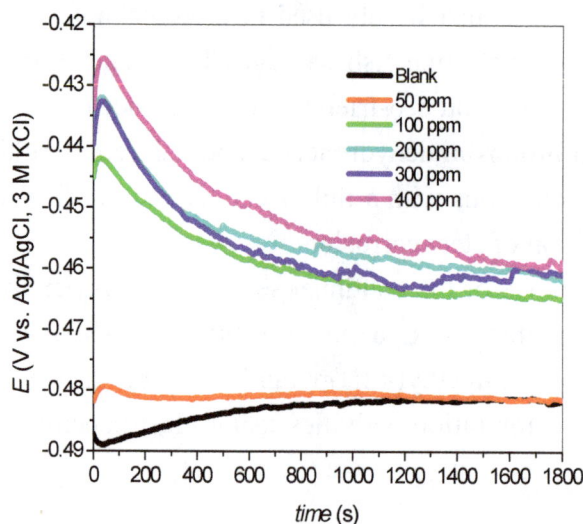

Figure 1. OCP scan for mild steel corrosion in 1 M HCl with and without various concentrations of SB.

The nature of the OCP in the presence of SB was different from that of the blank. This is due to differences in surface activities on the steel in the absence and presence of the inhibitor. The OCP values for the inhibited systems were generally more positive than that of the uninhibited blank system. This can be attributed to the formation of protective film of the SB inhibitor on the steel surface [18] and suggests the inhibition of anodic dissolution of the steel by the SB under open circuit conditions [19]. Also, the OCP values increase with increase in inhibitor concentrations. This may be due to the increased number of inhibitor molecules in the protective layer formed on steel surface leading to thicker protective films. The OCP profile at 50 ppm of SB though reaches plateau at almost the same potential as the uninhibited 1 M HCl system, the 50 ppm SB containing electrochemical system obviously has a more positive potential on the average compared to the uninhibited system.

2.2.2. Potentiodynamic Polarization Measurements

The polarization measurements were carried out on mild steel electrode immersed in 1 M HCl with and without various concentrations of SB after 30 min of immersion. Polarization curves were obtained as the plots of potential against logarithm of current density as presented in Figure 2. Electrochemical kinetic parameters including the corrosion current density (i_{corr}), corrosion potential (E_{corr}), anodic Tafel slope (β_a), cathodic Tafel slope (β_b) and percentage inhibition efficiency (%IE) were determined by linear extrapolations of Tafel lines within the straight-line regions of the polarization curves. These curves exhibit a shift to lower current density in the presence of inhibitor, which implies that SB reduces the rate of mild steel corrosion in 1 M HCl. The polarization curves are shifted to more positive (noble) values of E_{corr} in the presence of SB. This suggests the formation of protective layer of SB on the steel surface. This is also in agreement with the observation during the OCP monitoring. Table 2 shows the kinetic parameters obtained after the linear Tafel fitting. An inhibitor can be regarded as anodic or cathodic type inhibitor if the shift in E_{corr} value is greater than 85 mV [20]. As shown in Table 2, the maximum shift in E_{corr} value in this study was 24 mV, which suggests that SB is a mixed-type inhibitor. That is, it reduces the rate of anodic reaction comprising the mild steel oxidation as well as cathodic reaction, which is hydrogen gas evolution. A closer look at the polarization curves (Figure 2), however, reveals that the anodic inhibiting effect is more pronounced in 1 M HCl, especially between 50 and 300 ppm, while at 400 ppm the mix-type inhibition characteristics is more obvious. This fact is also reflected in the relative magnitudes of the differences between the βa of the inhibitor containing and blank systems as well as the corresponding differences in the β_c values as shown in Table 2. A maximum difference of 49 mV/dec was observed between the β_a values in the absence and presence of inhibitor. Such a slight change in the values of βa and β_c upon addition of the inhibitors when compared with the blank suggests that SB adsorbed onto the metal surface and inhibit the corrosion rate without changing the mechanism of the mild steel corrosion in hydrochloric acid [21].

The percentage inhibition efficiency (%IE_P) was calculated at different concentrations of SB according to the equation:

$$\%IE_P = \left(\frac{i^0_{corr} - i_{corr}}{i_{corr}} \right) \times 100 \tag{1}$$

where i^0_{corr} and i_{corr} are the corrosion current density with and without various concentrations of SB, respectively. The values of %IE_P at various concentrations of SB are presented in Table 2. It is clear that %IE_P increases with increase in concentration of SB with the highest value of 80.6% obtained at 400 ppm of the inhibitor.

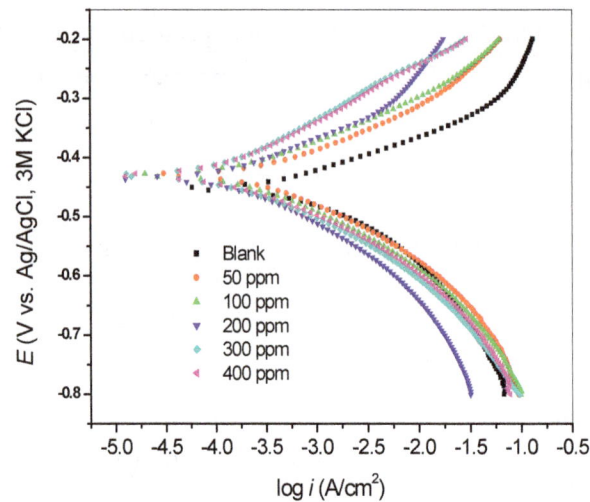

Figure 2. Polarization curves for the corrosion of mild steel in 1 M HCl with and without various concentrations of SB.

Table 2. Electrochemical kinetic parameters from potentiodynamic polarization experiment.

Inhibitor Concentration (ppm)	$-E$ (mV)	i_{corr} (μA/cm^2)	β_a (mV/dec)	β_c (mV/dec)	R_p (Ω)	%IE
Blank	452	529.7	58	102	4.832	-
50	428	310.9	71	94	9.34	41.31
100	428	214.6	72	97	14.16	59.49
200	434	189.6	82	108	20.41	64.21
300	428	172.0	107	101	27.5	67.53
400	429	102.7	86	73	26.44	80.61

2.2.3. Electrochemical Impedance Spectroscopy (Eis) Measurements

The Nyquist plots obtained from the EIS studies are presented in Figure 3. The Nyquist plots in Figure 3 show a single depressed capacitive arc over the frequency range studied. This is an indication that the dissolution of mild steel in 1 M HCl is controlled by a single charge transfer process [22]. It was observed that the diameter of the semicircles in the Nyquist plots increases with increase in concentration of the SB inhibitor. Impedance data fitted properly with Randles equivalent circuit of the form $R_s(R_{ct}Q)$, which consists of solution resistance (R_s), in series with the parallel combination of the constant phase element (CPE), denoted as Q, and a charge-transfer resistance (R_{ct}). Due to non-ideal capacitive behavior of the electrode/electrolyte systems investigated in the present study, the CPE was introduced in order to obtain a good agreement between experimental and simulated EIS data. The impedance (Z) of the CPE is defined as:

$$Z_{CPE} = Q^{-1}(j\omega)^{-n} \tag{2}$$

where Q is the CPE constant (in $\Omega^{-1}S^n cm^{-2}$); $j^2 = -1$ is the imaginary number; ω is the angular frequency (in rads^{-1}); and n is a CPE exponent, which can be used as a gauge of the heterogeneity or roughness of the surface.

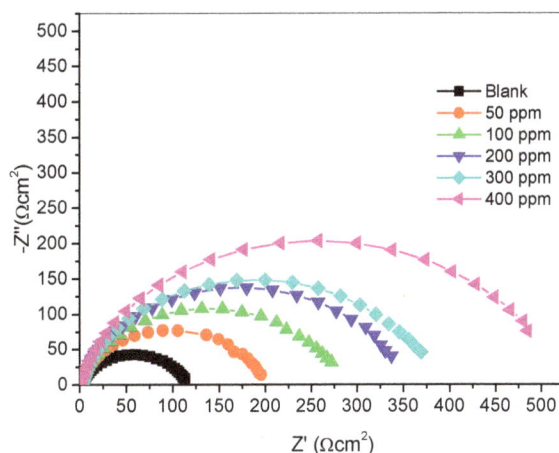

Figure 3. Nyquist plots for mild steel corrosion in 1 M HCl with and without different concentrations of SB.

The CPE can represent resistance (n = 0, Q = R), capacitance (n = 1, Q = C), inductance (n = −1, Q = L), or Warburg impedance (n = 0.5, Q = W). Thus, the closer the value of n to unity, the better the capacitive behavior of Q. The electrochemical kinetic parameters obtained from the fitting of impedance data are presented in Table 3. Since the CPE exponent, n, is close to 1 for the studied system, Q can be assumed to have some capacitive features and referred to as the double-layer capacitance. The percentage inhibition efficiency (%IE_{EIS}) was calculated by using the equation:

$$\%IE_{EIS} = \left(\frac{R_{ct} - R_{ct}^0}{R_{ct}} \right) \times 100 \qquad (3)$$

where R_{ct}^0 and R_{ct} are the charge transfer resistances with and without various concentrations of SB inhibitor, respectively. As shown in Table 3, %IE_{EIS} increases with increase in concentration of SB. This confirms the inhibition potency of SB against mild steel corrosion in 1 M HCl. There is an increase in the values of R_{ct} as the concentration of SB increases. This is attributed to an increase in the interface between the metal surface and the aggressive solution due to increase in the area of the adsorption film formed on the metal surface. According to the Helmholtz model [23], Q is expressed as:

$$Q = \frac{\varepsilon^0 \varepsilon}{d} S \qquad (4)$$

where ε^0 is the permittivity of the vacuum; ε is the dielectric constant of the medium; d is the thickness of the film and S is the surface area of the electrode. The decrease in values of Q as the concentration of SB increases may be due to an increase in the area of the adsorption film, which corresponds to the decrease in the exposed electrode surface area (S), or an increase in the thickness of the adsorbed protective layer (d) or a decrease in the medium dielectric constant (ε). One or more of these result in the observed decrease in the values of Q in accordance to Equation (4) above.

The variations of percentage inhibition efficiency with concentrations of SB are plotted in Figure 4. There is a good agreement between the values obtained for the percentage inhibition efficiency from the polarization and impedance techniques. It is worthy of mention that the studied compound (SB) when compared with some previously reported organic inhibitors exhibits similar or relatively better inhibition performance for mild steel corrosion in 1 M HCl [24–26].

Table 3. Fitted parameters from electrochemical impedance spectroscopy.

Inhibitor Concentration (ppm)	R_s (Ωcm^{-2})	R_{ct} (Ωcm^{-2})	Q_1 (Y_0) (μFcm^{-2})	n	$\% IE_{EIS}$
Blank	1.947	115.30	145.00	0.8027	-
50	2.570	193.00	85.58	0.8756	40.26
100	1.808	274.10	82.21	0.8651	57.94
200	1.937	343.00	71.67	0.8736	66.38
300	1.754	374.00	64.57	0.8707	69.17
400	1.504	506.00	61.56	0.8718	77.21

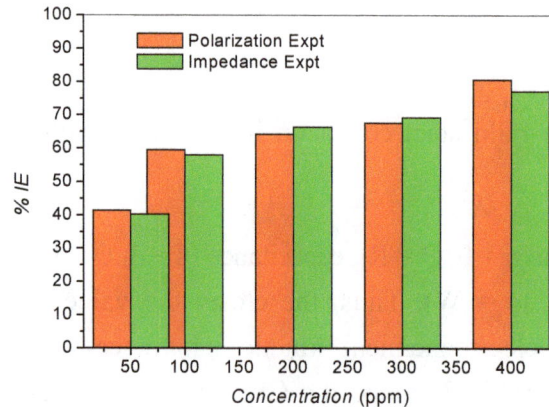

Figure 4. Variation of percentage inhibition efficiency (%*IE*) with concentration of SB for both polarization and impedance experiments.

2.3. Adsorption Isotherms

Important information on the adsorption behavior of an inhibitor on a metal surface can be obtained by fitting the experimental data into appropriate adsorption isotherms. The adsorption of inhibitor on metal/solution interface may occur through the displacement of water molecules by the inhibitor molecules [27] in accordance to the reaction equation:

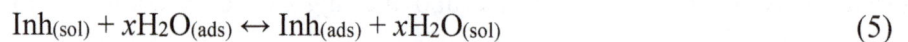

$$Inh_{(sol)} + xH_2O_{(ads)} \leftrightarrow Inh_{(ads)} + xH_2O_{(sol)} \tag{5}$$

where, *x*, the mole ratio, is the number of water molecules replaced by one molecule of organic inhibitor. The surface coverage, θ, was calculated from the percentage inhibition efficiency (θ = %*IE*/100), obtained from both polarization and impedance measurements. The experimental data obtained fitted well with the Langmuir adsorption isotherm represented by the equation:

$$\frac{C_{inh}}{\theta} = \frac{1}{K_{ads}} + C_{inh} \tag{6}$$

where C_{inh} is the concentration of the inhibitor and K_{ads} is the equilibrium adsorption constant. The plots of C_{inh}/θ vs C_{inh} (Figure 5) gave straight lines with strong linear correlation coefficients.

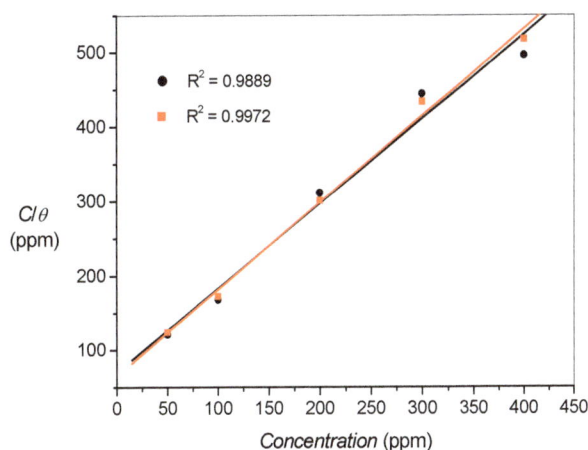

Figure 5. Langmuir adsorption isotherms of SB adsorption on mild steel surface from (▪) polarization and (●) impedance experiments.

The change in free energy of adsorption (ΔG_{ads}) was calculated from the relation:

$$\Delta G_{ads} = -RT \ln(55.5 K_{ads}) \tag{7}$$

where R is gas constant; T is absolute temperature and the constant 55.5 is the molar concentration of water. The values of K_{ads} and ΔG_{ads} are presented in Table 4. The negative value of ΔG_{ads} implies that the adsorption process is spontaneous. The magnitude of ΔG_{ads} is usually used to predict the nature of adsorption, whether it is physisorption or chemisorption. A value of ΔG_{ads} around −20 kJ/mol or less negative has been attributed to electrostatic interactions between the charged inhibitor molecules and the charged metal surface (physisorption), while values around −40 kJ/mol, or larger negative values, involve charge sharing or charge transfer from organic molecules to the metal surface to form coordinate bond (chemisorption) [28]. The magnitudes of ΔG_{ads} obtained in the present study suggest that the adsorption of SB on mild steel surface is a combination of both physisorption and chemisorption processes.

Table 4. Adsorption constant and change in free energy of adsorption of SB on mild steel surface from polarization and impedance experiments at 303 K.

	K_{ads} (×10^{-3})	$-\Delta G_{ads}$ (kJ/mol)
Polarization Expt.	6.47	31.71
Impedance Expt.	6.89	31.86

2.4. Ultraviolet-Visible and Raman Spectroscopic Studies

UV-vis and Raman spectroscopic analyses were carried out on pure SB and the resulting solution of SB after five days of mild steel immersion. Comparison of the spectroscopic features of pure SB and SB with mild steel can provide information about possible formation of Fe/SB complex. The UV-vis spectrum of SB solution (Figure 6) shows two absorption peaks, at 310 nm and 234 nm, corresponding to n→π* transition, π→π* transition and intermolecular charge transfer. After mild steel immersion, the maximum absorption peak seems to be unaffected as it re-appeared at 309 nm. However, there is a slight blue shift of the peak at 234 nm to 229 nm. This may be attributed to the formation of coordinate bonds between Fe and SB.

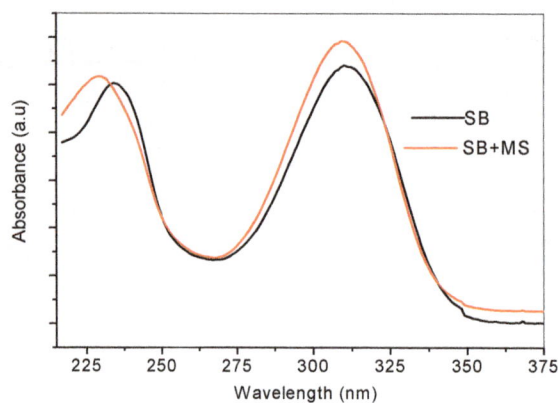

Figure 6. UV-Vis spectra of pure SB solution (black, SB) and SB solution after mild steel immersion (red, SB+MS).

The Raman spectroscopic analysis of possible formation Fe/SB complex was carried out by investigating the change in prominent Raman bands between 200 cm^{-1} and 2000 cm^{-1} after mild steel immersion in SB solution. As shown in Figure 7, in the pure SB, a strong band at 371 cm^{-1} is assigned to aliphatic CC chains in the molecule, while a medium peak, characteristic of C-O-C band appears at 978 cm^{-1}, coupled with its asymmetric band at 1074 cm^{-1}. Aromatic CC bands of medium intensity appear at 1173 cm^{-1} and 1451 cm^{-1}. Two strong bands appear at 1602 cm^{-1} and 1674 cm^{-1} which may be due to C=N functional group. The band at 1732 cm^{-1} is of medium intensity and it is attributed to C=O. After mild steel immersion, more bands appear in the region below 500 cm^{-1}. The strong bands at 382 cm^{-1} and 458 cm^{-1} are of special interest and can be assigned to Fe-O, suggesting bond formation between SB and Fe through the sp^2 oxygen. Other bands have also shifted correspondingly after mild steel immersion in SB. Most of the bands shifted to higher wavenumbers after mild steel immersion. This includes the bands for C=N functional groups.

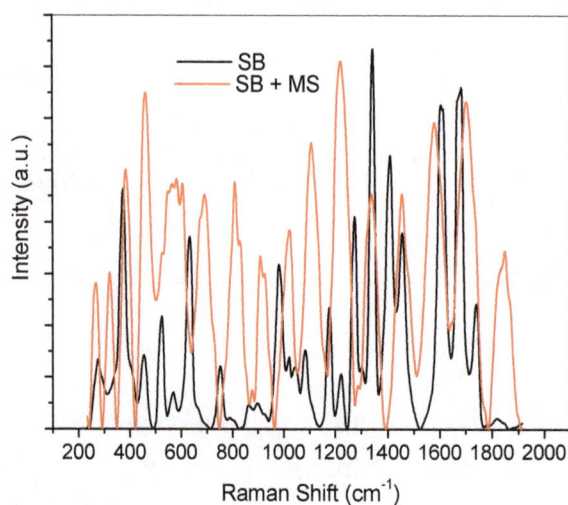

Figure 7. Raman spectra of pure SB solution (black, SB) and SB solution after mild steel immersion (red, SB + MS).

2.5. Quantum Chemical Calculations

The gas phase optimized structure of SB with atom labels is shown in Figure 8. SB is expected to be a symmetrical molecule but the optimized structure shows that there is extension of π-conjugation to the O27 atom of the –OH group, whereas this is not observed on the similar oxygen atom O13 of the –OH group on the other end of the molecule. This non-uniform distribution of electron density makes the molecule to be unsymmetrical. Relevant quantum chemical parameters of the optimized structure are listed in Table 5. Adsorption of an inhibitor on a metal surface is often explained based on donor-acceptor phenomenon between the inhibitor and metal atom. The energy of the highest occupied molecular orbitals (E_{HOMO}) is associated with the tendency of an inhibitor molecule to donate its least stable electron(s) to the appropriate vacant orbitals of the metal atom. On the other hand, the energy of the lowest unoccupied molecular orbitals (E_{LUMO}) informs the tendency of the inhibitor molecule to accept charges from the metal atom towards back-bonding. The higher the E_{HOMO}, the higher the possibility of forward donation of charges to the metal, and the lower the E_{LUMO}, the better the chances of back-donation of charges. The results in Table 5 show that the values of E_{HOMO}, E_{LUMO} and energy gap ($\Delta E_{LUMO-HOMO}$), the ionization potential ($I = -E_{HOMO}$) and the electron affinity ($A = -E_{LUMO}$) of SB are within the range of values that have been reported for some other Schiff bases that had also been adjudged good corrosion inhibitors [29]. SB has a reasonably high dipole moment, which suggests that it has the potential to interact with metal atom or ions in aqueous system. The results of the chemical potential (μ), hardness (η) and electrophilicity (ω) as shown in Table 5 reveal that SB has a potential to react with Fe in the mild steel thereby protecting the steel surface against corrosion. The value obtained for fraction of electrons transferred (ΔN) from the inhibitor molecule to the metal surface confirms the possibility of charge transfers from SB to the mild steel.

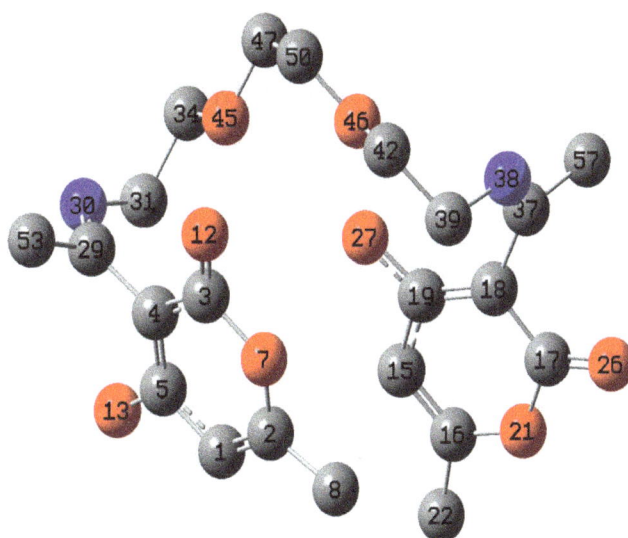

Figure 8. Optimized structure of SB showing only non-hydrogen atoms with atomic numbering.

Table 5. Some gas-phase quantum chemical parameters of SB calculated at B3LYP/6-31G(d) level.

Parameter	Value
Total energy (au)	−1566.4527
E_{HOMO} (eV)	−5.99
E_{LUMO} (eV)	−1.46
$\Delta E_{LUMO-HOMO}$	4.53
I	5.99
A	1.46
Dipole moment (Debye)	5.69
Chemical potential, μ (eV)	−3.72
Hardness, η (eV)	2.26
Electrophilicity, ω (eV)	3.06
ΔN	0.72
ΔE (eV)	-0.57

The change in energy (ΔE) following charge transfer implies that the charge transfer from SB to mild steel followed by back donation from mild steel to the inhibitor is energetically favorable. The HOMO and LUMO graphic surfaces (Figure 9) are distributed around the unsaturated ring of the dehydroacetic acid with the HOMO surface being around the ring with the unsaturated –OH group and the LUMO surface on the ring that has no π-conjugation to the –OH. The Mulliken atomic charges of SB are displayed on the atoms in the optimized structure (Figure 10). There are quite a number of atoms with negative Mulliken charges, which implies that SB has potential atoms that can interact with relatively positive centers on the mild steel surface.

The Fukui functions (f(r)) are often used as indices of local reactivity to analyze the active atomic sites in inhibitor molecules [30,31]. The Fukui functions (f (r)) measure the change in the electron density of an N electron system upon addition (f^+(r)) or removal (f^-(r)) of an electron [32]. Atom condensed Fukui functions using the Mulliken population analysis (MPA) and the finite difference (FD) approximations approach introduced by Yang and Mortier [33] were calculated using the equations:

$$f_k^+ = \rho_{k(N+1)}(r) - \rho_{k(N)}(r) \tag{8}$$

$$f_k^- = \rho_{k(N)}(r) - \rho_{k(N-1)}(r) \tag{9}$$

where $\rho_{k(N+1)}$, $\rho_{k(N)}$ and $\rho_{k(N-1)}$ are the electron densities of the (N+1)-, N- and (N-1)- electron systems, respectively, and approximated by Mulliken gross charges; f^+ and f^- are the Fukui indices condensed on atom k and measure its electrophilic and nucleophilic tendencies, respectively. The Fukui indices are listed in Table 6 for selected non-hydrogen atoms in SB. The atomic labels employed in Table 6 are the same as those displayed in Figure 8. The most susceptible sites for nucleophilic attacks are O45, C31, C47 and C50, while the most susceptible sites for electrophilic attack are C39, O45, C50, C47 and C31 in that order.

Figure 9. HOMO and LUMO surfaces of SB at an isosurface value of 0.02.

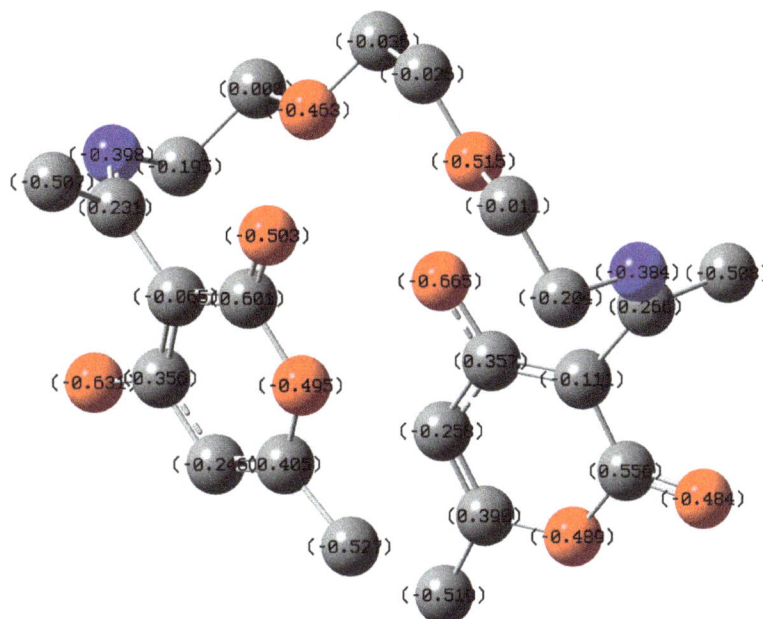

Figure 10. Mulliken atomic charges of atoms in the optimized structure of SB.

Table 6. Fukui indices for non-hydrogen selected atoms in SB molecule.

Atom	f^+	Atom	f^-
C1	0.0013	C8	0.0022
C8	0.0098	C22	0.0085
C15	0.0104	C29	0.0054
C22	0.0072	C31	0.0102
C31	0.0182	C34	0.0097
C34	0.0076	C37	0.0005
C39	0.0040	C39	0.0163
C42	0.0081	C42	0.0077
O45	0.0215	O45	0.0135
O46	0.0078	O46	0.0088
C47	0.0127	C47	0.0114
C50	0.0102	C50	0.0136
C53	0.0076	C53	0.0042
C57	0.0052	C57	0.0072

3. Experimental Section

3.1. Materials

All reagents and solvents were of analytical grade and were used without further purification. The experiments were performed on mild steel samples with the chemical composition (wt %) C = 0.17, Mn = 0.46, Si = 0.26, S = 0.017, P = 0.019, and balance Fe. For all electrochemical studies, mild steel coupons were cut into 1 cm × 1 cm dimensions and embedded in a Teflon holder using epoxy resin, exposing only 1 cm^2 surface area. Prior to each measurement, mild steel surface was mechanically abraded on Struers MD PianoTM 220 (size: 200 dia) mounted on Struers LaboPol-1 machine to remove traces of epoxy resin from the surface. The surface was then polished with SiC papers of various grit sizes ranging from 600 to 1200 to achieve a finely ground surface then washed with water followed by acetone and then water again, and finally wiped with clean tissue paper and air-dried. Mild steel specimens were used immediately after surface preparation.

3.2. Chemicals and Instrumentation

3-acetyl-4-hydroxy-6-methyl-(2H) pyran-2-one and 2,2'-(ethylenedioxy)diethylamine were used as supplied by Fluka. Elemental analyses of C, H and N were performed by using Carlo Erba Elemental analyzer EA 1108. Melting point was taken in open capillaries on a melting point apparatus model no 125. IR spectra were recorded on a Perkin Elmer Spectrum 100. 1H and 13C NMR spectra were obtained from a Bruker AV 500 MHz for 1H and 125 MHz for 13C using a 5 mm Quadra Nuclei Probe (QNP). UV-Vis spectra were recorded on Cary 300 UV-Vis by Agilent Technologies. Raman spectra were obtained from Xplora Raman spectrometer from Horiba Scientific.

3.3. Synthesis of 3-acetyl-4-hydroxy-6-methyl-(2H) pyran-2-one Schiff base (SB)

A solution of 3-acetyl-4-hydroxy-6-methyl-(2H) pyran-2-one (3.36 g, 0.02 mol) in 20 mL ethanol was mixed with a solution of 2,2'-(ethylendioxy) diethylamine in 20 mL ethanol. The equation of reaction is as shown in Equation (10):

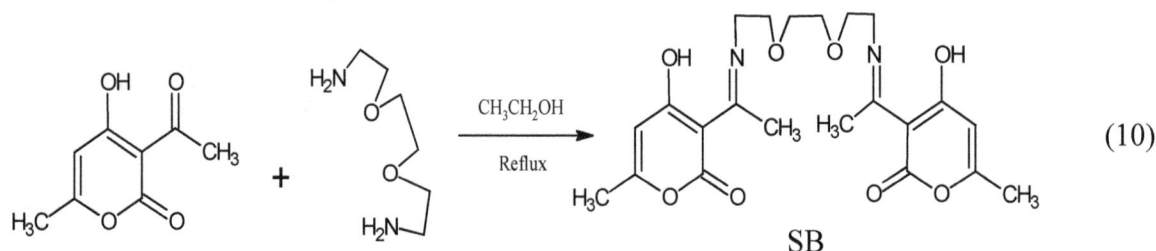

$$(10)$$

The mixture was refluxed for 3 h, and the resulting solution was chilled to −10 °C to obtain a whitish product which was filtered, dried and recrystallized in water.

3.4. Electrochemical Measurements

All electrochemical measurements were carried out on the Autolab PGSTAT 302N obtained from Metrohm, equipped with a three-electrode system. Ag/AgCl with 3 M KCl was used as the reference electrode, while platinum rod was used as the counter electrode.

The system was allowed to reach the steady open circuit potential (OCP) before each electrochemical measurement. The OCP measurements were carried out for 30 min in the aggressive solutions with and without various concentrations of SB. The systems were confirmed to have reached OCP before 30 min with less than ±10 mV change in potential. The potentiodynamic polarization tests were performed after 30 min of mild steel immersion in the aggressive solutions by sweeping the potential between –200 mV and –800 mV *versus* Ag/AgCl, 3 M KCl reference electrode potential at the scan rate of 1 mV/s. Electrochemical impedance spectroscopy measurements were carried out after 30 min of mild steel immersion in the aggressive solution with and without various concentrations of the inhibitor. The electrochemical impedance spectroscopy measurements were conducted at the OCP by analyzing the frequency response of the electrochemical system in the range of 100 kHz to 1 Hz at 5 mV amplitude. All electrochemical experiments were conducted under unstirred conditions at 303 K.

3.5. Quantum Chemical Calculations

All quantum chemical calculations were carried out using the DFT method involving the Becke 3-Parameter exchange functional together with the Lee-Yang-Parr correlation functional (B3LYP) [34,35] and 6-31G(d) basis set. SB was modeled with Gaussview 5.0 software to obtain the initial geometries. Gas phase geometry optimization was carried out using Gaussian 09W D.01 software [36].

4. Conclusions

The new Schiff base (SB) synthesized from the condensation of 3-acetyl-4-hydroxy-6-methyl-(2H)pyran-2-one and 2,2'-(ethylenedioxy)diethylamine showed corrosion inhibition potency for the protection of mild steel in 1 M HCl, as confirmed by potentiodynamic polarization and electrochemical impedance spectroscopy experiments. SB was found to be a mixed-type inhibitor and its inhibition property was associated with its spontaneous adsorption onto mild steel surface via both physisorption and chemisorption. The experimental data fitted Langmuir adsorption isotherm. UV-vis and Raman spectroscopic analyses revealed the possibilities of chemical interactions and bond formation between mild steel and SB. Quantum chemical calculations suggested the possible sites for nucleophilic and electrophilic attack on SB.

Acknowledgments

Eno E. Ebenso gratefully acknowledges the National Research Foundation (NRF) of South Africa for incentive funding for rated researchers. A.S.A. thanks the NRF/Sasol Inzalo foundation for financial support and NWU for postdoctoral fellowship. L.O.O. acknowledges NRF/Sasol Inzalo foundation for funding support towards his PhD studies.

Author Contributions

All the authors have equally contributed to the manuscript.

Conflicts of Interest

The authors declare no conflict of interest.

References

1. De la Fuente, D.; Diaz, I.; Simancas, J.; Chico, B.; Morcillo, M. Long-term atmospheric corrosion of mild steel. *Corros. Sci.* **2011**, *53*, 604–617.

2. Ulaeto, S.B.; Ekpe, U.J.; Chidiebere, M.A.; Oguzie, E.E. Corrosion inhibition of mild steel in hydrochloric acid by acid extracts of *Eichhornia crassipes*. *Int. J. Mat. Chem.* **2012**, *2*, 158–164.

3. Guzman-Lucero, D.; Olivares-Xometl, O.; Martinez-Palou, R.; Likhanova, N.V.; Dominguez-Aguilar, M.A.; Garibay-Febles, V. Synthesis of selected vinylimidazolium ionic liquids and their effectiveness as corrosion inhibitors for carbon steel in aqueous sulfuric acid. *Ind. Eng. Chem. Res.* **2011**, *50*, 7129–7140.

4. Sastri, V.S. *Corrosion Inhibitors Principles and Applications*; Jonh Wiley & Sons: New York, NY, USA, 1998.

5. Mishra, M.; Tiwari, K.; Singh, A.K.; Singh, V.P. Synthesis, structural and corrosion inhibition studies on Mn(II), Cu(II) and Zn(II) complexes with a Schiff base derived from 2-hydroxypropiophenone. *Polyhedron* **2014**, *77*, 57–65.

6. Quan, Z.; Chen, S.; Li, S. Protection of copper corrosion by modification of self-assembled films of Schiff bases with alkanethiol. *Corros. Sci.* **2001**, *43*, 1071–1080.

7. Battaini, G.; Monzani, E.; Casella, L.; Santagostini, L.; Pagliarin, R. Inhibition of the catecholase activity of biomimetic dinuclear copper complexes by kojic acid. *J. Biol. Inorg. Chem.* **2000**, *5*, 262–268.

8. Puerta, D.T.; Cohen, S.M. Examination of novel zinc-binding groups for use in matrix metalloproteinase inhibitors. *Inorg. Chem.* **2003**, *42*, 3423–3430.

9. Rao, P.V.; Narasaiah, A.V. Synthesis, characterization and biological studies of oxovanadium(IV), manganese(II), iron(II), cobalt(II), nickle(II) and copper(II) complexes derived from a quadridentate ligand. *Indian J. Chem. A* **2003**, *42*, 1896–1899.

10. Chalaca, M.Z.; Figueroa-Villar, J.D.; Ellena, J.A.; Castellano, E.E. Synthesis and structure of cadmium and zinc complexes of dehydroacetic acid. *Inorg. Chem. Acta* **2002**, *328*, 45–52.

11. Fouad, D.M.; Bayoumi, A.; El-Gahami, M.A.; Ibrahim, S.A.; Hammam, A.M. Synthesis and thermal studies of mixed ligand complexes of Cu(II), Co(II), Ni(II) and Cd(II) with mercaptotriazoles and dehydroacetic acid. *Natural Science* **2010**, *2*, 817–827.

12. Chitrapriya, N.; Mahalingam, V.; Zeller, M.; Jayabalan, R.; Swaminathan, K.; Natarajan, K. Synthesis, crystal structure and biological activities of dehydroacetic acid complexes of Ru(II) and Ru(III) containing PPh3/AsPh3. *Polyhedron* **2008**, *27*, 939–946.

13. Asegbeloyin, J.N.; Ujam, O.T.; Ngige, C.M.; Onwukeme, V.I.; Groutso, T. Crystal structure of N'-[(1E)-1-(6-methyl-2,4-dioxo-3,4-dihydro-2H-pyran-3-ylidene)ethyl]benzenesulfonohydr-azide. *Acta Cryst.* **2014**, *E70*, 01179–01180.

14. Kubaisi, A.A.; Ismail, K.Z. Nickel(II) and palladium(II) chelates of dehydroacetic acid Schiff bases derived from thiosemicarbazide and hydrazinecarbodithioate. *Can. J. Chem.* **1994**, *72*, 1785–1788.

15. Asegbeloyin, J.N.; Babahan, I.; Ukwueze, N.N.; Oruma, U.S.; Poyrazoglu, E.C.; Eze, U.F. Synthesis, characterization and antimicrobial activity of 3-acetyl-4-hydroxy-6-methyl-(2H)pyran-2-one Schiff base with 2,2'-(ethylenedioxy)diethylamine and its Co(II), Ni(II) and Cu(II) complexes. *Asian J. Chemistry* **2015**, *27*, in press.

16. Sudheer; Quraishi, M.A. 2-Amino-3,5-dicarbonitrile-6-thio-pyridines: New and effective corrosion inhibitors for mild steel in 1 M HCl. *Ind. Eng. Chem. Res.* **2014**, *53*, 2851–2859.

17. Rammelt, U.; Koehler, S.; Reinhard, G. Electrochemical characterization of the ability of dicarboxylic acid salts to the corrosion inhibition of mild steel in aqueous solutions. *Corros. Sci.* **2011**, *53*, 3515–3520.

18. Sheriff, E.M.; Erasmus, R.M.; Comins, J.D. *In situ* Raman spectroscopy and electrochemical techniques for studying corrosion and corrosion inhibition of iron in sodium chloride solutions *Electrochimica. Acta* **2010**, *55*, 3657–3663.

19. El Adnani, Z.; Mcharfi, M.; Sfaira, M.; Benzakour, M.; Benjelloun, A.T.; Ebn Touhami, M.; Hammouti, B.; Taleb, M. DFT study of 7-R-3methylquinoxalin-2(1H)-ones (R=H; CH3; Cl) as corrosion inhibitors in hydrochloric acid. *Int. J. Electrochem. Sci.* **2012**, *7*, 6738–6751.

20. Satapathy, A.K.; Gunasekaran, G.; Sahoo, S.C.; Kumar, A; Rodrigues, P.V. Corrosion inhibition by Justicia gendarussa plant extract in hydrochloric acid solution. *Corros. Sci.* **2009**, *51*, 2848–2856.

21. Abdallah, M.; Asghar, B.H.; Zaafarany, I.; Fouda, A.S. The inhibition of carbon steel corrosion in hydrochloric acid solution using some phenolic compounds. *Int. J. Electrochem. Sci.* **2012**, *7*, 282–304.

22. Daoud, D.; Douadi, T.; Issaadi, S.; Chafaa, S. Adsorption and corrosion inhibition of new synthesized thiophene Schiff base on mild steel X52 in HCl and H2SO4 solutions. *Corros. Sci.* **2014**, *79*, 50–58.

23. Benabdellah, M.; Ousslim, A.; Hammouti, B.; Elidrissi, A.; Aouniti, A.; Dafali, A.; Bekkouch, K.; Benkaddour, M.J. The effect of poly(vinyl caprolactone-co-vinyl pyridine) and poly(vinyl imidazol-co-vinyl pyridine) on the corrosion of steel in H3PO4 media. *Appl. Electrochem.* **2007**, *37*, 819–826.

24. Chakravarthy, M.P.; Mohana, K.N.; Kumar, C.B.P. Study of adsorption properties and inhibition of mild steel corrosion in hydrochloric acid media by water soluble composite poly (vinyl alcohol-omethoxy aniline). *J. Assoc. Arab Univ. Basic Appl. Sci.* **2014**, *16*, 74–82.

25. Chakravarthy, M.P.; Mohana, K.N.; Kumar, C.B.P. Corrosion inhibition effect and adsorption behaviour of nicotinamide derivatives on mild steel in hydrochloric acid solution. *Int. J. Ind. Chem.* **2014**, *5*, 1–21.

26. Abd-Elaal, A.A.; Aiad, I.; Shaban, S.M.; Tawfik, S.M.; Atef, S. Synthesis and Evaluation of Some Triazole Derivativesas Corrosion Inhibitors and Biocides. *J. Surfact. Deterg.* **2014**, *17*, 483–491.

27. John, S.; Kuruvilla, M.; Joseph, A. Adsorption and inhibition effect of methyl carbamate on copper metal in 1 N HNO3: An experimental and theoretical study. *RSC Adv.* **2013**, *3*, 8929–8938.

28. Murulana, L.C.; Singh, A.K.; Shukla, S.K.; Kabanda, M.M.; Ebenso, E.E. Experimental and quantum chemical studies of some bis(trifluoromethyl-sulfonyl) imide imidazolium-based ionic liquids as corrosion inhibitors for mild steel in hydrochloric acid solution. *Ind. Eng. Chem. Res.* **2012**, *51*, 13282–13299.

29. Ahamad, I.; Prasad, R.; Quraishi, M.A. Thermodynamic, electrochemical and quantum chemical investigation of some Schiff bases as corrosion inhibitors for mild steel in hydrochloric acid solutions. *Corros. Sci.* **2010**, *52*, 933–942.

30. Gomez, B.; Likhanova, N.V.; Dominguez-Aguilar, M.A.; Martinez-Palou, R.; Vela, A.; Gazquez, J.L. Quantum chemical study of the inhibitive properties of 2-pyridyl-azoles. *J. Phys. Chem. B.* **2006**, *110*, 8928–8934.

31. Yan, Y.; Wang, X.; Zhang, Y.; Wang, P.; Zhang, J. Theoretical evaluation of inhibition performance of purine corrosion inhibitors. *Mol. Sim.* **2013**, *39*, 1034–1041.

32. Echegaray, E.; Cárdenas, C.; Rabi, S.; Rabi, N.; Lee, S.; Zadeh, F.H.; Toro-Labbe, A.; Anderson, J.S.M.; Ayers, P.W. In pursuit of negative Fukui functions: Examples where the highest occupied molecular orbital fails to dominate the chemical reactivity. *J. Mol. Model.* **2013**, *19*, 2779–2783.

33. Yang, W.; Mortier, W.J. The use of global and local molecular parameters for the analysis of the gas-phase basicity of amines. *J. Am. Chem. Soc.* **1986**, *108*, 5708–5711.

34. Becke, A.D. Density-functional exchange-energy approximation with correct asymptotic behavior. *Phys. Rev. A* **1988**, *38*, 3098–3100.

35. Lee, C.; Yang, W.; Parr, R.G. Development of the Colle-Salvetti correlation-energy formula into a functional of the electron density. *Phys. Rev. B* **1988**, *37*, 785–789.

36. Frisch, M.J.; Trucks, G.W.; Schlegel, H.B.; Scuseria, G.E.; Robb, M.A.; Cheeseman, J.R.; Scalmani, G.; Barone, V.; Mennucci, B.; Petersson, G.A.; *et al. Gaussian 09, Revision D.01*; Gaussian, Inc.: Wallingford, CT, USA, 2009.

13

Corrosion Behavior of Cast Iron in Freely Aerated Stagnant Arabian Gulf Seawater

El-Sayed M. Sherif [1,2,*], Hany S. Abdo [1,3,†] and Abdulhakim A. Almajid [1,4,†]

[1] Deanship of Scientific Research (DSR), Advanced Manufacturing Institute (AMI),
King Saud University, P.O. Box 800, Riyadh 11421, Saudi Arabia;
E-Mails: enghany2000@yahoo.com (H.S.A.); aalmajid@ksu.edu.sa (A.A.A.)

[2] Electrochemistry and Corrosion Laboratory, Department of Physical Chemistry,
National Research Centre (NRC), Dokki, Cairo 12622, Egypt

[3] Mechanical Design and Materials Department, Faculty of Energy Engineering, Aswan University,
Aswan 81521, Egypt

[4] Department of Mechanical Engineering, College of Engineering, King Saud University,
P.O. Box 800, Riyadh 11421, Saudi Arabia

[†] These authors contributed equally to this work.

[*] Author to whom correspondence should be addressed; E-Mail: esherif@ksu.edu.sa;

Academic Editor: Steven L. Suib

Abstract: In this work, the results obtained from studying the corrosion of cast iron in freely aerated stagnant Arabian Gulf seawater (AGS) at room temperature were reported. The study was carried out using weight-loss (WL), cyclic potentiodynamic polarization (CPP), open-circuit potential (OCP), and electrochemical impedance spectroscopy (EIS) measurements and complemented by scanning electron microscopy (SEM) and energy dispersive X-ray (EDX) investigations. WL experiments between two and 10 days' immersion in the test electrolyte indicated that the weight-loss the cast iron increases with increasing the time of immersion. CPP measurements after 1 h and 24 h exposure period showed that the increase of time decreases the corrosion via decreasing the anodic and cathodic currents, as well as decreasing the corrosion current and corrosion rate and increasing the polarization resistance of the cast iron. EIS data confirmed the ones obtained by WL and CPP that the increase of immersion time decreases the corrosion of cast iron by increasing its polarization resistance.

Keywords: Arabian Gulf seawater; cast iron; corrosion; electrochemical measurements; weight-loss

1. Introduction

Cast iron is a durable and fire-resistant material that is used in the home and industry. It is a complex material with stable and meta-stable phases and has elements in the solution, which influence the extent and stability of the desirable properties not obtained by other alloys. Cast iron is primarily an alloy of iron that contains carbon content of 2%–5% and at least 1% silicon, in addition to traces of manganese, sulfur, and phosphorus. Cast iron has a wide variety of properties such as cast complex shape with low cost, low melting temperature, high fluidity when molten, it does not form undesirable surface film when poured due to less reactivity with molten materials and has slight to moderate shrinkage during solidification and cooling [1]. It is also hard, brittle, nonmalleable, and more fusible than steel. Cast irons are widely used materials for components handling seawater and brine such as large intake, recycling and blow down pumps for desalination and power plants, and other hydraulic machinery. In particular, diesel engine cylinder liners are manufactured almost exclusively from flake graphite grey irons [2–5].

Upon manufacture, cast iron develops a protective film or scale on the surface which makes it initially more resistant to corrosion than wrought iron or mild steel [6]. It has been reported [7] that alloying elements play an important role in the susceptibility of cast irons to corrosion attack. Where, its corrosion depends mainly on the percent of silicon existed in the alloy; the higher the silicon content, the higher the corrosion resistance [7]. The corrosion of cast iron in different environments has attracted few investigators. Yilbas *et al.* [8] have reported the improved corrosion resistance of cast iron surface treated with laser gas assisted of dual matrix structured. Al-Hashem *et al.* [9] studied the effect of microstructure of nodular cast iron on its cavitation corrosion behavior in seawater and found that the surface becomes very rough with large size cavity pits. Olawale *et al.* [10] have evaluated the corrosion behavior of grey cast iron (GCI) and low alloy steel (LAS) in cocoa liquor and well water and found that LAS has better corrosion resistance than GCI in both media, and cocoa liquor is more aggressive than well water. They [10] also recommended that LAS is a better material for piping and pumping systems in cocoa processing industries than GCI.

The present work aims at investigating the corrosion behavior of cast iron after different immersion periods of time in the naturally aerated stagnant AGS solution at room temperature using gravimetric weight-loss, varied electrochemical, and spectroscopic techniques. The weight-loss method was employed to report the dissolution behavior for the cast iron over 10 days' immersion. The electrochemical techniques such as CPP, EIS, and OCP and were for reporting the electrochemical and kinetic parameters for the cast iron under investigation after its immersion for 1 h and 24 h in AGS medium. Spectroscopic investigations using SEM and EDX profile analysis were to understand the formed corrosion products onto the surface of cast iron that was immersed for 10 days' immersion in AGS.

2. Results and Discussion

2.1. Weight-Loss Measurements and SEM/EDX Investigations

The variation of (a) the weight-loss (Δm, mg/in^2) and (b) the corrosion rate (R_{Corr}, mpy) *vs.* time for the cast iron coupons in 300 cm^3 of aerated stagnant solutions of AGS are shown in Figure 1. The values of Δm and R_{Corr} over the exposure time were calculated as reported in our previous work as following [11]:

$$\Delta m = \frac{m_1 - m_2}{A} \tag{1}$$

$$R_{Corr} = \frac{543 \, \Delta m}{Dt} \tag{2}$$

where, m_1 and m_2 are the weighs of the cast iron coupon per mg before and after its immersion in the test solution, A is the area of the cast iron coupon in inch^2, D is the density of cast iron ($D = 7.563$ g/cm^3), and t is the exposure time (h).

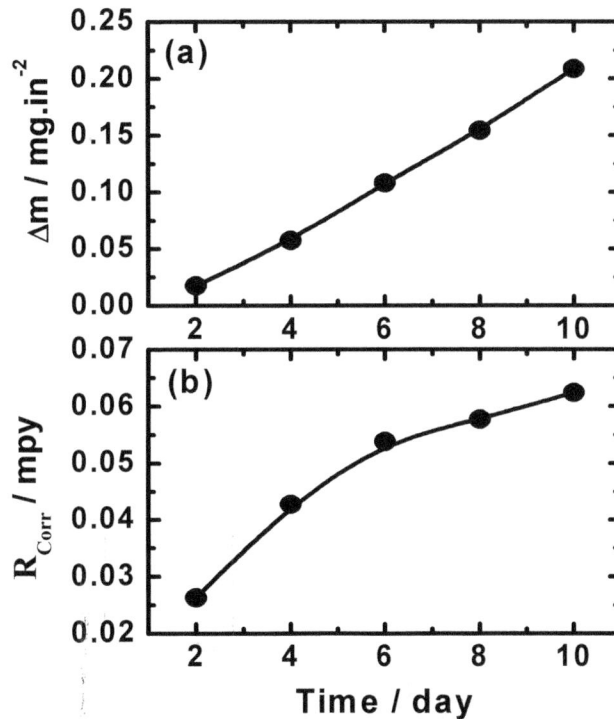

Figure 1. Change of (**a**) the dissolution rate (Δm, mg/in^2) and (**b**) the corrosion rate (R_{Corr}, mpy) with time for the cast iron coupons in the Arabian Gulf seawater (AGS) solutions.

One can see from Figure 1a that the values of Δm increased with time due to the aggressiveness attack of the corrosive ions present in the seawater toward the cast iron surface. It is well known that the cathodic reaction for metals and alloys in near neutral solutions is the oxygen reduction according to the following equation:

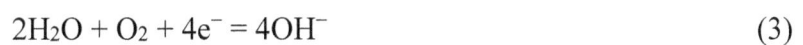

$$2H_2O + O_2 + 4e^- = 4OH^- \tag{3}$$

On the other hand, the anodic reaction of iron in aerated neutral solutions is the dissolution of metallic iron ($Fe°$) to ferrous cations (Fe^{2+}) as follows,

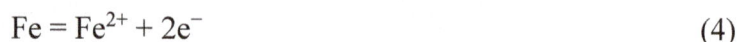

$$Fe = Fe^{2+} + 2e^-$$ (4)

The dissolution of iron gets activated due to the consumption of the produced electrons via the cathodic reaction at the anodic one, which could lead to the increased weight-loss with time, Figure 1a. Similar to the increase of Δm with time, the values of K_{Corr} also increased with increasing the immersion time, particularly at short periods, as can been seen from Figure 1b. This increment slightly decreased at longer immersion time due to the formation of thick corrosion products that decreases the attack of the corrosive species present in the seawater and thus decreases the corrosion rate.

In order to understand the morphology and the composition of the corrosion products formed on the cast iron coupon after its immersion in the seawater for 10 days, SEM/EDX investigations for the surface of the coupon were carried out. Figure 2a shows the SEM image obtained for the cast iron surface at different magnifications after 10 days' immersion in the aerated stagnant AGS solution, and Figure 2b represents the EDX profile analysis for the area of the surface shown in Figure 2a. It is clearly seen from the SEM image that the surface of the cast iron is fully covered with thick layers of corrosion products due to the immersion of the coupon in AGS for long time, 10 days. The atomic percentages of the components found on the surface were 56.87% O, 32.57% Fe, 9.64% Na, 3.50% Cl, 1.04% Mg, 0.29% Si, and 0.27% S. The very high content of oxygen and iron indicates that the layers formed on the cast iron surface after 10 days immersion in AGS were most probably iron oxide films. This can be explained on the light of the overall reaction that occurs on the iron surface as follows,

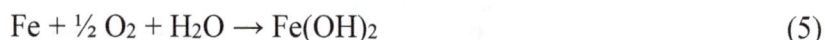

$$Fe + \tfrac{1}{2} O_2 + H_2O \rightarrow Fe(OH)_2$$ (5)

This ferrous hydroxide formed layer reacts, further in the presence of excess oxygen in the solution, to building up the final corrosion product Fe_3O_4 (magnetite) according to the following reaction [12];

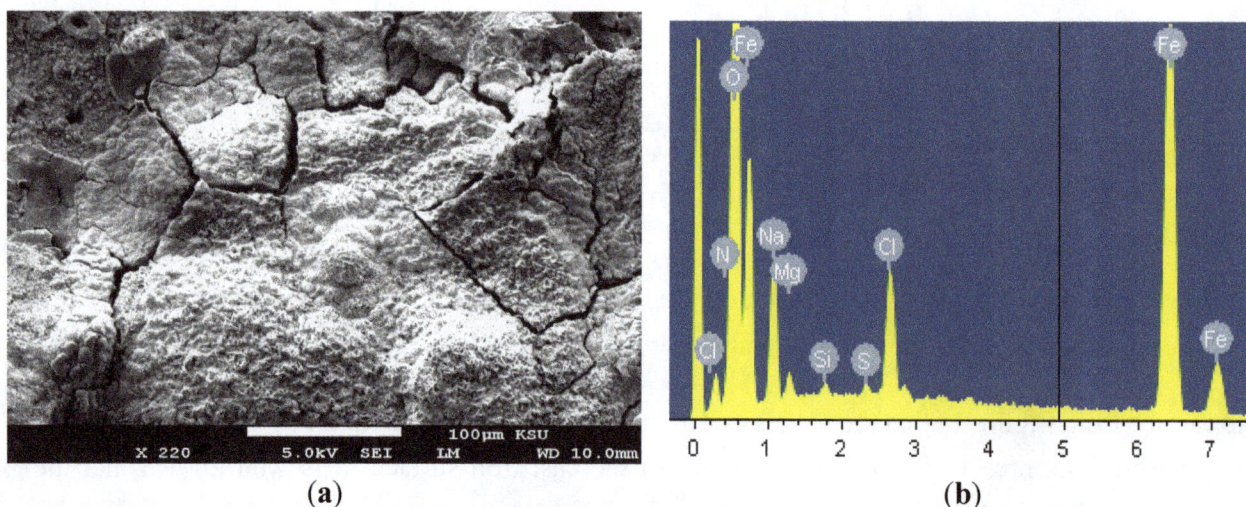

$$3Fe(OH)_2 + \tfrac{1}{2} O_2 \rightarrow Fe_3O_4 + 3H_2O$$ (6)

Figure 2. (**a**) SEM micrograph obtained for the cast iron coupons after its immersion in AGS for 10 days; and (**b**) the EDX profile analysis obtained for the surface shown in the SEM image.

According to Mohebbi and Li in a similar study [13], both Fe^{3+} and Fe^{2+} ion species existed in the corrosion product, which indicates that the electrons generated by oxidation of iron (cathodic reaction) can be readily consumed by the oxygen present in the electrolyte solution through the porous conductive corrosion layer. The presence of Cl ions gives indications on the increase of weight-loss and corrosion rate of cast iron with time due to possible reactions of Cl^- with the inner surface of iron [13]. This is due to the dissolution of iron as shown in Equation (4), which would lead to the formation of $FeCl_2$ and $FeCl_3$ in the solution. The concentration of $FeCl_2$ and $FeCl_3$ at this condition is at saturation and would precipitate to form a porous mixed film(s) of $FeCl_2$ and $FeCl_3$ on the iron electrode surface [14],

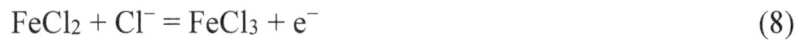

$$Fe + 2Cl^- = FeCl_2 + 2e^- \tag{7}$$

$$FeCl_2 + Cl^- = FeCl_3 + e^- \tag{8}$$

In addition to the iron oxides, the presence of Mg, Na and Si indicates that their oxides might have also formed on the surface.

2.2. Cyclic Potentiodynamic Polarization (CPP) Measurements

Cyclic potentiodynamic polarization testing were conducted to measure the corrosion parameters such as cathodic (β_c) and anodic (β_a) Tafel slopes, corrosion potential (E_{Corr}), pitting potential (E_{Pit}), protection potential (E_{Prot}), corrosion current (j_{Corr}), polarization resistance (R_P) and corrosion rate (R_{Corr}). Figure 3 shows the CPP curves obtained for the cast iron electrode after its immersion in AGS solutions for (a) 1 h and (b) 24 h, respectively. The corrosion parameters obtained from the CPP curves shown in Figure 3 are list in Table 1. The values of β_c and β_a were determined after at least 50 mV away from E_{Corr} and at least one decade of current densities (j_{Corr}). The values of E_{Corr} and j_{Corr} were obtained from the intersection of the extrapolation of anodic and cathodic Tafel lines located next to the linearized current regions. The values of polarization resistance, R_p, and corrosion rate, R_{Corr}, for the cast iron were calculated according to the following equations [12,15] corrosion as follows:

$$R_p = \frac{1}{j_{Corr}} \left(\frac{\beta_c \beta_a}{23 \left(\beta_c + \beta_a \right)} \right) \tag{9}$$

$$R_{Corr} = \frac{j_{Corr}}{d} \frac{k}{A} E_W \tag{10}$$

Where, k is a constant that defines the units for the corrosion rate ($=3272$ mm $amp^l cm^{-1} y^{-1}$); Ew the equivalent weight in grams/equivalent of cast iron alloy (Ew = 27.9 grams/equivalent); d the density in gcm^{-3} ($=7.563$); and A the area of the exposed surface of the electrode in cm^2.

It is clearly seen from Figure 3 that the anodic branch for the cast iron shows a passive region whether the measurement was taken after 1 h (Figure 3a) or after 24 h (Figure 3b). This passive region was formed due to the formation of an oxide film as depicted by Equations (5) and (6). Where, the formed ferrous hydroxide reacted with more oxygen to form the top layer of magnetite corrosion product, Fe_3O_4. The current then increased rapidly in the anodic side due to the breakdown of the formed oxide film and the occurrence of pitting corrosion. This was indicated by the higher current values in the backward direction of the scanned potential and the appearance of a hysteresis loop, the area of which decreased by increasing immersion time from 1h to 24 h. It is well known that AGS contains corrosive species such as chloride ions, which attacks the weak and flowed areas of the oxide film formed on the iron surface in the passive region causing its breakdown and corrosion via pitting by chloride ions attack as depicted in Equations (7) and (8).

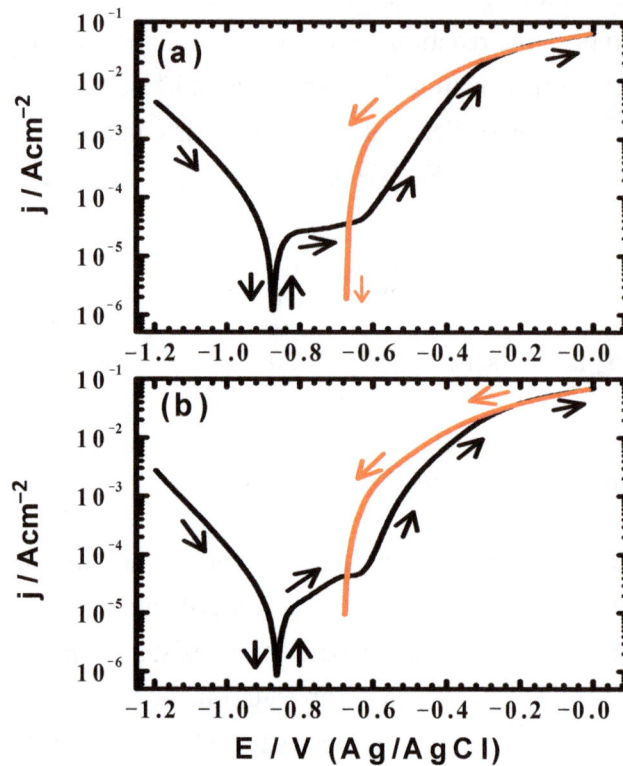

Figure 3. Cyclic potentiodynamic polarization curves obtained for the cast iron electrode after its immersion in AGS solutions for (**a**) 1 h and (**b**) 24 h, respectively.

Table 1. Polarization parameters obtained for the cast iron electrode after its immersion for 1 h and 24 h in the Arabian Gulf seawater (AGS).

Immersion time	Polarization Parameters							
	$\beta_c/$ mVdec^{-1}	$E_{Corr}/$ mV	$\beta_a/$ mV/dec^{-1}	$j_{Corr}/$ μA cm^{-2}	$E_{Pit}/$ mV	$E_{Prot}/$ mV	$R_p/$ Ω cm^2	$R_{Corr}/$ mmpy
AGS (1 h)	95	−860	230	20	−630	−665	1471	0.2414
AGS (24 h)	105	−825	210	11	−620	−680	2767	0.1328

The parameters shown in Table 1 show that the increase of immersion time from 1h to 24 h shifted E_{Corr} and E_{Pit} to the less negative values, while E_{Prot} was shifted to the more negative values. This indicates that both general and pitting corrosion decreased with increasing the time of immersion for the cast iron in AGS. Table 1 also shows that the increase of time before measurements decreased the values of j_{Corr} and R_{Corr} as well as increased the polarization resistance (R_p).

2.3. Open-Circuit Potential (OCP) and Electrochemical Impedance Spectroscopy (EIS) Measurements

Figure 4 shows the change of the OCP with time for the cast iron electrode in the AGS solution. It is seen from Figure 4 that the initial potential of iron rapidly increased towards the more negative values in the first few minutes due to the dissolution of a preformed air oxide film. The potential then slightly shifted in the more negative direction with the appearance of some fluctuations by increasing time for the first 12 h. This more negative shift might have resulted from the dissolution of iron by the corrosive ions attack such as chlorides. Finally, the potential very slightly decreases again towards the less negative direction till the end of the experiment as a result of the oxide and/or corrosion product layers on the surface.

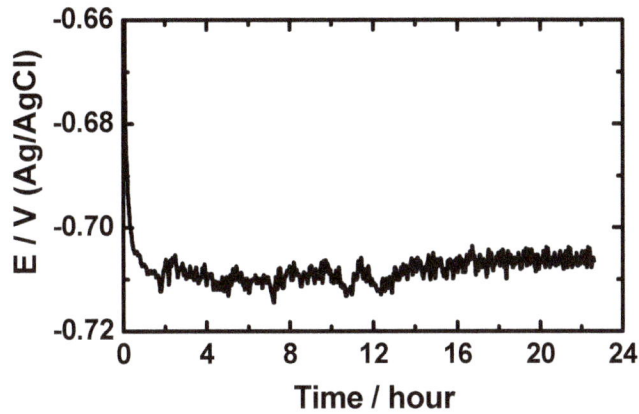

Figure 4. Change of the open-circuit potential with time for the cast iron electrode in the AGS solution.

EIS method is a powerful technique that has been used in studying corrosion and corrosion inhibition of metals and alloys in various corrosive media [14–21]. Typical Nyquist obtained for the cast iron electrode after (1) 1 h and (2) 24 h immersion in the AGS solution, respectively, are shown in Figure 5. The spectra represented in Figure 5 were analyzed by best fitting to the equivalent circuit model depicted in Figure 6. The EIS parameters obtained by fitting this circuit are listed in Table 2 and can be defined according to the usual convention as follows; R_S represents the solution resistance, Q is the constant phase elements (CPEs), C_{dl} is the double layer capacitance, R_{p1} is the polarization resistance for the solution/cast iron interface and may be defined as the charge transfer resistance of the cathodic reduction reaction of the cast iron, and R_{p2} is another polarization resistance for the corrosion product/cast iron interface [22].

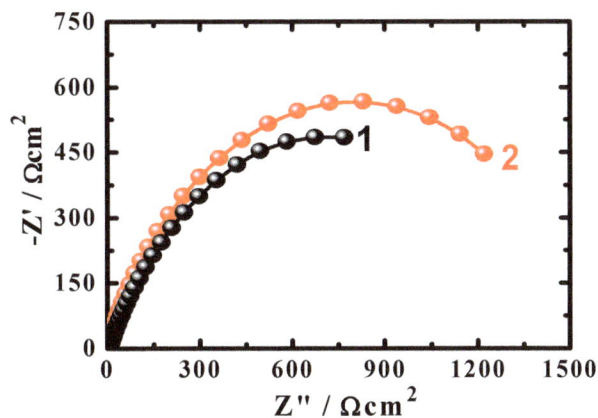

Figure 5. Typical Nyquist obtained for the cast iron electrode after (1) 1 h and (2) 24 h immersion in the AGS solution, respectively.

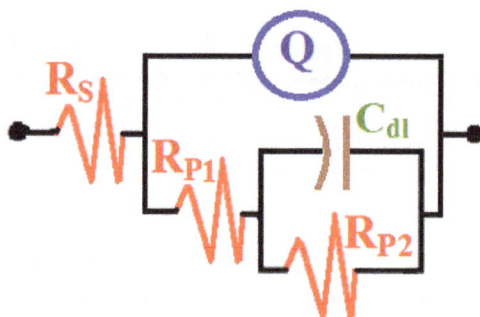

Figure 6. The equivalent circuit model used to the fit the EIS data shown in Figure 5; the symbols of the equivalent circuit are defined in the text and the values are listed in Table 2.

Table 2. Parameters obtained by fitting the EIS data shown in Figure 5 with the equivalent circuit shown in Figure 6 for the electrode after its immersion for 1 h and 24 h in AGS.

Immersion Time	Kinetic EIS Parameters					
	R_S/ Ω cm^2	Q		R_{P1}/ Ω cm^2	Cdl/ F cm^{-2}	R_{P2}/ Ω cm^2
		Y_Q/F cm^{-2}	n			
AGS (1 h)	5.814	0.000948	0.72	0.3977	5.232×10^{-5}	1047
AGS (24 h)	6.127	0.000539	0.72	40.36	2.374×10^{-5}	1571

It is seen from Figure 5 that the cast iron showed only one semicircle whether the immersion time for the cast iron before measurement was 1 h or 24 h. The diameter of the obtained semicircle got wider by increasing the immersion time to 24 h, which indicates that the resistance against corrosion increased by increasing the exposure period of time from 1h to 24 h. The values of R_S, R_{P1} and R_{P2} that are listed in Table 2 recorded higher values for the cast iron immersed in AGS for 24 h compared to those obtained after only 1 h. This is due to the increase of the corrosion resistance for the surface of the cast iron with increasing time. The constant phase elements (Q, CPEs) with its n values exactly 0.72 for the cast iron after 1 h and 24 h immersion in the AGS electrolyte represent double layer capacitors with some pores that allow the dissolution of iron [14–17], which agrees with the work of Mohebbi and Li [13]. Where and depending on the value of n, a CPE can represent resistance (Z(CPE) = R, n = 0), capacitance (Z(CPE) = Cdl, n = 1) or Warburg impedance for (n = 0.5). Therefore, the CPE for iron and steel is substituted for the capacitor to fit the semicircle more accurately. According to Zhang *et al.*, the admittance and the impedance of a CPE at this condition can be defined by the following equations, respectively [16].

$$Y_{CPE} = Y_0 (j\omega)^n \qquad (11)$$

$$Z_{CPE} = (1/Y_0) (j\omega)^{-n} \qquad (12)$$

where, Y_0 is the modulus; ω is the angular frequency; and n is the phase. The decrease of the CPEs and Cdl values byincreasing the immersion period to 24 h reveals also that elongating time lowers the corrosion of the cast iron [16,23].

Figure 7 shows the typical Bode (Figure 7a) impedance of the interface, |Z|, and (Figure 7b) phase angle plots obtained for the cast iron electrode after (1) 1 h and (2) 24 h immersion in the AGS solution, respectively. It is obvious that the increase of immersion time to 24 h before measurements increased the impedance $|Z|$ values (Figure 7a). It has been reported by Mansfeld *et al.* [24] that the surface is more protected when higher $|Z|$ values are shown, particularly at the low frequency region. The increase of immersion time also increased the maximum degree of the phase angle as can be seen from Figure 7b. This confirms that the better corrosion resistance of the cast iron with increasing time. The EIS Nyquist and Bode plots were consistent with each other and both are in good agreement with the data obtained by weight-loss and cyclic polarization measurements.

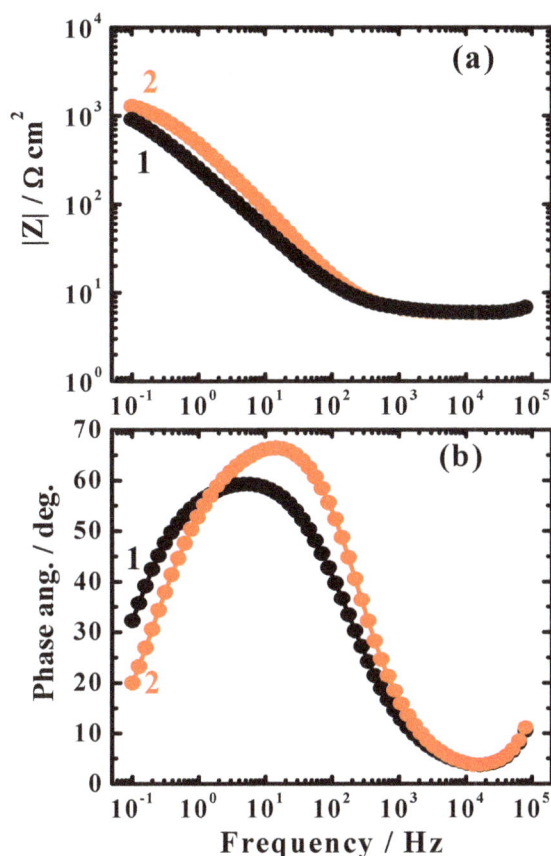

Figure 7. Typical Bode (**a**) impedance of the interface, $|Z|$; and (**b**) phase angle plots obtained for the cast iron electrode after (**1**) 1 h and (**2**) 24 h immersion in the AGS solution, respectively.

3. Experimental Section

Cast iron with the chemical compositions (in wt %) listed in Table 3 was employed in the present study. The Arabian Gulf seawater (AGS) was brought directly from the Arabian Gulf at the eastern region of Saudi Arabia. For electrochemical measurements, a conventional electrochemical cell accommodates only 200 cm^3 and a three-electrode configuration was used. The three electrodes were the cast iron, platinum foil, and an Ag/AgCl electrode (in 3.0 M KCl), which were used as working, counter, and reference electrodes, respectively. The cast iron electrode had a square shape with dimensions of 1 cm × 1 cm × 3 cm; the exposed surface area of the electrode to the electrolytic AGS solution was only 1 cm^2. Accordingly, the working electrode was prepared by welding an insulated copper wire to one face of the cast iron electrode and cold mounted in resin. The surface of the cast iron electrode to be exposed to the solution was first ground successively with metallographic emery paper of increasing fineness of up to 600 grits and further with 5 μm, 1 μm, 0.5 μm, and 0.3 μm alumina slurries

(Buehler). The electrode was then washed with doubly distilled water, degreased with acetone, washed using doubly distilled water again and finally dried with tissue paper.

Table 3. Chemical compositions of the cast iron that has be employed in the present study.

Element	C	Si	Mn	P	Mg	Ti	W	Cr	Cu	Zn	S	Ce	Fe
Wt %	4.58	2.13	0.27	0.08	0.07	0.04	0.03	0.02	0.02	0.02	0.02	0.01	Balance

The weight loss experiments were carried out using rectangular cast iron coupons that had the same chemical composition as cast iron electrodes. The coupons had a dimension of 4.0 cm length, 2.0 cm width, and 0.4 cm thickness and the exposed total area of 54 cm^2. The coupons were polished and dried as for the case of cast iron electrodes, weighed, and then suspended in 300 cm^3 solutions of AGS for different exposure periods (2–10 days).

The SEM investigation and EDX analysis were obtained on the surface of a cast iron samples after its immersion in open to air stagnant AGS solution for 10 days. The SEM/EDX analysis was collected on cast iron samples with dimensions of 1 cm × 1 cm × 0.4 cm that were cut from the coupons used in the weight-loss test. The SEM images were carried out by using a JEOL model JSM-6610LV (Japanese made) scanning electron microscope with an energy dispersive X-ray analyzer attached.

An Autolab Potentiostat-Galvanostat (Metrohm Autolab B.V., Amsterdam, The Netherlands, PGSTAT20 computer controlled) operated by the general purpose electrochemical software (GPES) version 4.9 (Metrohm, Amsterdam, The Netherlands) was used to perform the electrochemical experiments. Cyclic potentiodynamic polarization (CPP) curves were obtained by scanning the potential in the forward direction from −1.20 V to 0.0 V *vs.* Ag/AgCl at a scan rate of 0.001 V/s. The potential was scanned again in the backward direction at the scan rate till the end of the run. The open-circuit potential measurements were carried out for 24 h at room temperature. The electrochemical impedance spectroscopy (EIS) tests were performed at corrosion potentials over a frequency range of 100 kHz–100 mHz, with an ac wave of ±5 mV peak-to-peak overlaid on a dc bias potential, and the impedance data were collected using Powersine software at a rate of 10 points per decade change in frequency. Each experiment was carried out using fresh steel surface and new portion of the AGS solution. All CPP and EIS experiments were carried out after 1 h and 24 h immersion in the AGS electrolyte.

4. Conclusions

The corrosion of cast iron in freely aerated stagnant AGS electrolyte at room temperature using gravimetric and electrochemical measurements after varied exposure periods of time was reported. The results obtained by WL indicated that increasing the immersion time from 2 days to 10 days increases the weight-loss. Electrochemical (CPP, OCP, and EIS) measurements taken after 1 h and 24 h showed that the increase of immersion time decreases the corrosion of cast iron through decreasing its anodic, cathodic, and corrosion currents and corrosion rate, while increasing the polarization and solution resistances. The SEM image taken for the surface of the cast iron coupon that was immersed for 10 days in AGS showed that the presence of a thick layer of corrosion products fully covers the surface. The EDX profile analysis obtained for the cast iron surface after 10 days confirmed that the presence of high amount of oxygen and low amounts of iron compared to its content in the original coupon, which indicates that the corrosion product layer mainly consists of iron oxides. These results are consistent and confirm that the increase of immersion time decreases the corrosion of cast iron in AGS due to the formation of a thick layer of iron oxides that covers the surface and decreases its dissolution.

Acknowledgments

The authors would like to extend their sincere appreciation to the Deanship of Scientific Research at King Saud University for its funding of this research through the Research Group Project No. RGP-160.

Author Contributions

El-Sayed M. Sherif designed the work and participated in conducting the experiments and edited the final manuscript. Hany S. Abdo and Abdulhakim A. Almajid participated equally in conducting the experiments and wrote the initial draft of the manuscript.

Conflicts of Interest

The authors declare no conflict of interest.

References

1. Henkel, D.P.; Pense, A.W. *Structure and Properties of Engineering Materials*, 5th ed.; McGraw-Hill: New York, NY, USA, 2001.
2. Angus, H.T. *Cast Iron: Physical and Engineering Properties*, 2nd ed.; Batterworth: Guildford, UK, 1976.
3. Tomlinson, W.J.; Talks, M.G. Erosion and corrosion of cast iron under cavitation conditions. *Tribol. Int.* **1991**, *24*, 67–75.
4. Elliott, R. *Cast Iron Technology*; Butterworth: Guildford, UK, 1988.
5. Minkoff, I. *The Physical Metallurgy of Cast Iron*; Wiley: Chichester, UK, 1983.
6. Zahner, L.W. *Architectural Metals: A Guide to Selection, Specification, and Performance*; John Wiley & Sons: New York, NY, USA, 1995.
7. Rana, A.M.; Khan, A.; Amjad, S.; Abbas, T. Microstructural evaluation in heat-treated cast irons. *J. Res. Sci.* **2001**, *12*, 65–71.
8. Yilbas, B.S.; Toor, I.; Karatas, C.; Malik, J.; Ovali, I. Laser treatment of dual matrix structured cast iron surface: Corrosion resistance of surface. *Opt. Lasers Eng.* **2015**, *64*, 17–22.
9. Al-Hashem, A.; Abdullah, A.; Riad, W. Cavitation corrosion of nodular cast iron (NCI) in seawater—Microstructural effects. *Mater. Charact.* **2001**, *47*, 383–388.
10. Olawale, J.O.; Odusote, J.K.; Rabiu, A.B.; Ochapa, E.O. Evaluation of corrosion behaviour of grey cast iron and low alloy steel in cocoa liquor and well water. *Sci. Res.* **2013**, *1*, 44–48.
11. Oguzie, E.E.; Unaegbu, C.; Ogukwe, C.N.; Okolue, B.N.; Onuchukwu, A.I. Inhibition of mild steel corrosion in sulphuric acid using indigo dye and synergistic halide additives. *Mater. Chem. Phys.* **2004**, *84*, 363–368.

12. Sherif, E.-S.M. Effects of 5-(3-aminophenyl)-tetrazole on the inhibition of unalloyed iron corrosion in aerated 3.5% sodium chloride solutions as a corrosion inhibitor. *Mater. Chem. Phys.* **2011**, *129*, 961–967.

13. Mohebbi, H.; Li, C.Q. Experimental investigation on corrosion of cast iron pipes. *Int. J. Corros.* **2011**, *2011*, doi:10.1155/2011/506501.

14. Sherif, E.-S.M.; Erasmus, R.M.; Comins, J.D. *In situ* Raman spectroscopy and electrochemical techniques for studying corrosion and corrosion inhibition of iron in sodium chloride solutions. *Electrochim. Acta* **2010**, *55*, 3657–3663.

15. Sherif, E.-S.M. A comparative study on the electrochemical corrosion behavior of iron and X-65 steel in 4.0 wt % sodium chloride solution after different exposure intervals. *Molecules* **2014**, *19*, 9962–9974.

16. Zhang, Z.; Chen, S.; Li, Y.; Li, S.; Wang, L. A study of the inhibition of iron corrosion by imidazole and its derivatives self-assembled films. *Corros. Sci.* **2009**, *51*, 291–300.

17. Sherif, E.-S.M. Comparative study on the inhibition of iron corrosion in aerated stagnant 3.5 wt % sodium chloride solutions by 5-phenyl-1H-tetrazole and 3-amino-1,2,4-triazole. *Ind. Eng. Chem. Res.* **2013**, *52*, 14507–14513.

18. Gopi, D.; Sherif, E.-S.M.; Manivannan, V.; Rajeswari, D.; Surendiran, M.; Kavitha, L. Corrosion and corrosion inhibition of mild steel in ground water at different temperatures by newly synthesized benzotriazole and phosphono derivatives. *Ind. Eng. Chem. Res.* **2014**, *53*, 4286–4294.

19. Sherif, E.-S.M. Corrosion inhibition in 2.0 M sulfuric acid solutions of high strength maraging steel by aminophenyl tetrazole as a corrosion inhibitor. *Appl. Surf. Sci.* **2014**, *292*, 190–196.

20. Macdonald, J.R. *Impedance Spectroscopy*; Wiley & Sons: New York, NY, USA, 1987.

21. Sherif, E.-S.M.; Seikh, A.H. Effects of immersion time and 5-phenyl-1H-tetrazole on the corrosion and corrosion mitigation of cobalt free maraging steel in 0.5 M sulfuric acid pickling solutions. *J. Chem.* **2013**, *2013*, doi:10.1155/2013/497823.

22. Ma, H.; Chen, S.; Niu, L.; Zhao, S.; Li, S.; Li, D. Inhibition of copper corrosion by several Schiff bases in aerated halide solutions. *J. Appl. Electrochem.* **2002**, *32*, 65–72.

23. Okafor, P.C.; Liu, C.B.; Liu, X.; Zheng, Y.G.; Wang, F.; Liu, C.Y.; Wang, F. Corrosion inhibition and adsorption behavior of imidazoline salt on N80 carbon steel in CO_2-saturated solutions and its synergism with thiourea. *J. Solid State Electrochem.* **2010**, *14*, 1367–1376.

24. Mansfeld, F.; Lin, S.; Kim, S.; Shih, H. Pitting and surface modification of SIC/Al. *Corros. Sci.* **1987**, *27*, 997–1000.

Permissions

List of Contributors

Ahmed Abdel-Mohti
Civil Engineering Department, Ohio Northern University, Ada, OH 45810, USA

Alison N. Garbash
Mechanical Engineering Department, Ohio Northern University, Ada, OH 45810, USA

Saad Almagahwi
Mechanical Engineering Department, Ohio Northern University, Ada, OH 45810, USA

Hui Shen
Mechanical Engineering Department, Ohio Northern University, Ada, OH 45810, USA

Alexandra Muñoz-Bonilla
Departamento de Química Física Aplicada, Facultad de Ciencias, Universidad Autónoma de Madrid (UAM), C/ Francisco Tomás y Valiente 7, Cantoblanco, 28049 Madrid, Spain

Marta Fernández-García
Instituto de Ciencia y Tecnología de Polímeros (ICTP-CSIC), C/ Juan de la Cierva 3, 28006 Madrid, Spain

Guo Chen
Department of Biotechnology and Bioengineering, Huaqiao University, Xiamen 361021, China

Bin Zhang
Department of Biotechnology and Bioengineering, Huaqiao University, Xiamen 361021, China

Jun Zhao
Department of Biotechnology and Bioengineering, Huaqiao University, Xiamen 361021, China

Zarina Yahya
Center of Excellence Geopolymer and Green Technology, School of Materials Engineering, Universiti Malaysia Perlis (UniMAP), P.O. Box 77, D/A Pejabat Pos Besar, Kangar 01000, Perlis, Malaysia
Faculty of Engineering Technology, Uniciti Alam Campus, Universiti Malaysia Perlis, Sungai Chuchuh 02100, Padang Besar, Perlis, Malaysia

Mohd Mustafa Al Bakri Abdullah
Faculty of Engineering Technology, Uniciti Alam Campus, Universiti Malaysia Perlis, Sungai Chuchuh 02100, Padang Besar, Perlis, Malaysia

Faculty of Technology, Universitas Ubudiyah Indonesia, Jl. Alue Naga, Kec. Syiah Kuala Desa Tibang 23536, Banda Aceh, Indonesia

Kamarudin Hussin
Center of Excellence Geopolymer and Green Technology, School of Materials Engineering, Universiti Malaysia Perlis (UniMAP), P.O. Box 77, D/A Pejabat Pos Besar, Kangar 01000, Perlis, Malaysia
Faculty of Engineering Technology, Uniciti Alam Campus, Universiti Malaysia Perlis, Sungai Chuchuh 02100, Padang Besar, Perlis, Malaysia

Khairul Nizar Ismail
School of Environmental Engineering, Universiti Malaysia Perlis (UniMAP), P.O. Box 77, D/A Pejabat Pos Besar, Kangar 01000, Perlis, Malaysia

Rafiza Abd Razak
Center of Excellence Geopolymer and Green Technology, School of Materials Engineering, Universiti Malaysia Perlis (UniMAP), P.O. Box 77, D/A Pejabat Pos Besar, Kangar 01000, Perlis, Malaysia

Andrei Victor Sandu
Faculty of Materials Science and Engineering, Gheorghe Asachi Technical University of Iasi, Blvd. D. Mangeron 41, Iasi 700050, Romania

Ruisi Zhang
Department of Mechanical Engineering, Iowa State University, Ames, IA 50011, USA

Yuanfen Chen
Department of Mechanical Engineering, Iowa State University, Ames, IA 50011, USA

Reza Montazami
Department of Mechanical Engineering, Iowa State University, Ames, IA 50011, USA
Center for Advanced Host Defense Immunobiotics and Translational Comparative Medicine, Iowa State University, Ames, IA 50011, USA

Hong-Jiang Jiang
Wendeng Orthopaedic Hospital, No. 1 Fengshan Road, Wendeng 264400, Shandong, China

Jin Xu
Kangda College of Nanjing Medical University, No. 8 Chunhui Road, Xinhai District, Lianyungang 222000, Jiangsu, China

Zhi-Ye Qiu
School of Materials Science and Engineering, Tsinghua University, Haidian District, Beijing 100084, China
Beijing Allgens Medical Science and Technology Co., Ltd., No. 1 Disheng East Road, Yizhuang Economic and Technological Development Zone, Beijing 100176, China

Xin-Long Ma
Tianjin Hospital, No. 406 Jiefang South Road, Tianjin 300211, China

Zi-Qiang Zhang
Beijing Allgens Medical Science and Technology Co., Ltd., No. 1 Disheng East Road, Yizhuang Economic and Technological Development Zone, Beijing 100176, China

Xun-Xiang Tan
Wendeng Orthopaedic Hospital, No. 1 Fengshan Road, Wendeng 264400, Shandong, China

Yun Cui
Beijing Allgens Medical Science and Technology Co., Ltd., No. 1 Disheng East Road, Yizhuang Economic and Technological Development Zone, Beijing 100176, China

Fu-Zhai Cui
School of Materials Science and Engineering, Tsinghua University, Haidian District, Beijing 100084, China

Laura Fiocco
Department of Industrial Engineering, University of Padova, Via Marzolo 9, Padova 35131, Italy

Hamada Elsayed
Department of Industrial Engineering, University of Padova, Via Marzolo 9, Padova 35131, Italy

Letizia Ferroni
Department of Biomedical Sciences, University of Padova, Via Ugo Bassi 58/B, Padova 35131, Italy

Chiara Gardin
Department of Biomedical Sciences, University of Padova, Via Ugo Bassi 58/B, Padova 35131, Italy

Barbara Zavan
Department of Biomedical Sciences, University of Padova, Via Ugo Bassi 58/B, Padova 35131, Italy

Enrico Bernardo
Department of Industrial Engineering, University of Padova, Via Marzolo 9, Padova 35131, Italy

Jaecheol Ahn
Department of Architecture, Dong-A University, 550 Beon-gil Saha-gu, Busan 604-714, Korea

Takafumi Noguchi
Department of Architecture, the University of Tokyo, Hongo 7-3-1, Tokyo 113-8654, Japan

Ryoma Kitagaki
Department of Architecture, the University of Tokyo, Hongo 7-3-1, Tokyo 113-8654, Japan

Jesús Carmona
Civil Engineering Department, Universidad de Alicante, Ctra. San Vicente s/n, San Vicente del Raspeig 03690, Spain

Miguel-Ángel Climent
Civil Engineering Department, Universidad de Alicante, Ctra. San Vicente s/n, San Vicente del Raspeig 03690, Spain

Carlos Antón
Civil Engineering Department, Universidad de Alicante, Ctra. San Vicente s/n, San Vicente del Raspeig 03690, Spain

Guillem de Vera
Civil Engineering Department, Universidad de Alicante, Ctra. San Vicente s/n, San Vicente del Raspeig 03690, Spain

Pedro Garcés
Civil Engineering Department, Universidad de Alicante, Ctra. San Vicente s/n, San Vicente del Raspeig 03690, Spain

Cheng-Chih Fan
Institute of Materials Engineering, National Taiwan Ocean University, No. 2 Pei-Ning Road, Keelung 20224, Taiwan

Ran Huang
Department of Harbor and River Engineering, National Taiwan Ocean University, No. 2 Pei-Ning Road, Keelung 20224, Taiwan

Howard Hwang
Graduate Institute of Architecture and Sustainable Planning, National Ilan University, No.1, Sec. 1, Shen-Lung Road, I-Lan 26047, Taiwan

Sao-Jeng Chao
Department of Civil Engineering, National Ilan University, No.1, Sec. 1, Shen-Lung Road, I-Lan 26047, Taiwan

Younoussa Millogo
Unité de Formation et de Recherche en Sciences et Techniques (UFR/ST), Université Polytechnique de Bobo-Dioulasso, 01 BP 1091 Bobo 01, Burkina Faso
Laboratoire de Chimie Moléculaire et de Matériaux (LCMM), UFR/Sciences Exactes et Appliquées, Université de Ouagadougou, 03 BP 7021 Ouagadougou 03, Burkina Faso

Jean-Emmanuel Aubert
Université de Toulouse, UPS, INSA, LMDC (Laboratoire Matériaux et Durabilité des Constructions), 135 avenue de Rangueil, F-31077 Toulouse cedex 4, France

Erwan Hamard
Institut Français des Sciences et Technologies des Transports, de l'Aménagement et des Réseaux, Département Matériaux, GPEM, route de Bouaye, 44344 Bouguenais, CS4, France

Jean-Claude Morel
Ecole Nationale des Travaux Publics de l'Etat, Université de Lyon CNRS-LTDS, UMR 5513, LGCB, 3 rue Maurice Audin, Vaulx-en-Velin cedex, F-69120, France

Jonnie N. Asegbeloyin
Department of Pure and Industrial Chemistry, University of Nigeria, Nsukka 40001, Enugu State, Nigeria

Paul M. Ejikeme
Department of Pure and Industrial Chemistry, University of Nigeria, Nsukka 40001, Enugu State, Nigeria

Lukman O. Olasunkanmi
Material Science Innovation and Modelling (MaSIM) Research Focus Area, Faculty of Agriculture, Science and Technology, North-West University (Mafikeng Campus) Private Bag X2046, Mmabatho 2735, South Africa
Department of Chemistry, Faculty of Science, Obafemi Awolowo University, Ile-Ife 220005, Nigeria

Abolanle S. Adekunle
Material Science Innovation and Modelling (MaSIM) Research Focus Area, Faculty of Agriculture, Science and Technology, North-West University (Mafikeng Campus) Private Bag X2046, Mmabatho 2735, South Africa
Department of Chemistry, Faculty of Science, Obafemi Awolowo University, Ile-Ife 220005, Nigeria

Eno E. Ebenso
Material Science Innovation and Modelling (MaSIM) Research Focus Area, Faculty of Agriculture, Science and Technology, North-West University (Mafikeng Campus) Private Bag X2046, Mmabatho 2735, South Africa

El-Sayed M. Sherif
Deanship of Scientific Research (DSR), Advanced Manufacturing Institute (AMI), King Saud University, P.O. Box 800, Riyadh 11421, Saudi Arabia
Electrochemistry and Corrosion Laboratory, Department of Physical Chemistry, National Research Centre (NRC), Dokki, Cairo 12622, Egypt

Hany S. Abdo
Deanship of Scientific Research (DSR), Advanced Manufacturing Institute (AMI), King Saud University, P.O. Box 800, Riyadh 11421, Saudi Arabia
Mechanical Design and Materials Department, Faculty of Energy Engineering, Aswan University, Aswan 81521, Egypt

Abdulhakim A. Almajid
Deanship of Scientific Research (DSR), Advanced Manufacturing Institute (AMI), King Saud University, P.O. Box 800, Riyadh 11421, Saudi Arabia
Department of Mechanical Engineering, College of Engineering, King Saud University, P.O. Box 800, Riyadh 11421, Saudi Arabia

www.ingramcontent.com/pod-product-compliance
Lightning Source LLC
Chambersburg PA
CBHW080635200326
41458CB00013B/4638